GEOLOGICAL SOCIETY MEMOIR NO. 18

A Reassessment of the Southern Ocean Biochronology

By

ANTHONY T. S. RAMSAY
Marine Geoscience Research Group
Department of Earth Sciences
Cardiff University
Cardiff CF1 3YE, UK

and

JACK G. BALDAUF
Department of Oceanography and Ocean Drilling Program
Texas A&M University
College Station, Texas 77845
USA

1999
Published by
The Geological Society
London

Geological Society Memoirs
Series Editor
A. J. FLEET
R. E. HOLDSWORTH
A. C. MORTON
M. S. STOKER

It is recommended that reference to this book should be made in the following way.

RAMSAY, A. T. S. & BALDAUF, J. G. 1999. *A Reassessment of the Southern Ocean Biochronology*. Geological Society, London, Memoir **18**.

Contents

Preface	v	Hole 704B	57
Acknowledgements	v	Hole 737A	63
		Hole 744A	67
Introduction	1	Hole 744B	71
Previous work	3	Hole 755B	76
Methods	16	Hole 746A	80
Results	22	Hole 747A	83
Hole 514	23	Hole 747B	91
Hole 689B	26	Hole 748B	95
Hole 690B	33	Hole 751A	101
Hole 695A	40	Discussion	106
Hole 696A	43	Conclusion	115
Hole 697B	46	References	116
Hole 699A	49	Index	119
Hole 701A	54		

Preface

Numerous biochronological models have been developed for the Southern Ocean based on coring results from the various Deep Sea Drilling Project and Ocean Drilling Program expeditions to this region. These models are based on combining micropalaeontological datums, including the first and last occurrences and sometimes the first common or abundant occurrences of individual species with the record of magnetic polarity reversals.

Previous biostratigraphical age models represent an amalgamation of cruise-specific micropalaeontological datums derived from specific geographical sectors of the Southern Ocean and from other ocean basins. In the development of these models previous workers were constrained by limited sedimentary sequences containing an integrated biostratigraphical (siliceous and calcareous microfossils) and magnetostratigraphical events.

The results presented here integrate data sets generated through previous studies into a Southern ocean-wide perspective. The biostratigraphical (diatoms, radiolarians, and calcareous nannofossils) and magnetostratigraphical data sets are reviewed, integrated, calibrated, and reinterpreted to the Geopolarity Timescale of Cande & Kent (1995) providing a comprehensive Southern Ocean data base. In doing so, this study identifies inconsistent chronostratigraphical ranges of many previously used biostratigraphical markers thereby questioning the usefulness of these events within the Southern Ocean. The reassessment of previous data sets results in the recognition of 13 primary events which are chronostratigraphically useful throughout the Southern Ocean for the Late Eocene to Recent. Twenty additional events are identified as potentially stratigraphically useful; however further refinement of their spatial and/or temporal distribution is required. The limited number of primary events suggests that the biostratigraphy in the Southern Ocean is less sophisticated than currently assumed.

The integrated approach applied in this study provides a reassessment of the stratigraphy previously established for the 18 holes re-examined. The revised stratigraphy indicates typically more complete sequences compared to previous interpretations.

Acknowledgements

This study was completed with financial support from the National Environmental Research Council, Grant No. NERC-GST/02/698 to the Paleoceanography of the Southern Ocean Working Group (PALSOS) at the University of Wales, Cardiff. The group consisted of Rob Kidd, Tony Ramsay, Tim Sykes (University of Wales, Cardiff), Jack Baldauf and Tom Davies (Texas A&M University, USA) and J.-Y. Royer (Villefranche-sur-Mer, France). Numerous individuals assisted in completing this study and we specifically thank Rosie Sada, Doris Cooley and Tom Parmesen for their dedication, patience and thoroughness. The manuscript was reviewed by John Barron, Lloyd Burckle and Sherwood Wise, and we are grateful for their their critical reviews and constructive comments. We also thank our families who have endured the endless hours focused on this project and who, at times, lived with the flow of the Southern Ocean and the chill of the Antarctic.

Rob Kidd died in June 1996. He was an enthusiastic advocate for the Ocean Drilling Program and for synthesis investigations of ocean drilling data. This study is dedicated to Rob who was a valued colleague and a good friend.

Introduction

Understanding Earth's climatic history has lead researchers to explore the glacial history of Antarctica and its role in the Earth's climate system. Scientists have used the information recorded both on the continent and from the surrounding marine sediments. Attention to the deep sea record has focused on the use of proxy indicators such as ice-rafted detritus, sedimentological variations, stable-isotope records and changes in faunal and floral assemblages to identify fluctuating ice sheets and changes in continental ice volume. Dating of specific geological events has relied on microfossil biostratigraphy with calibration of specific biostratigraphical datums (e.g. first and last occurrences) to polarity stratigraphies. In the absence of polarity records, dating of biostratigraphical events was based on known calibrations and/or stratigraphical successions of specific species from regions other than the Southern Ocean. For example, Miocene diatom events in the Southern Ocean were typically assigned ages based on the known ranges of the species in the North Pacific. The chronological framework, based on imported microfossil datums, was used to calibrate the ages of datums specific to the Southern Ocean, i.e. a process involving secondary correlation. These approaches assumed isochroneity of events between oceans.

The biochronological framework in this paper is based on the results of the numerous shipboard and shorebased scientists who completed detailed analyses of the sediments recovered at the 59 Deep Sea Drilling Project (DSDP) and Ocean Drilling Program (ODP) Southern Ocean sites (Fig. 1; Table 1). Their contributions have made significant advances in the understanding of species biogeography, evolutionary succession and provide the cornerstone for developing a biochronological framework for the Southern Ocean.

Following 23 years of scientific ocean drilling, it is critical that the current biochronological model be evaluated for overall reliability and usefulness throughout the Southern Ocean and the assumption of isochroneity of specific biostratigraphical events between oceans is tested. This current study incorporates diatom, calcareous nannofossil, radiolarian and palaeomagnetic data sets from numerous Southern Ocean researchers (Table 2) to test the ocean wide isochroneity of currently used biostratigraphical events and ensure a consistent and coherent biochronological model for the Southern Ocean.

Table 1. *Latitude and longitude of sites occupied by the Deep Sea Drilling Project and the Ocean Drilling Program position south of 40° southern latitude*

Site	Latitude	Longitude
514	46°02.77'S	26°51.29'W
689	64°31.01'S	03°06.00'E
690	65°09.63'S	01°12.30'E
691	70°44.54'S	13°48.66'W
692	70°43.45'S	13°49.21'W
693	70°49.89'S	14°34.41'W
694	66°50.83'S	33°26.79'W
695	62°23.48'S	43°27.10'W
696	61°50.95'S	42°55.98'W
697	61°48.63'S	40°17.27'W
698	51°27.51'S	33°05.96'W
699	51°32.54'S	30°40.62'W
700	51°31.99'S	30°16.70'W
701	51°59.08'S	23°12.74'W
702	50°56.79'S	26°22.13'W
703	47°03.04'S	07°53.68'E
704	46°52.76'S	07°25.25'E
736	49°24.12'S	71°39.61'E
737	50°13.67'S	73°01.97'E
738	62°42.54'S	82°47.25'E
739	67°16.57'S	75°04.91'E
740	68°41.22'S	76°43.25'E
741	68°23.16'S	76°23.02'E
742	67°32.98'S	75°24.27'E
743	66°54.99'S	74°41.42'E
744	61°34.66'S	80°35.46'E
745	59°35.71'S	85°51.60'E
746	59°32.82'S	85°51.78'E
747	54°48.68'S	76°47.64'E
748	58°26.45'S	78°58.89'E
749	58°43.03'S	76°24.45'E
750	57°35.54'S	81°14.42'E
751	57°43.56'S	79°48.89'E

Table 2. *Previous shipboard and shorebased studies contributing to the development of the biochronological model presented in this volume*

Diatoms	
Baldauf & Barron (1991)	Leg 119
Ciesielski (1983)	Leg 71
Ciesielski (1991)	Leg 114
Fenner (1991)	Leg 114
Gersonde et al. (1990)	Leg 113
Gombos & Ciesielski (1983)	Leg 71
Gombos (1976)	Leg 36
Harwood & Maruyama (1992)	Leg 120
McCollum (1975)	Leg 28
Schrader (1976)	Leg 35
Weaver & Gombos (1981)	Other
Radiolarians	
Abelmann (1990)	Leg 113
Abelmann (1992)	Leg 120
Caulet (1991)	Leg 119
Lazarus (1990)	Leg 113
Lazarus (1992)	Leg 120
Ling (1991)	Leg 114
Takemura (1992)	Leg 120
Weaver (1976)	Leg 35
Weaver (1983)	Leg 71
Calcareous nannofossils	
Aubry (1992)	Leg 120
Chen (1975)	Leg 28
Crux (1991)	Leg 114
Gard & Crux (1991)	Leg 114
Madile & Monechi (1991)	Leg 114
Pospichal & Wise (1990a)	Leg 113
Pospichal & Wise (1990b)	Leg 113
Wei & Thierstein (1991)	Leg 119
Wei & Wise (1990)	Leg 113
Wei & Wise (1992a,b)	Leg 120
Wise & Wind (1977)	Leg 36
Wise (1983)	Leg 71
Magnetics	
Clement & Hailwood (1991)	Leg 114
Hailwood & Clement (1991a)	Leg 114
Hailwood & Clement (1991b)	Leg 114
Heider et al. (1992)	Leg 120
Inokuchi & Heider (1992)	Leg 120
Keating & Saka (1991)	Leg 119
Ledbetter (1983)	Leg 71
Salloway (1983)	Leg 71
Sakai & Keating (1991)	Leg 119
Others	
Barron et al. (1991)	Leg 119
Harwood et al. (1992)	Leg 120

INTRODUCTION

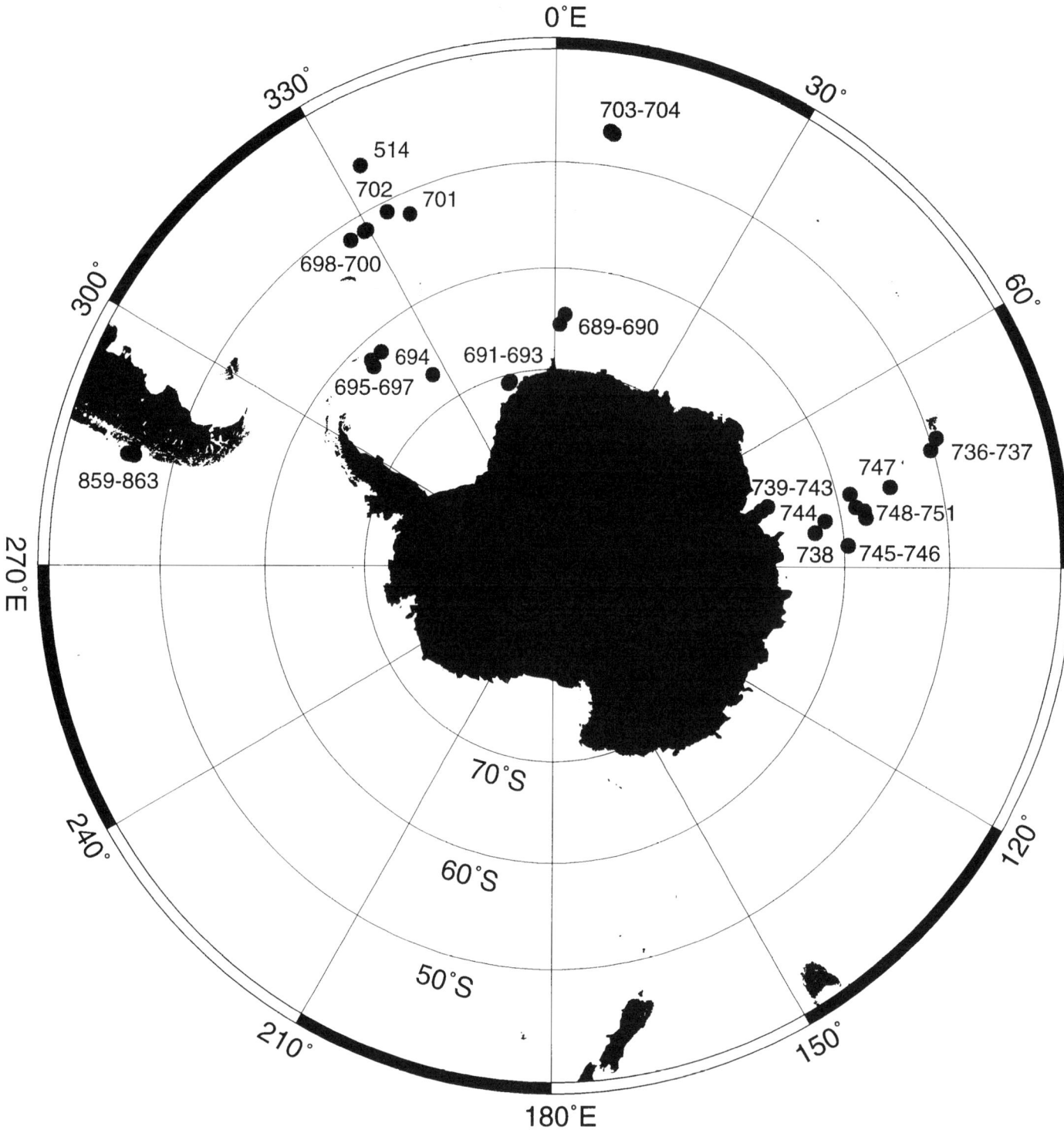

Fig. 1. Geographic location of sites occupied by the Deep Sea Drilling Project and the Ocean Drilling Program South of 40° Southern latitude. Sites used in this study contain both microfossils (diatoms, radiolarians and/or calcareous nannofossils) and a polarity record.

Previous work

Biochronological models for the Southern Ocean have developed since the mid 1960s and early 1970s and have accompanied the completion of numerous biostratigraphical studies during the last 30 years. The voluminous bibliographic literature available for discussion has been limited here to selected articles that focus primarily on the development of diatom and radiolarian stratigraphies and the calibration of these stratigraphies to a Geopolarity Timescale (GPTS). This focused discussion reflects the emphasis of these stratigraphies in this current study.

Prior to the DSDP and the ODP expeditions to the Southern Ocean, calibration of microfossil datums to a magnetostratigraphy was limited to stratigraphical successions obtained from relatively short piston cores (generally less than 10 m). This practice established age constraints for microfossil datums within the uppermost sedimentary sequence, i.e. approximately the last 4 million years. Age determination for older sequences was based on estimates derived from secondary and indirect correlations to a magnetostratigraphy. The introduction of DSDP and ODP cruises to the Southern Ocean, commencing with Leg 28 (Hays *et al.* 1975) and continuing through Leg 120 (Wise *et al.* 1992) has both expanded the geographical coverage and the stratigraphical extent of sequences in which direct calibration of biostratigraphical data is possible.

The current Cenozoic biochronological framework results directly from the eight DSDP and ODP Southern Ocean expeditions (DSDP Legs 28, 35, 36, 71; ODP Legs 113, 114, 119, and 120). The biochronological framework has been refined with each expedition, as additional sedimentary sequences with integrated magnetostratigraphical and biostratigraphical records were recovered. Refinement of this framework also continued with advances in species concepts, floral and faunal diversity, and an improved understanding of species biogeography. The Southern Ocean biochronological framework is based predominantly on diatom and radiolarian biostratigraphies for middle Miocene and younger sediments and diatom, radiolarian and calcareous nannofossil biostratigraphies for the lower Miocene and older sediments.

DSDP Leg 28

Deep Sea Drilling Project Leg 28 (Hays *et al.* 1975) was designed to explore the long-term glacial, climatic, biostratigraphical and geological history of Antarctica. Fourteen holes were completed at 11 sites, and Oligocene to Holocene sediments were recovered. Biostratigraphies for most microfossil groups were completed independently of obtaining a magnetostratigraphy. The results of McCollum (1975) and Chen (1975) are pertinent to this study. McCollum (1975) building on the previous work of Jousé *et al.* (1963), Donahue (1970) and Abbott (1972) established a diatom zonation (Fig. 2) consisting of 12 middle Miocene and younger zones based on the first occurrence (FO) or last occurrence (LO) of specific species. Age assignments for specific datums were based on previous calibrations of these datums to a magnetostratigraphy from piston cores (see Donahue 1970; McCollum 1975; Weaver & Gombos 1981) which was applicable for the youngest approximately 4.5 million years interval. The Pliocene and younger diatom zones established were based on the stratigraphical ranges of species endemic to the Southern Ocean. The Miocene zones were based on the stratigraphical ranges of specific species known from the North Pacific and relied on the studies by Burckle (1972), Koizumi (1972) and Schrader (1973). For example, McCollum (1975) equated the LO of *Denticulopsis lauta* (previously referred to as *Denticula lauta*) with the 'A' event of Magnetic Epoch 7 based on a similar placement by Schrader (1973) at DSDP Site 173 which is positioned in the California Current. McCollum (1975) also used the comparative ranges of Miocene diatoms, calcareous nannofossils and foraminifera zonal markers.

Chen (1975) used the previously defined radiolarian stratigraphical framework established by Hays (1965) and Hays & Opdyke (1967) based on piston core data. Zonal definitions, however, were modified by Chen (1975) to use the last occurrence rather than the last common occurrence (LCO) of specific species. In addition, Chen (1975) established eight new zones based on the stratigraphical ranges of selected species. This stratigraphy was correlated to the Geomagnetic Polarity Time Scale (GPTS) of Opdyke *et al.* (1966) and provided a framework for the youngest *c.* 4.5 million years.

Burns (1975) described the distribution of nannofossil assemblages, but did not assign these to recognized zones. This omission was deliberate and served to emphasize the sporadic occurrence of low latitude zonal markers defined by Martini (1970), Martini & Worsley (1970) and Bukry (1973*b*). Bukry (1975) used the results obtained by integrating calcareous nannofossil, silicoflagellate and occasionally diatom data to assign calcareous nannofossil assemblages from Leg 28 samples within a modified version of the low-latitude coccolith zonation which he had established for the Caribbean (Bukry 1973). Bukry attributed the paucity of sphenoliths and discoasters in the calcareous nannofossil assemblages to the reduced tolerance of these forms to cooler conditions.

Leg 35

Leg 35 (Hollister *et al.* 1976) occupied four sites in the Pacific sector of the Southern Ocean to address both the tectonic and palaeoceanographic history of the region. Biostratigraphical studies were completed for most microfossil groups, but unfortunately a palaeomagnetostratigraphy was not obtained. A diatom biostratigraphy (Fig. 2) was established by Schrader (1976), and Weaver (1976) developed a radiolarian biostratigraphy. The diatom stratigraphy developed by Schrader (1976) integrated the Pliocene and Pleistocene zonations of McCollum (1975), but established new or modified zones for the Miocene and Oligocene. Modification of the Miocene zones resulted from differences between McCollum (1975) and Schrader (1976) in the stratigraphical ranges of selected species. The lack of independent age control for the Miocene sediments required that age determination was based on secondary correlations, including known species ranges from non-Antarctic regions. For example, Schrader (1976) defined the Miocene zones based on the known ranges of *Coscinodiscus lewisianus*, *Denticula hustedtii* (now referred to as *Denticulopsis hustedtii*), *Denticula nicobarica* (now referred to as *Crucidenticula nicobarica*) and *Coscinodiscus yabei* (now referred to as *Thalassiosira yabei*) in the North Pacific.

Weaver (1976) used the radiolarian zonation defined by Chen (1975) for DSDP Leg 28. This zonation was modified slightly, based in part on the sporadic occurrence of nominated taxa. For example, Weaver (1976) noted the sporadic occurrence of *Helotholus vema* and replaced this zonal indicator with the FO of *Desmospyris spongiosa*. The absence of radiolarian datums and generally poor preservation of Miocene radiolarians prevented the development of a detailed biostratigraphy for the Leg 35 sites.

Haq (1976) described calcareous nannofossil assemblages for Leg 35 sites. Poor recovery of calcareous sediments, coupled with the poor to moderate preservation of calcareous nannofossils assemblages prohibited the assignment of these assemblages within a zonal framework.

Leg 36

Leg 36 (Barker *et al.* 1976), on the Falkland Plateau examined the gateway history of the Drake passage and the history of the Antarctic bottom water production. The six sites occupied resulted in recovery of Cretaceous through Quaternary sediments. Neogene and Palaeogene microfossil stratigraphies were completed postcruise for most major microfossil groups, with the exception of radiolarians. Gombos (1976) completed a diatom biostratigraphy and established eight late Palaeocene to early Miocene zones (Fig. 2). A palaeomagnetic stratigraphy was not available for independent age control and Gombos (1976) had to rely on previously

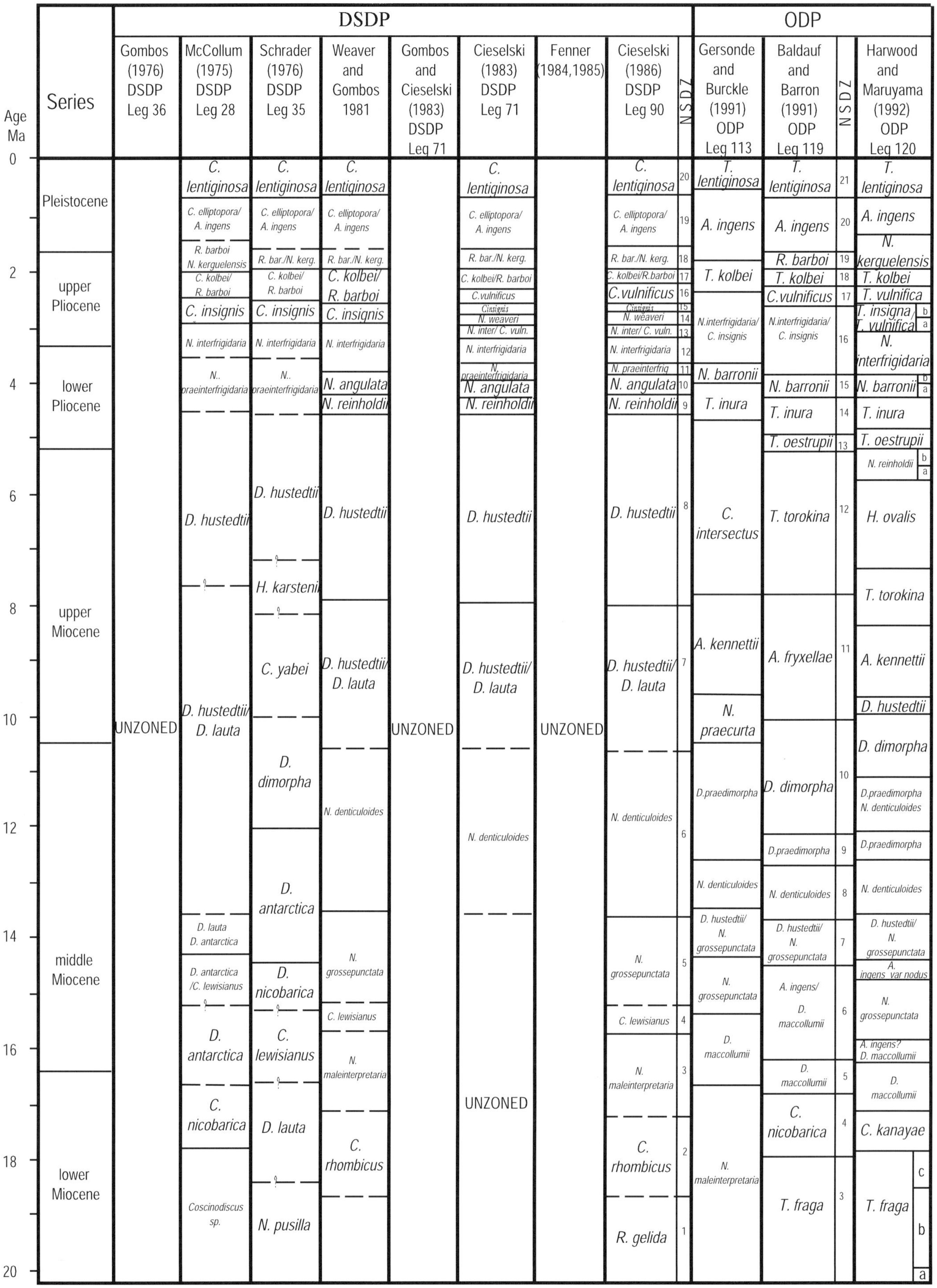

Fig. 2. Comparison of the numerous Palaeogene to Quaternary diatom zonations defined for the Southern Ocean (modified from Baldauf & Barron 1991).

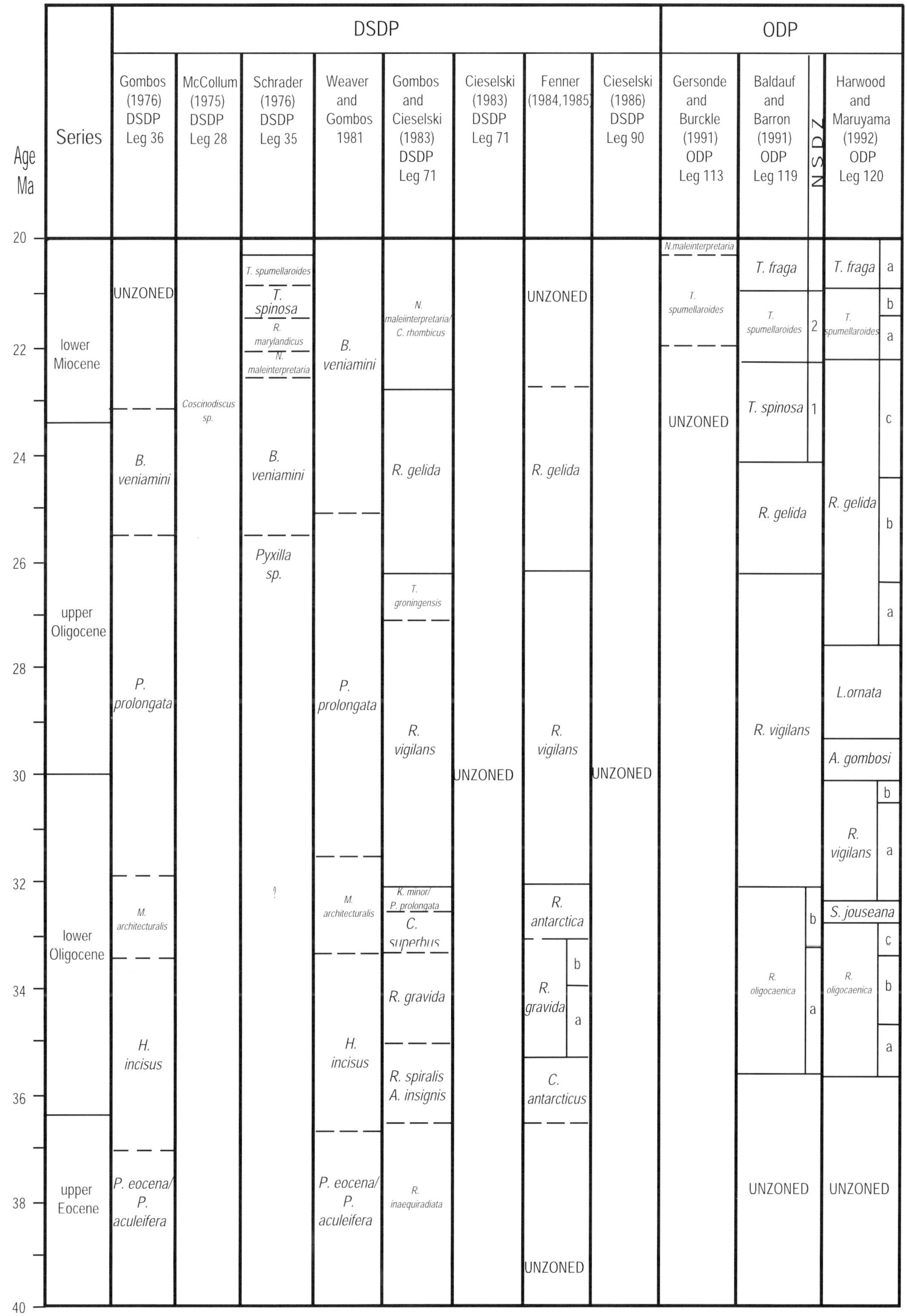

Fig. 2. (*continued*)

described ranges and secondary correlations to establish age estimates. He integrated diatom assemblages data with data derived from shipboard studies of calcareous nannofossils, radiolarians and silicoflagellate stratigraphies.

Wise & Wind (1976) used the standard low- or middle-latitude zonations of Bukry (1973a, b) for DSDP Leg 36. These workers discussed the influence of cooler climates on the geographical distribution of tropical guide species and the impact of their paucity/ absence on biostratigraphical resolution in the higher latitudes. Wise & Wind (1976) recorded the highly restricted nature of Oligocene calcareous nannofloras and the increased restriction in the diversity of Miocene assemblages. Consequently they questioned the usefulness of low-latitude calcareous nannofossil zonations for Oligocene and post-Oligocene Leg 36 material. In addition to using the cooler water form *Isthmoithus recurvus* as a high-latitude, late Eocene zonal marker Wise & Wind (1976) proposed the establishment of two new high-latitude provincial zones (*Fasciculithus involutus* Zone and *Heliolithus universus* Zone) for the late Palaeocene. These zones were considered to be of regional importance only.

The completion of DSDP Leg 36 culminated two seasons (Legs 28–29, 35–36) of Southern Ocean exploration and provided the opportunity to synthesize the established biochronological framework. Weaver & Gombos (1981), focusing on diatom biostratigraphical results, synthesized the results from the previous four cruises and re-examined sediments from DSDP Sites 266 and 328 to resolve taxonomic and stratigraphical differences between McCollum (1975) and Schrader (1976). They concluded that these differences, in part, reflected the occurrence of a hiatus that McCollum (1975) did not recognize. Based on these analyses, Weaver & Gombos (1981) established a new diatom zonation (Fig. 2) that incorporated portions of the zonation identified by McCollum (1975) for the Miocene through Pleistocene interval. Portions of these zones were modified, for example the *Nitzschia interfrigidaria* zone was changed because the last occurrence (LO) of *Nitzschia praeinterfrigidaria* was difficult to recognize due to the co-occurrence of several closely related taxa. They replaced this datum with the first occurrence (FO) of *Nitzschia interfrigidaria*. Similarly, Weaver & Gombos (1981) modified the Oligocene and early Miocene zones of Schrader (1976) based on stratigraphical ranges of diatoms from DSDP Site 328 in the Malavinas Outer Basin. Age estimates for the Miocene and older datums were based on secondary correlation or on extrapolation of average sedimentation rates at DSDP Site 278. This later method was used to derive ages for the LOs of *Denticulopsis lauta*, *Nitzschia denticuloides*, and *Nitzschia grossepunctata*.

Leg 71

Leg 71 (Ludwig *et al.* 1983) occupied two sites on the Falkland Plateau within the present-day Antarctic Convergence and two sites on the Mid-Atlantic Ridge, north of the present-day Antarctic Convergence. A good palaeomagnetic stratigraphy was obtained (Ledbetter 1983; Salloway 1983) for the last 4.0 million years, i.e. the interval represented by C4A-C5 (Opdyke *et al.* 1974) or the end of Anomaly 5a (La Breque *et al.* 1977). Identification of the polarity sequence was based primarily on the FO of the planktonic foraminifer *Globorotalia acostaesis* in DSDP Hole 512. Diatom biostratigraphies (Fig. 2) were completed by Ciesielski (1983) and Gombos & Ciesielski (1983) and a radiolarian biostratigraphy was completed by Weaver (1983).

Gombos & Ciesielski (1983) improved the Eocene to Miocene diatom biostratigraphy for the Southwest Atlantic and provided initial comparisons of the stratigraphical ranges of species between the high southern and lower latitudes. In this study, Gombos & Ciesielski (1983) identified 12 early Miocene to late Eocene zones with age estimates inferred from other microfossil biostratigraphies. Fenner (1984) proposed an alternative Palaeogene diatom zonation for the Southern Ocean in order to resolve inconsistent stratigraphical ranges between her study and that of Gombos & Ciesielski (1983). These differences reflect the different the methodologies employed. Correlation of Fenner's (1984) zonation with that of the low-latitudes is based on the first occurrences of cosmopolitan species (Fenner 1985).

Ciesielski (1983) substantially redefined the Neogene and Quaternary diatom zonation previously used by McCollum (1975) and Weaver & Gombos (1981). He established 15 diatom zones, of which five remained unchanged from previous definitions (Fig. 2). The new Pliocene zones were identified in order to improve the overall chronostratigraphical resolution, clarify previous zonal definitions or resolve inconsistent stratigraphical ranges. Ciesielski (1983) introduced several new Pliocene datums, including the LO of *Rhizosolenia barboi* (now referred to as *Simonseniella barboi*) and the LO of *Coscinodiscus vulnificus* (now referred to as *Thalassiosira vulnifica*). Age estimates were generally based on direct calibration to a magnetostratigraphy. Age estimates of datums in intervals lacking a polarity record relied on secondary correlations and age estimates derived at DSDP Site 513 (Ciesielski 1983). One exception was the use of a dated ash layer in Hole 513A which provided an age of 8.7 Ma for the LO of *Denticula lauta* (now referred to as *Denticulopsis lauta*). This age is approximately 0.8 my younger than that obtained by Koizumi (1977) for this event in the North Pacific.

The DSDP Leg 71 radiolarian biostratigraphy was completed by Weaver (1983) using a modified version of the Neogene biostratigraphy established by Chen (1975). Modification to the Pliocene portion consisted of subdividing the *Helotholus vema* zone of Chen (1975) into three zones. With the exception of the *Stichocorys peregrina* zone, datums used to identify these zones were directly calibrated to a polarity record. The age of the FO of *Stichocorys peregrina* was determined based on the relationship of this event to the first occurrence of the diatom *Thalassiosira praeconvexa* calibrated at that time to the upper reversed interval of Chonozone 6 in the low latitudes (see Weaver 1983; Haq *et al.* 1980). The late to mid-Miocene portion of Chen's (1975) zonation was modified to adjust for the inconsistent range of *Actinomma tanyacantha*. Age determinations for the Eocene and Oligocene interval are based on calibration of species ranges from low latitudes.

Wise (1983) described the calcareous nannofossil zonation for DSDP Leg 71. In this report he extends the use of provincial zones to Oligocene sequences and defines six new zones for this time interval. These are: the *Cyclicargolithus abisectus*, *Reticulofenestra bisecta*, *Chiasmolithus altus*, *Reticulofenestra daviesii*, *Clausicoccus fenestratus* and *Blackites spinosus* zones. These zonal units are correlated with the standard low- or middle-latitude Cenozoic zonations of Bukry (1973a, b) and Okada & Bukry (1980) and were erected to be used either in place of or in conjunction with the standard low-latitude zonal schemes. In addition the *Reticulofenestra oamaruensis* zone of Stradner & Edwards (1968) was recorded.

Leg 113

Ocean Drilling Program Leg 113 (Barker *et al.* 1988, 1990) commenced the second generation of exploration of the Southern Ocean and the first expedition by the ODP into the southern high latitudes. Nine sites were occupied in order to investigate the palaeoceanographical history of the Weddell Sea and the glacial history of Antarctica. Shipboard biochronology relied predominantly on diatoms and radiolarians for the Neogene and Quaternary, and calcareous microfossil stratigraphies for Miocene and older sequences. The GPTS of Berggren *et al.* (1985a, b) was used for chronostratigraphical analysis. ODP Leg 113 recovered Neogene sediment at eight sites and detailed palaeomagnetostratigraphies were recovered from Sites 689 and 690 (Spiess 1990). The recovery of a polarity record together with diatoms (Gersonde & Burckle 1990) and radiolarians (Abelmann 1990; Lazarus 1990) provided an opportunity to test and refine the previously established biochronology. Gersonde *et al.* (1990), in their synthesis of the siliceous biostratigraphy for the Leg 113 sites, focused on Holes 689B and 690B.

Gersonde & Burckle (1990) identified 16 diatom zones (Fig. 2) for the Neogene; six of these zones are modified from previous work based on the inconsistent ranges of *Simonseniella barboi, Thalassiosira elliptopora, Nitzschia interfrigidaria* and *Nitzschia reinholdii*. They replaced these datums with species such as *Thalassiosira kolbei* and *Thalassiosira inura*. Gersonde & Burckle (1990) also noted that several events such as the FO of *Thalassiosira oestrupii*, the FO of *Denticulopsis praedimorpha* and the LO of *Denticulopsis hustedtii* have similar ages between the Southern Ocean and the North Pacific. Specific events are calibrated directly to the polarity record of Spiess (1990) for the entire Neogene with the exception of the interval representing the late Miocene *Cosmiodiscus intersectus* and the late early to early mid-Miocene *Denticulopsis maccollumnii* zones. Ages within these intervals and for the early Miocene interval rely, in part, on the radiolarian biostratigraphy of Abelmann (1990).

Radiolarian biostratigraphies for Leg 113 sediments were completed by Abelmann (1990) for the Oligocene to mid-Miocene, and by Lazarus (1990) for the mid-Miocene to Pleistocene. Abelmann (1990) identified ten new zones, five of which were modified from those of Chen (1975). Modification and/or the replacement of Chen's zones was necessary to accommodate inconsistencies in stratigraphical ranges and differences in species concepts. In addition, Abelmann's (1990) zonation is based only on the first occurrence of radiolarian species, thus eliminating the inconsistent placement resulting from possible reworking of specimens in last occurrences. Datums used as zonal indicators were directly calibrated to the polarity record obtained by Spiess (1990). The mid-Miocene to Pleistocene radiolarian zonation used by Lazarus (1990) was also new and revised the zonations of Hays (1965) and Chen (1975). This new zonation also incorporated some of the datums previously defined by Weaver (1976) and Keany (1979).

The calcareous nannofossil stratigraphy for ODP Leg 113 was completed by Pospichal & Wise (1990) and Wei & Wise (1990). Pospichal & Wise were able to apply the low-latitude, number coded coccolith zonation of Okada & Bukry (1980) to interpret the stratigraphical distribution of Palaeocene to mid-Eocene calcareous nannofossil assemblages. They were, however, unable to achieve the higher biostratigraphical resolution that is obtainable by subzone recognition.

Wei & Wise (1990) described the mid-Eocene to Pleistocene calcareous nannofossil assemblages. The absence of calcareous nannofossil assemblages for most of the Neogene sequence prohibited the establishment of a formal calcareous nannofossil zonation for this interval. The calcareous nannofossil zonations proposed for the sub-Antarctic Palaeogene of New Zealand (Edwards 1971) and for the Eocene and Oligocene of the Falkland Plateau (Wise 1983) were shown to have limited value in the Weddell Sea sites. Wei & Wise (1990) recorded a succession of nine calcareous nannofossil zones from the middle Eocene to upper Oligocene sequence and correlated these with the calcareous nannofossil zones used by Okada & Bukry (1980) and Wise (1983). The recorded zones include the *Reticulofenestra umbilica* and *Reticulofenestra reticulata* Zones, which were new; the *Discoaster saipanensis* and *Chiasmolithus oamaruensis* zones of Martini (1970), which were adopted without amendment as is the *Chiasmolithus altus* Zone of Wise (1983). The *Isthmolithus recurvus* Zone (Martini 1970), *Reticulofenestra oamaruensis* Zone (Stradner & Edwards 1968) *Blackites spinosus* Zone (Wise 1983) and *Reticulofenestra daviesii* Zone (Wise 1983) were emended to eliminate the use of species with variable ranges, time transgressive ranges and uncertain taxonomy. Wei & Wise (1990) also correlated calcareous nannofossil datums (LO *Reticulofenestra umbilica*; LO and FO *Isthmolithus recurvus*; FO and LO *Reticulofenestra oamaruensis*; FO and LO *Reticulofenestra reticulata*; FO and LO *Chiasmolithus oamaruensis*; LO *Chiasmolithus solitus*) with the magnetic time scale of Berggren (1985a, b).

Leg 114

Leg 114 (Ciesielski *et al.* 1988, 1991) investigated the tectonic basement and palaeoceanographic history of the Atlantic sector of the Southern Ocean. Partial polarity records were obtained from the Miocene and Pliocene section at Site 701, the Eocene sediments at Site 702, Miocene to Eocene interval at Site 703 and the Pleistocene to lower Miocene section at Site 704 (see Hailwood & Clement 1991a, b; Clement & Hailwood 1991). These polarity sequences were correlated with the GPTS of Berggren *et al.* (1985a, b). Calibration of the Pliocene diatom biostratigraphy with the polarity record was completed by Fenner (1991). A radiolarian biostratigraphy for Leg 114 sediments was completed by Ling (1991).

Fenner (1991) completed a quantitative analysis of diatoms from Sites 699, 701 and 704 for the Pleistocene and Pliocene (Bruhnes through Gauss equivalent). Diatom age determinations were based on direct calibration to polarity records. Results from this study questioned the reliability of several previously used datums, including the LO of *Thalassiosira vulnifica*, the range of *Cosmiodiscus insignis*, and the LO of *Simonseniella barboi*. Fenner (1991) suggested that the use of the first abundant appearance (FAAD) and last abundant appearance (LAAD) of species provided more reliable datums than the true first or last occurrences. The isochroneity of FAAD and LAAD datums between Sites 699, 701 and 704, during the late Matuyama magnetic epoch, was attributed to migrations of the Polar Front Zone that affected all three sites. Fenner postulated that the diachroneity of the FAAD of *Thalassiosira kolbei* and the LAAD of *Thalassiosira vulnifica* between Sites 699 and 701 (isochronous) and 704, during the early Matuyama, resulted from varying palaeoceanographical conditions with stable antarctic conditions at Sites 699 and 701 and stable subantarctic conditions at Site 704.

Ciesielski (1991) investigated and compiled the stratigraphical ranges and relative abundances of selected diatom and silicoflagellate species for Holes 699A and 704A and 704B. His preliminary data for the Site 704 sequences revealed the completeness of the Neogene succession.

Crux (1991) used the calcareous nannofossil zonal scheme of Martini (1970) to interpret the Palaeogene to earliest Miocene succession in Leg 114. In the absence of datums used by Martini (1970), additional datums were imported from the zonal schemes of Edwards (1971), Wise & Constans (1976), Romein (1979), Varol (1981) and Wise (1983). No formal zonal scheme was proposed for the Miocene to Quaternary interval. Madile & Monechi (1991) focused on the upper Eocene to lower Oligocene sequences of Sites 699 and 703. These authors applied the zonal scheme for calcareous nannofossil assemblages proposed by Wise (1983) for the Falkland Plateau region. They assigned calcareous nannofossil nannofossil asemblages to the *Chiasmolithus oamaruensis* and *Isthmolithus recurvus* Zones.

Wei (1991) described and compared the calcareous nannofossil magnetobiochronology for Sites 699A and 703A and compared these data with results from South Atlantic mid-latitude (Leg 72, Site 516) and other Southern Ocean (Leg 113, Sites 689 and 690; Leg 119, Site 744; Leg 120, Site 748) sites. He recognized the LO of *Reticulofenestra bisecta*, the LO of *Chiasmolithus altus* and the LO of *Isthmolithus recurvus* as consistent, reliable biostratigraphical markers from low to high latitudes. The LOs of *R. bisecta* and *C. altus* are influenced by the presence of hiatuses in the extremely high-latitude Sites 689, 690 and 744 and are therefore unreliable. The LO of *Reticulofenestra oamaruensis* is recorded as isochronous between four Southern Ocean Sites (689, 699, 744 and 748). Wei (1991) recorded an age of 32.9 Ma, i.e. 2.9 million years younger than the age assignment of Berggren *et al.* (1985a) based on mid-latitude data.

Leg 119

Ocean Drilling Program Leg 119 (Barron *et al.* 1989, 1991) addressed the palaeoceanographic history of the Indian Ocean sector of the Southern Ocean, the glacial history of Antarctica, and the tectonic history of the Kerguelen plateau. Eleven sites were occupied and recovered excellent Neogene siliceous microfossils and older siliceous and calcareous microfossils. The high-quality

Table 3. *Calculated ages (Ma) for previously used biostratigraphical events using a calibration to the GPTS of Cande & Kent (1995)*

Event	514t	514b	689Bt	689Bb	690Bt	690Bb	695At	695Ab	696At	696Ab	697Bt	697Bb	699At	699Ab	701At	701Ab
LO *Pterocanium trilobium*																
LO *Triceraspyris antarctica*																
LO *Pseudoemiliana lacunosa*																
FO *Emiliana huxleyii*																
LO *Thalassiosira elliptopora*	0.72	0.84											0.55	0.66		
LO *Hemidiscus karstenii*													0.00	0.03	0.23	0.26
LO *Simonseniella barboi*	1.82	1.83			2.93	3.09					3.86	4.05	1.68	1.92	1.48	1.58
FO *Nitzschia miocenica*																
FO *Nitzschia kerguelensis*	2.40	2.42			?	?	1.17	1.70	5.13	5.65						
LO *Actinocyclus ingens*	0.59	0.72	?	?	?	?	0.20	0.64	?	?	0.79	1.10	0.55	0.66	0.81	0.83
LO *Thalassiosira kolbei*	**1.92**	**2.25**	?	?	?	?					**1.74**	**1.95**	**1.93**	**2.10**	**1.73**	**1.98**
LO *Thalassiosira inura*			?	?	?	?	3.20	3.30	?	?	2.89	2.94				
LO *Thalassiosira vulnifica*	**2.37**	**2.40**					**0.64**	**1.95**			**2.17**	**2.40**	**2.43**	**2.49**	**2.12**	**2.26**
LO *Thalassiosira insigna*	3.20	3.26					2.11	2.77			2.94	3.08	2.68	2.72	2.47	?
LO *Nitzschia interfrigidaria*	2.91	2.93	?	?	?	?	2.89	3.11			3.08	3.11	2.72	2.82	2.47	?
LO *Denticulopsis dimorpha*													?	?		
LO *Nitzschia barronii*			?	?	?	?	3.11	3.12	?	?	1.10	1.25				
LO *Helicosphaera sellii*																
LO *Discoaster browerii*																
FO *Thalassiosira elliptopora*													?	?		
LO *Rouxia diploneides*																
LO *Nitzschia cylindrica*																
FO *Actinocyclus actinochilus*																
LO *Antarctissa cylindrica*			?	?	?	?	0.00	2.27								
LO *Cycladophora pliocenica*			?	?	?	?	0.00	2.27								
LO *Eucyrtidium calvertense*			?	?	?	?										
LO *Desmospyris spongiosa*	2.42	2.50	?	?	?	?	2.11	2.89			2.39	2.82	2.44	3.26	2.12	2.46
LO *Helotholus vema*	2.42	2.50	?	?	?	?	2.11	2.89			2.39	2.82	2.44	3.26	2.12	2.46
LO *Cycladophora davisiana*	2.59	2.63	?	?	?	?	2.11	2.89			3.08	3.34			2.12	2.46
LO *Nitzschia weaveri*	2.74	2.76					3.11	3.12					2.72	2.89	?	?
LO *Nitzschia reinholdii*	2.91	2.93											2.53	2.64		
LO *Cosmiodiscus intersectus*							4.80	4.89								
LO *Rouxia heteropolara*																
LO *Thalassiosira torokina*																
LO *Rouxia californica*																
LO *Stephanopyxis grunowii*																
LO *Nitzschia praeinterfrigidaria*			?	?	3.09	?	2.77	2.92	4.52	4.64	3.50	3.66	3.43	3.57		
FO *Stephanopyxis grunowii*																
FO *Thalassiosira lentiginosa*	2.91	2.93			?	?					3.40	3.49	?	?		
FO *Thalassiosira kolbei*	3.24	3.36	4.69	4.75							2.61	2.82				
FO *Thalassiosira inura*			4.84	4.89	4.76	4.81			6.18	6.77	?	?				
FO *Thalassiosira vulnifica*							2.89	3.11			2.61	2.82				
FO *Nitzschia praeinterfrigidaria*	**3.24**	**3.36**	**6.13**	**6.54**	**4.82**	**4.85**	**4.83**	**4.86**	**4.76**				**5.25**	**5.52**		
FO *Navicula wisei*																
FO *Nitzschia weaveri*	3.37	3.44					4.54	4.60					3.43	3.70		
LO *Lampromitra coronata*																
FO *Nitzschia interfrigidaria*	?	?	?	?	?	?	**3.30**	**3.71**			**4.05**	**4.35**	**3.82**	**4.10**		
LO *Prunopyle titan*			4.27	4.34	?	?	4.60	4.68					3.26	4.64		
LO *Nitzschia fossilis*																
LO *Rhizosolenia miocenica*																
FO *Helotholus vema*			**4.44**	**4.49**	?	?										
FO *Nitzschia barronii*			**4.47**	**4.55**	**4.52**	**4.63**	**4.42**	**4.48**	**4.51**	**4.52**	?	?				
LO *Thalassiosira complicata*									4.52	4.52	3.52	3.53				
LO *Denticulopsis hustedtii*							4.60	4.68	?	?	2.61	2.82	4.71	4.92		
FO *Thalassiosira complicata*																
LO *Neobrunia miraibilis*																
FO *Nitzschia praecurta*									?	?	?	?				
FO *Nitzschia angulata*													4.64	4.64		
FO *Thalassiosira oestrupii*			?	?					?	?	?	?				
LO *Thalassiosira burckliana*																
LO *Thalassiosira miocenica*																
LO *Hemidiscus cuneiformis*													4.98	5.25		
LO *Amphymenium challengerae*																
LO *Stichocorys peregrina*																
LO *Nitzschia marina*																

704Bt	704Bb	737At	737Ab	744At	744Ab	744Bt	744Bb	745Bt	745Bb	746At	746Ab	747At	747Ab	747Bt	747Bb	748Bt	748Bb	751At	751Ab
								0.00	0.09			0.84 ?	0.95 ?			?	?	?	?
0.00	0.15																		
						0.11	1.76	0.79	1.27			0.38 0.85	1.70 0.95	0.75	0.80			?	?
0.00	0.07	3.81	3.84			1.76	2.32	0.26	0.44	7.99	8.01			?	?	?	?		
2.12	2.13	3.68	3.76			2.32	3.46	1.57	1.75			1.80	?			?	?	?	?
										6.70	6.74								
		?	?			1.76	2.32	2.12	2.18			?	?	?	?			?	?
0.58	0.64	7.17	7.35			0.11	1.76	0.44	0.61	8.27	8.32	0.37	0.53	0.67	0.75				
2.30	**2.34**					**1.76**	**2.32**	**1.75**	**1.97**			?	?	?	?	?	?	?	?
												0.69	0.85	?	?	?	?	?	?
3.00	**3.13**					**1.76**	**2.32**	**2.23**	**2.33**			?	?	?	?	?	?	?	?
4.42	4.57	3.28	3.60			1.76	2.32	2.18	2.23			?	?	?	?	?	?	?	?
		?	?			1.76	2.32	1.97	2.12			?	?	?	?	?	?	?	?
9.05	9.37					1.76	2.32	0.61	0.79	6.60	6.65	?	?	?	?	?	?	?	?
												1.17	1.32	3.60	?	?	?	?	?
2.47	2.65																		
2.96	3.13																		
						2.32	3.46	1.57	1.75			1.32	1.48	?	?	?	?	?	?
												?	?						
								2.23	2.33										
												?	?	?	?			?	?
								0.61	0.65			0.84	0.95					?	?
						1.33	1.69	1.57	1.58			1.64	1.80			?	?	?	?
						1.76	3.46	0.61	0.62			?	?			?	?	?	?
		3.24	3.25			1.76	3.46	2.50	2.53			?	?			?	?	?	?
						1.76	3.46	2.48	2.50			?	?			?	?	?	?
								2.66	2.72			?	?			?	?	?	?
		?	?			3.46	?	1.75	1.97			?	?					?	?
4.42	4.57							2.71	2.83			?	?						
4.52	4.57																		
								2.18	2.23										
								2.12	2.18										
								2.33	2.43										
								2.71	2.83										
		?	?			2.32	3.46	2.71	2.83			?	?	3.30	?	?	?	?	?
		4.08	4.13			1.76	2.32					?	?	?	?				
		3.12	3.23			2.32	3.46	2.90	2.96			?	?	3.30	?	?	?	?	?
		4.72	4.77			2.32	3.46	4.80	4.89			5.89	6.14	4.17	?	?	?	?	?
3.31	3.32					2.32	3.46	2.71	2.83			?	?			?	?	?	?
		4.70	**4.72**			?	?	**4.72**	**4.87**			?	?	?	?	?	?	?	?
3.48	3.66	3.32	3.60			?	?	3.06	3.13			?	?	0.75	0.80			?	?
												?	?	3.30	?	?	?	?	?
		3.98	**4.03**			?	?	**3.71**	**3.80**			?	?	3.30	?	?	?	?	?
						1.76	3.46	3.44	3.52			?	?			?	?	?	?
4.57	4.63							3.95	4.08										
								3.95	4.08										
		4.32	**4.66**					**4.57**	**4.60**			?	?	3.75	?	?	?		
		4.32	**4.57**			?	?	**4.22**	**4.27**			?	?	4.17	?	?	?	?	?
4.18	5.42																		
												?	?			?	?		
												?	?			?	?		
												?	?					6.49	6.65
4.05	4.18																		
5.99	6.04	5.01	5.08			3.46	?	5.38	5.56			?	?	?	?	?	?	?	?
		6.93	7.08							6.81	6.86								
4.18	4.25	5.41	5.12			?	?	?	?					?	?			6.47	6.50
4.18	4.25							1.57	1.75										
5.42	6.42																		
5.42	6.42	5.07	5.12																
4.57	4.63																		

(*continued*)

Table 3. (*continued*)

Event	514t	514b	689Bt	689Bb	690Bt	690Bb	695At	695Ab	696At	696Ab	697Bt	697Bb	699At	699Ab	701At	701Ab
LO *Thalassiosira praeconvexa*																
LO *Thalassiosira convexa v. aspinosa*																
LO *Lamprocyclas aegles*																
FO *Thalassiosira praeconvexa*																
LO *Nitzschia porteri*																
FO *Lamprocyclas aegles*																
LO *Diartus hughesi*																
FO *Stichocorys peregrina*																
FO *Nitzschia reinholdii*			4.67	4.69										?	?	
FO *Thalassiosira insigna*							3.13	3.16								
FO *Nitzschia marina*																
LO *Denticulopsis lauta*													?	?		
FO *Navicula wisei*													?	?		
FO *Thalassiosira inura*																
FO *Rouxia heteropolara*									6.18	6.77	?	?				
LO *Cycladophora spongothorax*			8.28	8.51	4.88	?										
LO *Hemiaulus triangularis*																
FO *Simonseniella barboi*													?	?		
LO *Raphidodiscus marylandicus*																
FO *Cestodiscus peplum*																
LO *Thalassiosira convexa*																
FO *Thalassiosira convexa v. aspinosa*																
LO *Trinacria excavata*																
FO *Hemidiscus karstenii*													?	?		
LO *Asteromphalus kennettii*			7.02	7.68	?	?										
LO *Reticulofenestra pseudoumbilica*																
LO *Amphymenium challengeri*																
LO *Thalassiosira burckliana*																
LO *Hemidiscus ovalis*																
LO *Actinocyclus ingens v. nodus*			7.68	8.07												
LO *Acrosphaera australis*																
FO *Cosmiodiscus intersectus*			8.23	8.32	?	?					?	?				
FO *Acrosphaera labrata*			8.28	8.51												
LO *Reticulofenestra gelida*			8.88	9.66	12.12	12.51										
FO *Acrosphaera australis*			10.02	10.13												
LO *Calcidiscus macintyrei*																
FO *Thalassiosira torokina*													?	?		
LO *Cycladophora humerus*			10.02	10.13	4.89	4.98										
FO *Eucyrtidium pseudoinflatum*			10.31	10.39	10.00	10.58										
FO *Nitzschia fossilis*																
LO *Diartus hughesi*																
LO *Hemidiscus cuneiformis*	2.60	2.62											?	?		
LO *Thalassiosira miocenica*																
LO *Denticulopsis dimorpha*			4.34	4.38	4.63	4.68										
LO *Actinomma golownini*			11.86	11.87	10.00	10.58										
FO *Asteromphalus kennettii*			10.10	10.20	10.17	10.29										
FO *Actinocyclus fryxellae*																
LO *Cyrtocapsella japonica*																
LO *Denticulopsis praedimorpha*			12.51	12.76	10.88	?										
LO *Nitzschia denticuloides*			11.88	11.92	?	?							?	?		
LO *Reticulofenestra hesslandii*			11.91	11.93	12.12	12.51										
FO *Reticulofenestra hesslandii*																
FO *Cyrtocapsella japonica*																
FO *Denticulopsis dimorpha*			11.91	11.92	12.72	12.88										
LO *Dendrospyris megalocephalis*			11.88	11.92												
FO *Cycladophora spongothorax*			11.90	11.91	12.12	12.93										
LO *Nitzschia grossepunctata*			12.15	12.51	13.68	14.00										
FO *Denticulopsis praedimorpha*			**12.51**	**12.76**	**13.09**	**13.25**										
LO *Crucidenticula nicobarica*			11.95	12.15	12.51	12.72										
FO *Actinomma golownini*			13.06	13.28	13.41	13.82										
LO *Nitzschia denticuloides*			**13.43**	**13.54**	**13.68**	**14.00**							?	?		
FO *Antarctissa deflandrei*			13.38	13.44	13.75	14.51										
LO *Denticulopsis maccollumnii*			13.99	14.24	14.24	14.44										
FO *Denticulopsis hustedtii*			**13.99**	**14.24**	**14.24**	**14.44**					?	?				
LO *Cycladophora humerus*			14.13	?	13.75	14.51										

704Bt	704Bb	737At	737Ab	744At	744Ab	744Bt	744Bb	745Bt	745Bb	746At	746Ab	747At	747Ab	747Bt	747Bb	748Bt	748Bb	751At	751Ab
5.69	5.92					?	?												
6.43	6.69																		
6.70	6.78																		
6.97	7.01									6.60	6.65								
8.04	8.12																		
8.10	8.25																		
8.39	8.46																		
8.39	8.46																		
8.79	8.81	?	?			?	?							5.19	5.49	?	?	6.49	6.65
8.79	8.81							2.90	2.96					3.30	?	?	?	?	?
8.79	8.81									8.12	8.18								
7.12	7.31					10.76	10.88									?	?	?	?
		6.36	6.41					3.95	4.08			?	?			?	?	?	?
										9.11	9.13	?	?	5.38	6.45			9.23	9.30
												?	?						
16.77	?																		
?	?			20.69	?														
								5.98	6.09										
								6.09	6.18										
								6.09	6.18										
										?	?								
						?	?			9.17	?	?	?	5.49	?	?	?	6.42	6.47
3.83	4.00					10.33	10.45					16.51	16.75			18.02	18.29	10.14	10.21
6.70	6.78																		
										9.17	?								
												?	?			11.51	11.54	6.49	6.65
11.90	12.18					?	?			6.60	6.65	?	?					12.17	?
						10.41	10.65												
8.79	8.81	?	?							8.44	8.48								
												6.20	?	5.19	5.38			?	?
														10.73	10.93	?	?	10.54	10.61
2.96	3.13					10.45	10.56					9.68	9.85			?	?		
		?	?			?	?			8.38	8.39	6.14	?	9.77	9.87	?	?	?	?
				8.23	8.27					8.23	8.27					?	?	10.79	10.84
						16.45	17.20			?	?	?	?	10.82	10.93	?	?	10.79	10.84
9.16	9.25																		
9.25	9.96																		
10.05	10.15					10.52	10.64					?	?						
10.15	10.30	6.57	6.68			?	?			?	?			5.19	5.49			6.49	6.65
				?	?	2.36	3.46			6.60	6.65			?	?	?	?	?	?
						10.76	11.12					?	?	10.53	10.62			10.90	10.97
10.51	10.60					10.13	10.24			?	?			8.88	9.77	?	?	9.63	9.69
		?	?			10.99	11.05												
10.57	10.62																		
						11.38	11.44					?	?	?	?	11.33	11.40	11.72	11.84
						?	?					?	?			?	?	11.83	11.89
						17.55	17.92					?	?					6.49	6.55
												?	16.73						
11.30	11.76																		
11.90	12.18					?	?	0.79	1.27			?	?			?	?	12.17	?
						?	?												
						11.32	?					?	?	?	?	?	?	12.03	12.08
14.16	15.05					?	?					?	?			?	?	12.78	?
						?	?					?	?			?	?	12.78	?
12.18	?																		
						?	?					?	?			?	?	13.66	14.40
15.06	**15.13**					**15.21**	**15.55**					**13.41**	**13.60**			?	?	**13.67**	**13.84**
?	15.05					?	?					14.31	14.49			?	?	?	?
?	?	?	?			?	?					**14.13**	**14.31**			?	?	**14.08**	**14.15**
						?	?			?	?								

(continued)

Table 3. (*continued*)

Event	514t	514b	689Bt	689Bb	690Bt	690Bb	695At	695Ab	696At	696Ab	697Bt	697Bb	699At	699Ab	701At	701Ab
FO *Dendrospyris megalocephalis*			14.13	?												
LO *Synedra jouseana*			?	?	14.51	14.74										
FO *Actinocyclus ingens v. nodus*			?	?	14.51	14.74										
LO *Coscinodiscus lewisianus*			15.07	15.11												
FO *Simonseniella barboi*																
LO *Coscinodiscus rhombicus*			15.11	15.16												
FO *Eucyrtidium punctatum*			15.16	?												
LO *Nitzschia maleinterpretaria*			?	?	14.51	14.74										
LO *Crucidenticula kanayae*			?	?	?	?										
FO *Denticulopsis lauta*																
FO *Nitzschia grossepunctata*			?	?	?	?										
FO *Nitzschia maleinterpretaria*					?	?										
LO *Thalassiosira fraga*			?	?	?	?										
FO *Calcidiscus macintyrei*																
FO *Actinocyclus ingens*	2.35	2.47	?	?	?	?							3.21	3.34		
FO *Denticulopsis maccollumnii*			?	?	?	?										
FO *Calcidiscus leptoporus*																
FO *Crucidenticula kanayae*			?	?	?	?										
FO *Cycladophora golli regipileus*			18.01	18.15	?	?										
FO *Crucidenticula nicobarica*			18.18	18.49	?	?										
FO *Cyrtocapsella tetrapera*			18.02	18.31	?	?										
FO *Cyrtocapsella longithorax*			18.86	19.41	?	?										
LO *Thalassiosira aspinosa*					?	?										
FO *Coscinodiscus lewisianus*																
LO *Thalassiosira spumellaroides*			24.91	25.05	?	?										
FO *Raphidodiscus marylandicus*			19.51	19.74	?	?										
FO *Thalassiosira fraga*			19.51	19.74	?	?										
LO *Chiasmolithus altus*			19.76	20.29	?	?									26.54	26.80
LO *Reticulofenestra bisecta*			19.76	20.29	?	?									24.22	24.30
FO *Thalassiosira aspinosa*					?	?										
LO *Rocella gelida*			20.05	?	?	?									23.49	23.51
LO *Rocella schraderi*																
LO *Rocella vigilans*			24.91	25.05	?	?										
LO *Bogorovia veniamini*															?	?
LO *Lisitzina ornata*					?	?										
FO *Thalassiosira spumellaroides*			?	24.91	?	?										
FO *Rocella vigilans*			26.07	26.13	25.93	26.01										
FO *Rocella gelida*			26.07	26.13	25.93	26.01									25.33	25.71
LO *Thalassiosira primalabiata*																
FO *Lisitzina ornata*					26.34	26.43										
LO *Reticulofenestra umbilica*			30.74	31.03	?	?									31.75	31.87
LO *Isthmolitus recurvus*			31.83	32.35	?	?									32.31	32.38
LO *Reticulofenestra oamaruensis*			33.58	33.78											33.48	?
FO *Reticulofenestra oamaruensis*			34.55	34.59												
FO *Isthmolitus recurvus*			34.83	34.88	36.36	36.39										
FO *Cyclicargolithus abisectus*					30.75	?									30.99	31.49
LO *Reticulofenestra reticulata*																
LO *Rhizosolenia oligocenica*																
LO *Synedra jouseana*																

Emboldened events are primary stratigraphical indicators based on this study. ? indicates the occurrence of undated specific biostratigraphical events. Dating these events was prohibited either by the absence of magnetic data and/or the presence of an unconformity.

palaeomagnetostratigraphical record recovered from Sites 744–746, represents a nearly continuous sequence for the last 39 million years (see Keating & Sakai 1991; Sakai & Keating 1991). Barron *et al.* (1991) summarize the biochronological and magnetostratigraphical results correlated with the GPTS of Berggren *et al.* (1985*a*, *b*). The Leg 119 diatom biostratigraphy was completed by Baldauf & Barron (1991) and the radiolarian biostratigraphy was completed by Caulet (1991). Baldauf & Barron (1991) identified 24 diatom zones (Fig. 2). The Pliocene and Quaternary portion of this zonal scheme integrates the previous zones of Ciesielski (1983) and Gersonde & Burckle (1990). The Miocene portion is modified from the zonation of Gersonde & Burckle (1990). The Oligocene zones are new. In this study, Baldauf & Barron (1991) questioned the reliability of *Asteromphalus kennettii* as a stratigraphical marker. These workers considered that the valves of this species were prone to dissolution and that its low preservation potential would impact the reliable stratigraphical placement of the FO and LO datums for this species. Baldauf & Barron (1991) elected to use the FO of *Denticulopsis dimorpha* rather than the FO of *Asteromphalus kennettii* as a zonal marker.

The radiolarian biostratigraphy for Leg 119 (Caulet 1991) incorporated the zones of Chen (1975), Weaver (1983) and Caulet (1982, 1985 and 1986). Although this zonation differs from zonal schemes previously used in the Southern Ocean, the principal

704Bt	704Bb	737At	737Ab	744At	744Ab	744Bt	744Bb	745Bt	745Bb	746At	746Ab	747At	747Ab	747Bt	747Bb	748Bt	748Bb	751At	751Ab
						?	?									?	?		
						?	?					14.49	14.80					14.52	14.59
14.16	15.05			15.55	15.89					8.07	8.12	14.49	14.68					14.29	14.36
?	15.05			1.76	2.32							?	?			?	?		
11.90	11.99																		
17.09	17.16																		
						16.40	16.80									?	?	16.42	16.53
12.18	?											14.56	14.80			11.51	11.54	16.53	16.56
?	?					16.95	17.02					?	?			11.51	11.54	?	?
?	15.05					16.95	17.02					?	?			?	?	?	?
18.68	19.28			25.98	26.22											?	?	17.64	17.94
?	?					16.80	16.95					17.97	18.38			?	?	?	?
?	16.84					?	?					17.17	17.38			11.51	11.54	16.03	16.53
?	15.05					15.89	16.40	2.23	2.33			?	?			?	?	?	?
16.77	16.85					16.95	17.02					16.74	17.14			?	?	16.53	16.66
						?	?					17.52	17.68			?	?	17.58	17.67
17.03	**17.09**					**17.49**	**17.69**					**17.43**	**17.51**			?	?	**17.64**	**17.94**
						17.20	?												
17.51	17.6					17.49	17.69											14.41	14.45
11.89	12.04																		
				19.61	19.78														
17.17	17.21																		
17.21	17.51																		
20.05	20.45			19.61	19.78											18.85	?		
				20.69	?	17.02	17.18					21.68	22.19						
20.75	21.54			19.85	20.02							20.33	20.57			?	?		
				26.04	26.11							25.96	25.97			25.93	?		
				25.93	25.76							23.80	?			?	?		
19.86	20.05																		
18.68	19.28			20.55	20.70							22.68	23.50						
18.68	19.28			20.01	20.24														
18.68	19.28							5.38	5.56			?	?			?	?		
20.75	21.54			20.69	?							20.79	21.26			?	?		
				25.80	25.86							23.41	23.78			?	?		
				20.55	20.70											?	?		
				30.69	30.83			?	?			26.53	?			30.29	30.52		
?	?			26.23	26.38							25.89	25.92			?	?		
				25.53	25.60							23.78	24.14			?	?		
				26.29	26.44							26.53	?						
				31.21	31.38														
				31.72	31.93											32.11	32.94		
				33.64	33.75											33.74	33.76		
				35.86	35.92														
				36.03	**36.15**														
				31.39	31.50											30.89	31.25		
				36.03	**36.15**														
				33.56	33.66														
				30.27	30.56														

stratigraphical datums recognized in the Leg 119 sediments can be correlated to the earlier stratigraphies including those of Abelmann (1992) and Lazarus (1992). Pliocene and Pleistocene radiolarian datums were directly calibrated to the polarity records obtained from Sites 744–746.

The upper Cretaceous and Cenozoic calcareous nannofossil biostratigraphy was completed by Wei & Thierstein (1991). These authors record the Neogene calcareous nannofossil zonation as unworkable. Zonal assignments for the middle Eocene to Oligocene sequence were based on the standard low-latitude zonation of Okada & Bukry (1980) including marker species that are recorded in the zonation of Martini (1970). In addition the zonation adopted for the Weddell Sea (Leg 113) was modified to accommodate the common occurrence of some cool water marker species and the absence of some marker species in the zonations of (Martini 1970) and Okada & Bukry (1981).

Leg 120

Leg 120 (Wise et al. 1989, 1992) returned to the Kerguelen Plateau region to improve the understanding of the tectonic history of the plateau. Secondary objectives were to obtain continuous

PREVIOUS WORK

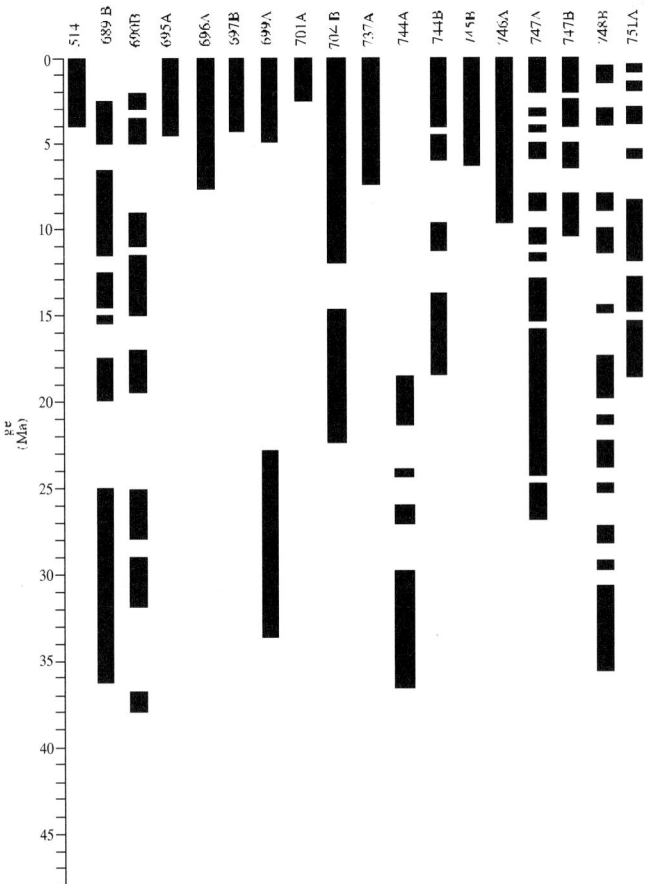

Fig. 3. Chronostratigraphical assignment of sediments recovered from the DSDP and ODP holes based on previous work (Table 2). The age-model used for each hole is based on the shipboard and subsequent shorebased investigations (see text for discussion). Solid bars represent intervals containing diatoms, radiolarians, or calcareous nannofossils and a polarity record.

stratigraphical sequences for biostratigraphical and palaeoceanographic analyses. This leg cored 12 holes at five sites and recovered an excellent polarity sequence from Sites 747, 748 and 751 (Heider et al. 1992; Inokuchi & Heider 1992) The biochronological results for the leg are discussed in Harwood et al. (1992) using the GPTS of Berggren et al. (1985a,b). The diatom biostratigraphy was completed by Harwood & Maruyama (1992) and the radiolarian stratigraphy by Takemura (1992), Abelmann (1992) and Lazarus (1992). The involvement of the same radiolarian workers with Legs 113, 119 and 120 minimized the potential for applying different species concepts.

Harwood & Maruyama (1992) defined 35 Oligocene to Pleistocene diatom zones (Fig. 2) based on the comparison of Leg 120 results with previous stratigraphies of Weaver & Gombos (1981), Ciesielski (1983, 1986), Gersonde & Burckle (1990) and Baldauf & Barron (1991). Diatom datums were calibrated directly to the polarity sequence of Heider et al. (1992) and Inokuchi & Heider (1992) for the last 31 million years (see also Harwood et al. 1992). The Pleistocene to mid-Miocene portion of the diatom zonation represented an integration between the zonations of Gersonde & Burckle (1990) and Baldauf & Barron (1991). The late Miocene diatom zones were newly defined and the early Miocene and Oligocene zones reflected a modified version of that presented in Baldauf & Barron (1991). Harwood & Maruyama (1992) expressed concern with several taxa based on their diachronous occurrences (*Thalassiosira vulnifica*, *Thalassiosira torokina*) and taxonomic uncertainties (*Thalassiosira inura*, *Thalassiosira jacksonii*, and *Thalassiosira gracilis*).

The Leg 120 radiolarian studies include an analysis of the Palaeogene (Takemura 1992), Miocene (Abelmann 1992) and Neogene (Lazarus 1992). Takemura (1992) elected not to use the previous late Oligocene zonations of Abelmann (1990) based on the taxonomic uncertainty of the *Stylosphaera* group and stratigraphical uncertainties in the previously recorded species ranges. Instead, Takemura (1992) established three new zones for the upper Eocene and Oligocene sediments. Abelmann (1992) and Lazarus (1992) used previously established Leg 113 zonations for the Neogene sediments. Modifications to these zones including both redefinition of zonal boundaries, replacing zones or re-calibrating datums to the GPTS, were completed based on comparisons of Leg 113 and Leg 120 data. Lazarus (1990) made minor modifications to his previous zones and established three new zones. These modifications were completed based on the rare occurrence of the nominated taxa (e.g. *Acrosphaera australis*). The *A. australis* Zone of Lazarus (1990) is subdivided into three zones (*Amphymenium challengerii*, *Siphonosphaera vesuvius* and *Acrosphaera australis*).

The stratigraphy of the Palaeogene calcareous nannofossils was described by Aubry (1991). Calcareous nannofossil assemblages were assigned within Martini's (1970) standard zonation scheme, which was chosen in preference to Bukry's biozonation (1973a, 1975) for its supposed broader geographical applicability.

Wei & Wise (1992a) considered the Oligocene to Quaternary calcareous nanoplankton assemblages. They were unable to assign Neogene assemblages to a zonal scheme, but suggested that non-traditional datums, which include the LO of *Reticulofenestra gelida*, the LO of *Reticulofenestra hesslandii* and the FO of *Calcidiscus leptoporus/C. macintyrei*, were potentially useful datums in Leg 120 sites. Neogene Zones (defined tentatively) are correlated with the zonation of Okada & Bukry (1980). Wei & Wise (1992a) applied the high-latitude zonation of Wei & Thierstein (1991) to interpret the upper Oligocene sequence. Wei & Wise (1992b) correlated selected calcareous nannofossil datums (LO and FO *Reticulofenestra perplexa*; LO *Cyclicargolithus floridanus*; and FO *C. leptoporus/ C. macintyrei*) from Legs 113, 119 and 120 sites with the palaeomagnetic time scale of Berggren et al. (1985a).

Wei et al. (1992) investigated, in detail, the relationship between the Eocene to Oligocene calcareous nannofossil datums at Site 748B and the preliminary magnetostratigraphy of Schlich et al. (1989). This investigation resulted in a reinterpretation of the preliminary magnetostratigraphy which deviates significantly from the magnetostratigraphy presented by Harwood et al. (1992)

Wei (1992) addressed the problem of using numerical ages, derived from calibrating calcareous nannofossil and planktonic foraminiferal biostratigraphical datums with magnetostratigraphy at mid-latitudes, to interpret the magnetostratigraphy and construct age–depth curves for Southern Ocean sites. Many of these microfossil datums are shown to be diachronous from mid- to high latitudes. Wei (1992) used direct biomagnetostratigraphical correlations for Southern Ocean sites to eliminate the affects of diachroneity and provided an update on the Palaeogene chronology of eight ODP sites (690, 699, 700, 702 703, 737, 738 and 748) in the Southern Ocean. The published magnetostratigraphies for some intervals from Sites 690, 699, 700, and 703 are reinterpreted.

The above discussion outlines the evolution of the Southern Ocean siliceous (diatom and radiolarian) and calcareous nannofossil biochronology and provides an understanding of the limitations of the current biochronological model. Age constraints (standardized to the GPTS of Cande & Kent 1995) of the datums used by each worker to construct this model are shown in Table 3. Of the 59 sites drilled south of 40° south, 18 holes at 16 sites contain sediment with both a polarity record and siliceous microfossils. Nine of these holes contain calcareous nannofossils. The chronostratigraphical assignment of sedimentary sequences, used in this study, based on previously published age models is shown in Fig. 3. The Pliocene–Pleistocene interval is represented at all sites and eight holes contain a continuous sequence for the last 5 million years; three holes have a continuous sequence for the last 10 million years. The interval from 10 to 20 Ma is recorded in the sediments of 11 sites and only seven sites contain sediment representing the interval older than 20 Ma. Figure 3 also illustrates the discontinuous nature and limited geographic coverage of the currently available data used to develop the biochronological framework presented here.

The development of the current biochronological model benefited from the improved core recovery achieved by the ODP. The limitations inherent in the model include:

(1) the geographically and temporally limited nature of continuous stratigraphical sections containing both well preserved biota and a polarity record;
(2) differences in taxonomic concepts used by different workers;
(3) the lack of a consistent biostratigraphical framework for each microfossil group that is uniformly applied to all Southern Ocean sites (see Fig. 2 as an example);
(4) inconsistencies in chronozonal nomenclature (see discussion below);
(5) the use of qualitative versus quantitative data in the definition of datums (e.g. the last occurrence compared with the last common occurrence of a species);
(6) problems in defining species ranges introduces uncertainty concerning the reliability of datums used to interpret the polarity record.

The advances made in the Southern Ocean biochronology during the last 30 years are significant and reflect the ever increasing data sets and refinement in stratigraphical resolution. The requirement to evaluate the current biochronological framework has been noted for sometime with initial efforts being completed by Thierstein *et al.* (1995) for the Neogene world ocean and by Wei (1992) and Wei *et al.* (1992) for the Palaeogene Southern Ocean. Thierstein *et al.* (1995) integrated published biostratigraphical data for Neogene intervals from 94 DSDP and ODP sites. Data sets obtained for diatoms, radiolarians, calcareous nannofossils and planktonic foraminifers were reviewed and compared with palaeomagnetic data sets correlated to the GPTS of Berggren *et al.* (1985*a,b*). The age model derived for each hole generally assumed that the original assignment of the polarity reversals to specific chronozones was accurate and the original assignment of palaeomagnetic reversals was rarely reassessed. The global approach employed focused on numerous microfossil assemblages with significant compositional differences between high and low latitudes and between the primary ocean basins. The stratigraphical potential of previously used microfossil datums was re-evaluated and compared between sites to determine the most reliable datums. Once this was completed, a new age-depth model was generated for each hole.

Wei (1992) used data sets for calcareous nannofossils and planktonic foraminifera and Wei *et al.* (1992) utilized data sets for calcareous nannofossils. These data sets were reviewed, compared with palaeomagnetic data sets, and also correlated to the GPTS of Berggren *et al.* (1995*a,b*). Wei (1992) and Wei *et al.* (1992) reinterpreted the magnetostratigraphies and constructed new age models for some Southern Ocean sites.

This study has used the pioneering studies of Wei (1992), Wei *et al.* (1992) and Thierstein *et al.* (1995) as its foundation and has adopted several of the same premises used by these workers. The approach used here differs from the studies of Wei (1992) and Wei *et al.* (1992) by including Palaeogene and Neogene data sets for calcareous and siliceous plankton and by the use of the GPTS of Cande & Kent (1995). The methodology, however, is similar to that employed by these investigators. This investigation differs from that of Thierstein *et al.* (1995) in scale and in methodology. This study develops a revised age model for data sets derived only from the Southern Ocean. The reassessment of the age model is based only on diatom, radiolarian and calcareous nannofossil data sets which are integrated with reinterpreted polarity records correlated to the GPTS of Cande & Kent (1995).

Methods

Site selection

Data sets obtained from 59 DSDP and ODP sites located south of 40°S were examined and 16 sites containing palaeomagnetic, diatom and calcareous nannofossil stratigraphies were selected for this study (see Table 1). All sites except DSDP Site 514 (Leg 71) were occupied during two campaigns (ODP Legs 113–114 and 119–120) into the Southern Ocean during 1987 and 1988. The majority of the sites are located between 46°02.77S and 65°9.63S degrees latitude (see Fig. 1). The one exception is Site 693 at 70°49.892S. The quality of the magneto-polarity history varies at each site as a consequence of either core recovery, disturbance, sampling or the scattered nature of the magnetization directions.

Calcareous nannofossils are common to abundant in sediments of Oligocene to Miocene age and are less common, rare or absent from the Pliocene through Pleistocene sequences. Exceptions include Holes 695A, 697B, 701A, 746A in which calcareous nannofossils are extremely rare or not recorded. Calcareous nannofossils are recorded in the upper Miocene sediments in Hole 696A. Diatoms are generally present throughout the Oligocene to Pleistocene sequences.

Palaeontological data sets

This study concentrates on the spatial and temporal distributions of the phytoplankton, diatoms and calcareous nannofossils. Both of these groups belong to the botanical Division Chrysophyta (Tappan 1980) and inhabit the photic zone of the world's ocean. These organisms respond to changing surface water conditions directly linked to changes in climate, surface ocean circulation and upwelling. The difference in the skeletal composition (diatoms – opaline silica; calcareous nannofossils – calcite) affects their preservation and fossilization. The temporal distribution of these organisms in the Southern Ocean is directly related to global cooling associated with increased ice volume and intensified thermal gradients during the late Neogene and Quaternary.

The siliceous zooplankton, represented by the radiolarians, live throughout the water column (Funnell & Riedel 1971) and respond to changes in surface, intermediate and deep water conditions. Their spatial and temporal distributions were used to test the biochronological model developed in this study. The preservation of their skeletal remains and their abundance in the Southern Ocean Pliocene and Pleistocene sediments confers a distinct advantage when compared to the sporadic distribution of the calcareous zooplankton (planktonic foraminifera) within sediments of similar age. In addition, the limited number of radiolarian stratigraphers analysing the faunas from the selected sites minimizes errors introduced by differences between specialists in taxonomic concepts or methodologies. For these reasons we have incorporated radiolarian rather than planktonic foraminiferal data sets into this study.

This study relies on data sets derived from the literature (Table 2) and no additional samples were examined. This approach was adopted to evaluate current biochronologies and to integrate these into a single Southern Ocean biochronology. In adopting this approach we recognize the following limitations.

Species concept

The data sets included in this study result from exploratory drilling spanning nine years between 1983 (Leg 71) and 1992 (Leg 120). Results from four of these expeditions were published between July 1990 and April 1992. Species concepts for the microfossil groups used in this study have advanced during this time resulting in refinement of species definition and identification of outliers within a population. These advances have resulted in the modification of previous species concepts by segregation into several species, the identification of varieties or grouping of species.

The application of a species concept is influenced by an individual's interpretation and consistency in applying that concept. Difficulties arise when different concepts are applied to the same species. This issue may be magnified by an increased number of researchers applying non-uniform concepts. Differences in species concepts pertaining to the microfossil groups used in this study from the Southern Ocean were potentially minimized by the few specialists completing the biostratigraphies. Several of the later Southern Ocean studies, i.e. Wei (1991, 1992) and Wei & Wise (1992) for calcareous nannofossils; Baldauf & Barron (1991) and Harwood & Maruyama (1992) for diatoms; Lazarus (1992) for radiolarians have incorporated the species concepts used in earlier studies.

In our own effort the potential consequences of differing species concepts is reduced by the typical traits of stratigraphically useful indicators. These include: ease of identification, wide distribution, skeletal size and robustness and minimal intra-specific variation.

Taxonomic nomenclature

Modifications to taxonomic nomenclature have evolved as new information concerning species similarities and differences are identified. Advances in this field result primarily from use of the Scanning Electron Microscope (SEM) to define primary and secondary morphological characteristics used to identify genera and species. This particularly applies to diatoms and is best exemplified by subdivision of the genus *Denticula* into *Denticulopsis*, *Crucidenticula*, and *Neodenticula* (Yanagisawa & Akiba 1985) and by the incorporation of several forms of *Coscinodiscus* into *Thalassiosira* (Akiba & Yanagisawa 1985) or *Azpeitia* (Fryxell et al. 1986). We have applied the currently used nomenclatural concepts to data sets in this study.

Data quality

Current biochronologies rely on datums, i.e. the first or last occurrence (FO or LO) of a species which has been calibrated to an independent age control such as magnetostratigraphy. Accuracy in the placement of an event (FO/LO) and the age calibration of the event dictates the accuracy and reliability of any given datum. Factors controlling the placement of each event are summarized in Table 4 which is modified from Gradstein et al. (1985). We use the stratigraphical placement of each event as identified by each worker and are unable to address most of the factors outlined in Table 4. This study, however, is especially sensitive to potential errors introduced by each worker and has required the careful scrutinizing of each data set to eliminate any inconsistencies.

We have eliminated informally designated species such as *Rouxia* species 1 of Gersonde & Burckle (1990) or *Rhizosolenia* species B of Harwood & Maruyama (1992). An exception to this is our use of the radiolarian species *Acrosphaera labrata* (Lazarus 1992), which is equivalent to *Acrosphaera* sp. 'conical pore' collosphaerid (Lazarus 1990). We have also eliminated any uncertain assignments such as *Nitzschia* aff. *porteri* (Gersonde & Burckle 1990) and *Nitzschia* cf. *praeinterfrigidaria* (Baldauf & Barron 1991). Group assignments were also removed from the data sets, for example the *Azpeitia tabularis* group of Gersonde & Burckle (1990). In cases where morphological forms of species were segregated, only the primary species designation was incorporated into the data set. Thus the stratigraphical occurrences of *Thalassiosira oliverana* (coarse and umbonatus), Harwood & Maruyama 1992, were not incorporated.

Typically we adhered to the stratigraphical occurrences of species as assigned by the investigators including their assessment

Table 4. *Factors controlling the stratigraphical placement of biostratigraphical events (modified from Gradstein* et al. *1985)*

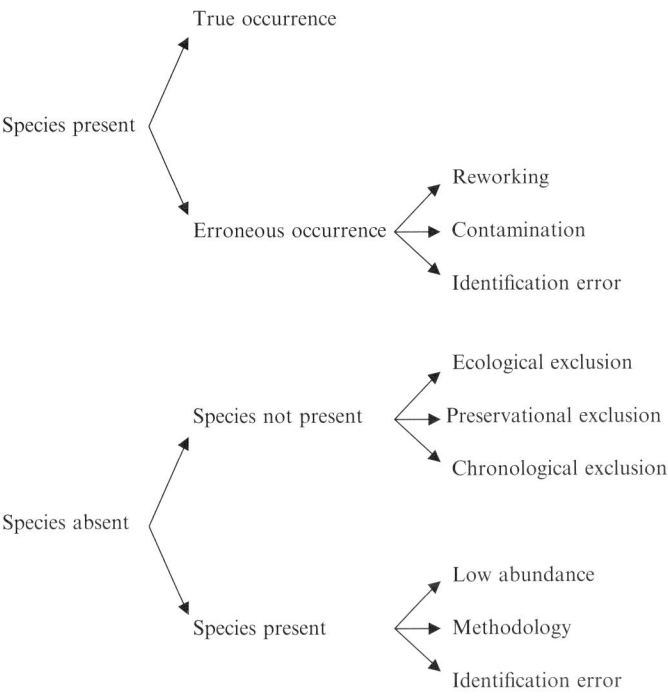

of reworking of specimens. In rare cases we have adjusted placement of first and last occurrences of selected species (e.g. when a single occurrence is isolated by numerous samples from its continuous stratigraphical range).

Previous workers have recognized the stratigraphical usefulness of abundance data and have defined datums based either on the first or last common occurrence or on the first or last abundance appearance of selected species. In regional settings such datums may prove reliable. We have not considered datums based on the first or last common occurrence or first or last abundance appearance in this study in order to minimize the subjectivity of the data set and also to avoid the influence of environmental factors on relative species abundance. Quantitative studies of diatom species distribution in the Southern Ocean indicate that relative abundances of diatom species were particularly sensitive to water-mass properties. Fenner (1991) speculated that temporal changes in water-mass dynamics of the Southern Ocean, particularly with respect to the position of the Polar Front Zone, influenced the isochroneity/diachroneity of quantitatively determined first and last abundance appearances of diatom species during the Matuyama magnetic epoch. Barron (1996) demonstrated that the relative percentages of diatom taxa, during a Pliocene warm interval (3.1–2.9 Ma) was related to a shift of the Polar Frontal Zone to a position 6° of latitude further to the south.

Calculation of the depths, in hole, of the first and last occurrence of a species represents the range between the sample in which the species is present and the adjacent sample.

Biostratigraphical zonations

Biostratigraphical zonations were continually refined with the addition of new data sets from the Southern Ocean. This holds true for the calcareous nannofossil, diatom and radiolarian biostratigraphies. Zonations are a convention that allow the placement of specific stratigraphical sequences into discrete arbitrary intervals and convey critical biochronological information to specialists and non-specialists. This study does not focus on zones, but uses FOs and LOs for stratigraphical evaluation and the establishment of an age model.

Magnetic data sets

Magnetic data sets obtained from the sites used in this study exhibit varying degrees of quality. These range from records with clear normal and reverse polarity, with directions close to the expected geocentric axial dipole directions, without excessive scatter (Category 1 of Clement *et al.* 1996), to records with a coherent and apparently stable remnant magnetization (Category 4 of Clement *et al.* 1996). In this latter category the dispersion is generally too great to allow polarity assignment. At most Southern Ocean sites, the correlation of the polarity directions is complicated by core recovery, drilling disturbance or sampling. Most Southern Ocean polarity records equate to Category 3 or 4 of Clement *et al.* (1996) as polarities appear to have been recorded, with unusually large amounts of scatter. In some cases, short polarity chrons may not be identified. An exception to this is the Palaeogene to Quaternary records in Holes 689B and 690B (Spiess 1990) and the Pliocene–Pleistocene record from Site 745 (Barron *et al.* 1991), which exhibit excellent polarity sequences.

Data collection. In the Southern Ocean the existence of few sedimentary records with recorded polarity histories is the primary limiting factor in the selection of sites. Of the potential factors influencing polarity history we have focused on inconsistencies in the magnetic history resulting from core disturbance, core recovery and sample frequency. We re-examined polarity records from each site concentrating on variations in the natural remnant magnetization (NRM) intensity and inclination. This was implemented in order to assess the placement of polarity reversals as identified by the various investigators. We are aware that other factors such as alternating field (AF) demagnetization levels, remagnetization of the drillstring, lithology and diagenesis may affect the quality of the magnetic data. With few exceptions, we have adhered to the stratigraphical placement previously identified polarity reversals.

Polarity calibration. Polarity timescales have evolved during and following the period of time represented by the acquisition of the data sets. On Leg 71, Salloway (1983) used the geomagnetic polarity timescale of La Brecque *et al.* (1977). Spiess (1990), Leg 113, applied the GPTS of Berggren *et al.* (1985a,b). During Leg 114, Hailwood & Clement (1991a; Site 699), Clement & Hailwood (1991; Site 701) and Hailwood & Clement (1991b; Site 704) also used the GPTS of Berggren *et al.* (1985a,b). Sakai & Keating (1991) and Barron *et al.* (1991) used the Berggren *et al.* (1985a,b) GPTS during Leg 119, but incorporated modifications proposed by Mankinen & Cox (1988). During Leg 120, Heider *et al.* (1992) and Inokuchi & Heider (1992) used the GPTS of Berggren *et al.* (1985a,b).

Nomenclature, identification of chronozones and sub chronozones were all in a state of flux and workers who applied the Berggren *et al.* (1985a,b) geomagnetic time scale often modified the chronozone nomenclature of Berggren *et al.* (1985a,b) to more readily identify specific chrons or sub-chrons (Fig. 4). One example is that of Clement & Hailwood (1991) who used the GPTS of Berggren *et al.* (1985a,b) and the zonal nomenclature of La Brecque *et al.* (1983).

Cande & Kent (1992, 1995) proposed a new GPTS for the late Mesozoic and Cenozoic. This time scale reassesses the sea-floor anomalies from profiles obtained throughout the ocean and establishes a polarity sequence using a composite approach. The identification of new events and the assignment of new ages for polarity intervals required re-examination of all data sets used in this study for standardization of these data sets to the Cande & Kent (1995) time scale. The application of this new GPTS has required reassessing previous zonal interpretations to address the critical chronostratigraphical control, and to adjust for changes in the polarity pattern introduced by increased or decreased rates of sediment accumulation. In addition, the polarity pattern was modified by the newly interpreted continuity of the stratigraphical record.

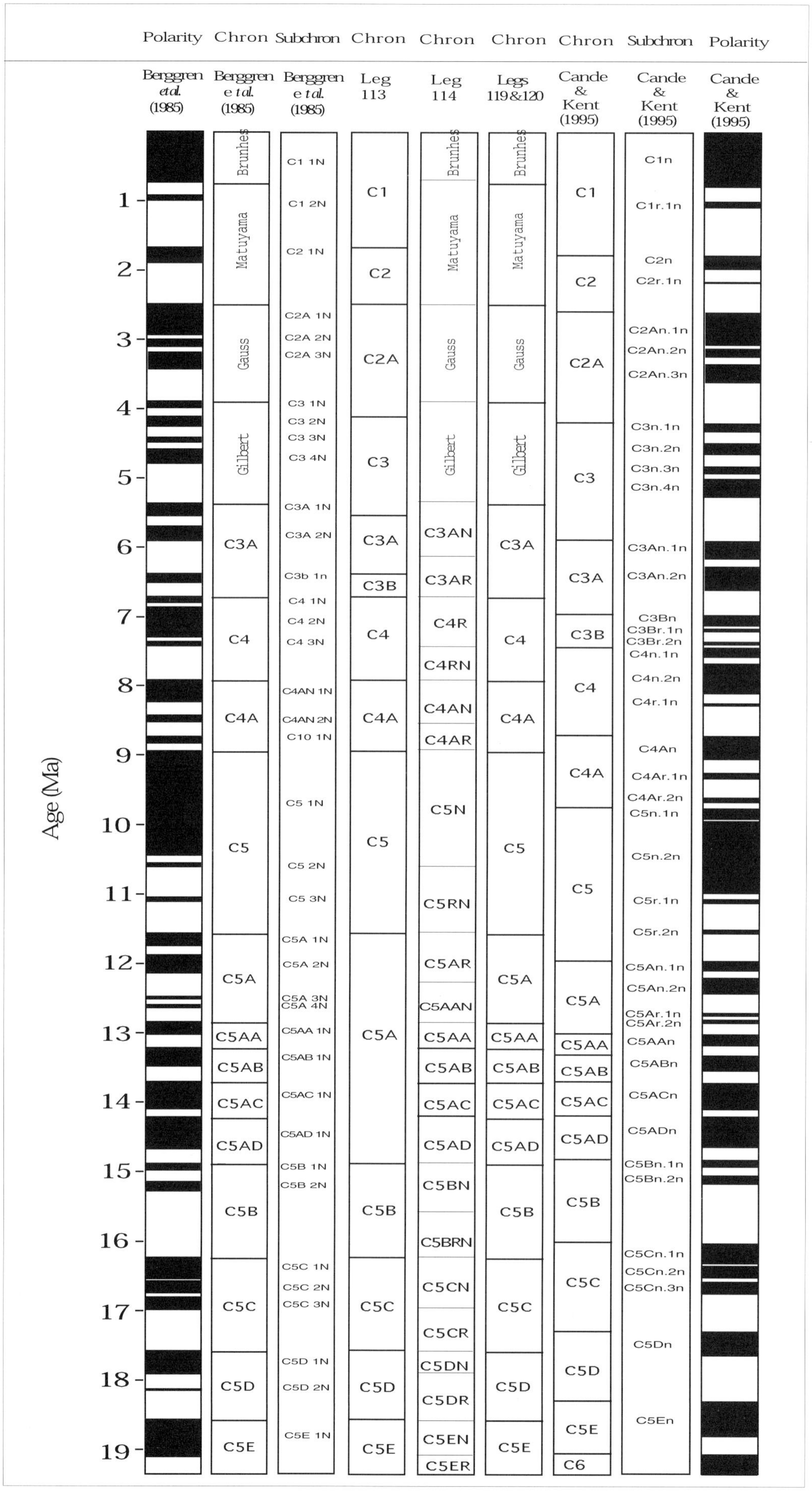

Fig. 4. Comparison of the Geopolarity Timescale (GPTS) of Berggren *et al.* (1985*a*, *b*) with the GPTS of Cande & Kent (1995). Shipboard and shorebased studies associated with the DSDP and ODP legs generally used the GPTS of Berggren *et al.* (1985*a*, *b*), but elected to modify the chronozonal nomenclature. This study uses the GPTS of Cande & Kent (1995).

METHODS

Chronostratigraphical control

A significant objective of this study was to develop a robust biochronological framework in which primary stratigraphical datums were validated and applied consistently throughout the Southern Ocean, when possible. The chronostratigraphical control used in this study is based on the integration of magnetostratigraphical and biostratigraphical data. The assignment of any given chronozone requires accurate identification of the polarity sequence which is achieved by using primary biostratigraphical species in conjunction with the polarity pattern represented in the geopolarity time scale. Each investigation in the Southern Ocean has reassessed previously identified primary datums and augmented these datums to accommodate the biostratigraphical results. This has a minimal impact on the accuracy of chronozonal assignments when the stratigraphical data sets contain numerous events. In the case where few events are present the selection of the primary events will have a significant impact on chronozonal identification.

Sedimentation rates

The reversal sequence provides a polarity pattern that can be compared to the GPTS. The polarity pattern observed at any given site will be affected by temporal variations in the sedimentation rates at that site. Therefore, independent age control (in this study biostratigraphy) is required for identification of the chronozones and to determine the sediment accumulation record. A constant sedimentation rate between chronostratigraphic control points is assumed throughout this study.

Continuity of the stratigraphical record

Correlation of any polarity and biostratigraphical data sets to any GPTS will result in the recognition of continuous stratigraphical sequences or in the interpretation of a stratigraphical break in sedimentary successions. The age model developed in this study was initially formulated using an assumption that the sedimentary sequences were continuous. Previously identified unconformities were not taken into consideration during the development of this age model. Stratigraphical breaks were interpreted following careful review of the stratigraphical data sets and an iterative reassessment of the age model.

Data processing

Data sets from previous workers (Table 2) were obtained for calcareous nannofossils, diatoms, radiolarians, and palaeomagnetic events from each site used in this study. The microfossil data sets contain information concerning the first occurrence and last occurrence of all species identified and the stratigraphical position of each event in the sedimentary sequence. Palaeomagnetic data was obtained for each hole and its depth, chronozone identification, and age of the polarity reversals were recorded using the authors calibration to a specific GPTS. These data sets were than combined to provide a composite record for each hole evaluated. Table 3 and Fig. 4 summarize the previously used biostratigraphical and chronostratigraphical indicators calibrated to the GPTS of Berggren et al. (1985a, b).

Previously identified chronozones were adjusted directly to the GPTS of Cande & Kent (1995) for use in this study. In some cases

Table 5. *Methodology used in this study to translate the previously generated datasets from the prior biochronological model to the model presented in this study*

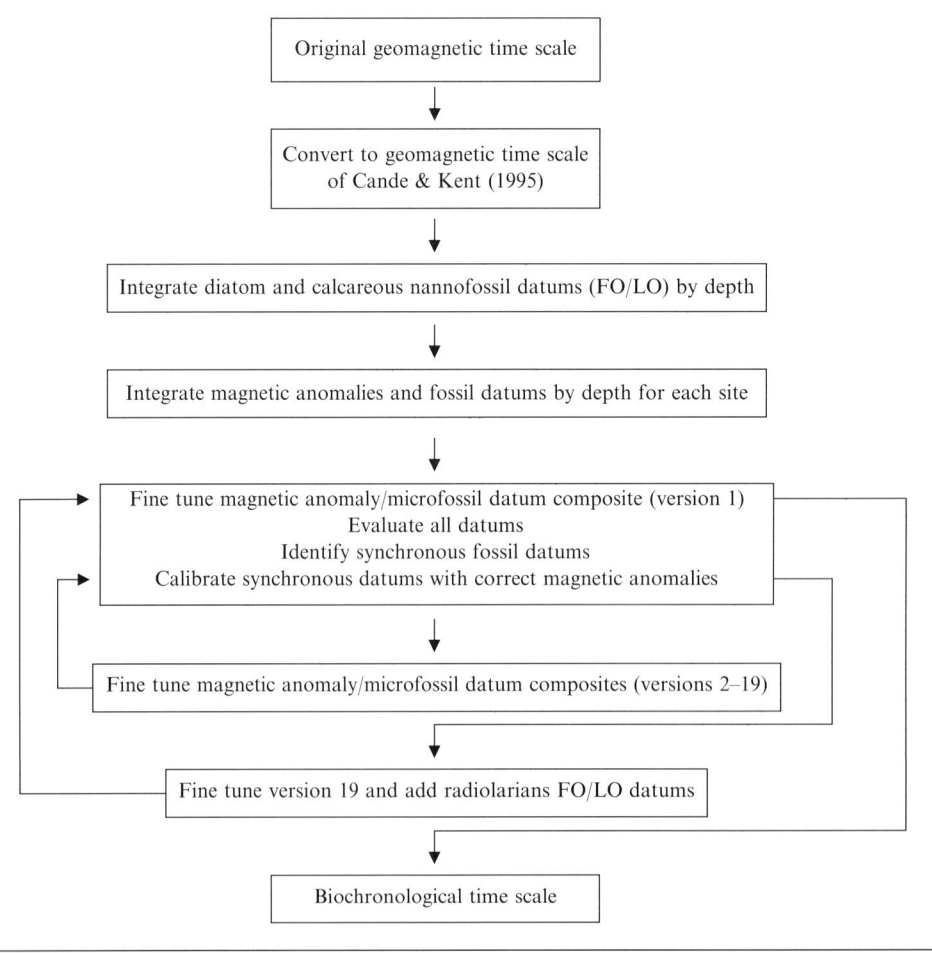

this required combining or subdividing polarity zones and chronozones. As a result, previously determined ages for specific datums were no longer valid and required re-calibration to the Cande & Kent (1995) GPTS and a new age assessment. The previously recognized biostratigraphical events were re-calibrated using the Cande & Kent (1995) GPTS applied to each data set and age-depth plots (designated original) were generated to establish a framework (Table 3).

At this stage we evaluated potentially useful biostratigraphical events and eliminated hole specific events, informally defined taxa and groups or events that failed to meet the criteria discussed above. We also identified polarity intervals requiring reassignment of chrono- or sub-chronozonations. Following this, we initiated an iterative process in which polarity assignments were reassessed and re-calibrated and the ages of microfossil datums were recalculated. Each step in the iterative reassessment of datums involved all sites and proceeded from the initial assumption that the sedimentary sequences at these sites were continuous. During the iterative process the initial age assignments for specific datums, within an individual site, were applied to all other sites in which these datums occurred. When a consistent result was obtained this was accepted. When a result was inconsistent we eliminated the interpretation and instigated a new iteration in which an age from a different hole was selected. We continued the process until either a consistent result was achieved or the datum was eliminated. The influence of this procedure on the assignment of polarity chrons and, on stratigraphical continuity within the sites was observed. where the existence of unconformities was clearly the product of applying a specific datum, we have chosen to maintain stratigraphical continuity at the expense of generating unconformities in order to retain the integrity of a biostratigraphical event. Therefore it was inevitable that some unconformities, identified on the basis of earlier age models, were eliminated. This process (Table 5) continued until we were confident that primary events were consistent at all sites in which they occur and at this stage, Age Model 1 was established representing the last 65 million years. Model 1 was used to generate age–depth plots for each hole and these were compared graphically to identify inconsistencies in the magnetic and biostratigraphical records from each hole. Biostratigraphical events older than about 36 Ma were eliminated due to the paucity of data. Subsequent iterations, similar to that described above were completed and resulted in establishing Age Model 2.

We applied the radiolarian stratigraphies developed by previous workers and synthesized by Lazarus (1992) as an independent assessment of Age Model 2. Following one additional iteration, diachronous biostratigraphical events were eliminated from the data sets and age–depth plots (designated revised) were reproduced to establish the final age model presented in this paper.

Primary events identified in this study are recorded from the Atlantic and Indian Ocean sectors of the Southern Ocean and are defined as typically having having an age constraint of less than 0.4 million years. This constraint is an artefact of sampling by previous workers and additional samples would have to be examined to refine the age constraint for a specific event. Although 0.4 million years constitutes our threshhold we recognize this as an upper limit. In rare instances we have allowed exceptions to accommodate outliers from otherwise consistent data. This occurred occasionally in cases where a specific event occurred in the majority of holes within our established criteria. In addition to the age constraint primary events were identified when the range of a specific event occurred within a less than 0.25 million years tolerance limit in the age ranges of the event in the majority of holes.

Results

The biochronological framework generated by this study is based on re-evaluation of the age model for the holes presented in this report. The results focus on a hole by hole discussion of the age model with a comparison between previous studies and the current results.

Legend

○   Diatom (top)

□   Radiolarian (top)

◇   Nannofossil (top)

△   Magnetic (top)

·   Diatom (base)

·   Radiolarian (base)

·   Nannofossil (base)

·   Magnetic (base)

Fig. 5. Legend of symbols used for age–depth graphs presented in this manuscript.

Hole 514

Thirty-five cores were recovered from a total penetration of 150.8 m below seafloor (mbsf) representing a recovery of 92% of the cored interval. Sediments recovered consist of Quaternary to lower Pliocene diatom ooze with a variable clay component (Ludwig *et al.* 1983). A good polarity record was recovered for the entire stratigraphical sequence. Diatoms and radiolarians are abundant throughout the section. Radiolarians are generally well preserved. Calcareous nannofossils are typically sparse exhibiting poor to moderate preservation (Ludwig *et al.* 1983).

Table 6 presents the integrated biostratigraphical and magnetostratigraphical events previously used (palaeomagnetics: Salloway 1983; diatoms: Ciesielski 1983; calcareous nannofossils: none; radiolarians: Weaver 1983) and calibrated to the GPTS of Berggren *et al.* (1985a, b). An age–depth graph for Hole 71-514 using these data is shown in Fig. 6.

Figure 7 compares the polarity chronozones and diatom zones assigned by previous workers with the chronozonal assignments in this study. The stratigraphical ranges of diatoms and radiolarians reflect the placement of the FO and LO of stratigraphically useful species. The revised chronozones illustrated are based on reassessing the data sets of these workers (Table 7). Comparison of the previous chronozonal assignments (compare Tables 6 and 7) with those of this study indicates that only minimal revisions were necessary. Modifications reflect in part the differences between the Berggren *et al.* (1985a, b) and Cande & Kent (1995) GPTS.

An age–depth graph for Hole 71–514 using the stratigraphical markers determined in this study is shown in Fig. 8. The section is interpreted to be continuous based on both palaeomagnetic and biostratigraphical control. Palaeomagnetic control consists of a continuous sequence of polarity reversals extending from the base of C1n at 8.43 mbsf to the base of C3n.1n at 133.51 mbsf. Biostratigraphical markers used to calibrate this polarity sequence include the LO of *Thalassiosira kolbei* at 22.26 mbsf, the LOs of *Helotholus vema* and *Desmospyris spongiosa* at 40.8 mbsf and the FO of *Nitzschia praeinterfrigidaria* at 105.96 mbsf. One primary biostratigraphical control point, the FO of *Nitzschia interfrigidaria* occurs at 133.79 mbsf. Unfortunately the absence of palaeomagnetic control points below this depth prevent an age assignment for this biostratigraphical event in this hole.

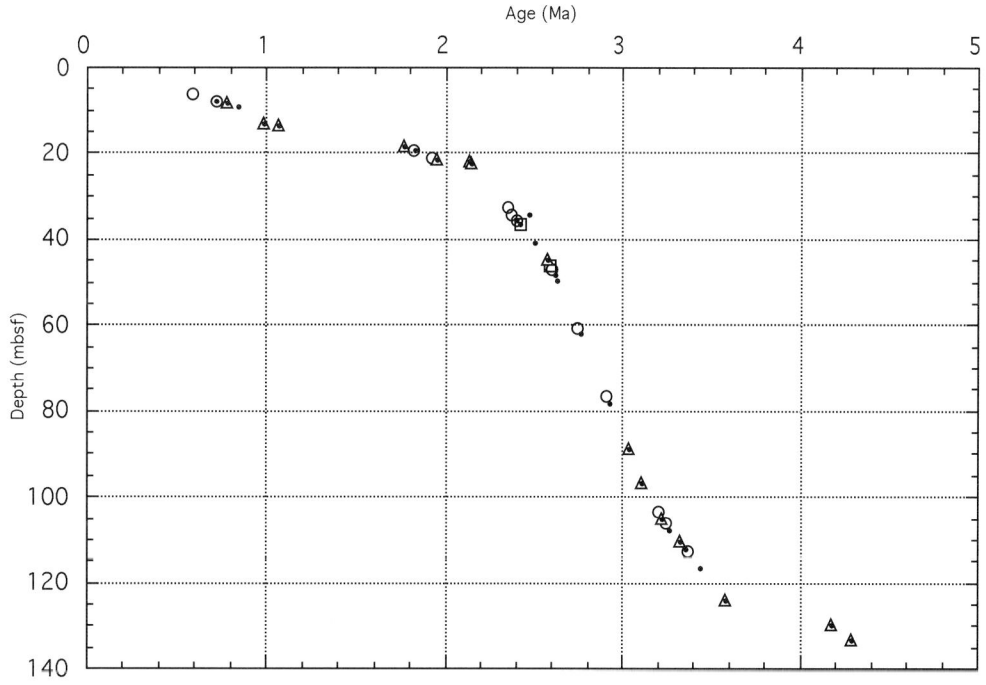

Fig. 6. An age–depth plot for DSDP Hole 514 utilizing the previously used biostratigraphical datums and polarity reversal interpretation calibrated to the GPTS of Berggren *et al.* (1985a, b). See Fig. 5 for legend of symbols.

Table 6. *Composite dataset for DSDP Hole 514 containing the diatom, radiolarian and palaeomagnetic data calibrated to the GPTS of Cande & Kent (1995)*

	Event	Sample (top)	Sample (bottom)	Depth (t)	Depth (b)	Age (t)	Age (b)
LO	*Actinocyclus ingens*	3-1, 72-74	3-2, 70-72	6.34	7.82	0.59	0.72
LO	*Thalassiosira elliptopora*	3-2, 70-72	3-3, 49-51	7.82	9.11	0.72	0.84
B	C1n			8.43	8.43	0.78	0.78
T	C1r.1n			13.06	13.06	0.99	0.99
B	C1r.1n			13.55	13.55	1.07	1.07
T	C2n			18.52	18.52	1.77	1.77
LO	*Simonseniella barboi*	5-4, 54-56	6-1, 74-79	19.45	19.59	1.82	1.83
LO	*Thalassiosira kolbei*			21.12	22.26	1.92	2.15
B	C2n			21.60	21.60	1.95	1.95
T	C2r.1n			22.00	22.00	2.14	2.14
B	C2r.1n			22.50	22.50	2.15	2.15
FO	*Actinocyclus ingens*	9-1, 86-88	9-2, 70-72	32.74	34.22	2.35	2.47

(continued)

Table 6. (*continued*)

	Event	Sample (top)	Sample (bottom)	Depth (t)	Depth (b)	Age (t)	Age (b)
LO	*Thalassiosira vulnifica*	9-2, 70-72	9-3, 63-65	34.22	35.65	2.37	2.40
FO	*Nitzschia kerguelensis*	9-3, 63-65	10-1, 33-35	35.65	36.75	2.40	2.42
LO	*Desmospyris spongiosa*			36.40	40.80	2.42	2.50
LO	*Helotholus vema*			36.40	40.80	2.42	2.50
T	C2An.1n			45.00	45.00	2.58	2.58
FO	*Cycladophora davisiana*			46.20	49.60	2.59	2.63
FO	*Hemidiscus cuneiformis*	35-1, 73-75	35-2, 73-75	47.15	48.63	2.60	2.62
LO	*Nitzschia weaveri*	15-2, 70-72	15-3, 70-72	60.62	62.12	2.74	2.76
LO	*Nitzschia interfrigidaria*	19-1, 72-74	19-1, 72-74	76.74	78.24	2.91	2.93
LO	*Nitzschia reinholdii*	19-1, 72-74	19-1, 72-74	76.74	78.24	2.91	2.93
FO	*Thalassiosira lentiginosa*			76.74	78.24	2.91	2.93
B	C2An.1n			89.00	89.00	3.04	3.04
T	C2An.2n			97.00	97.00	3.11	3.11
FO	*Thalassiosira insigna*	25-1, 95-97	26-1, 98-100	103.37	107.80	3.20	3.26
B	C2An.2n			105.00	105.00	3.22	3.22
FO	*Thalassiosira kolbei*			105.96	112.06	3.24	3.36
FO	*Nitzschia praeinterfrigidaria*			105.96	112.06	3.24	3.36
T	C2An.3n			110.57	110.57	3.33	3.33
FO	*Nitzschia weaveri*	27-2, 84-86	28-1, 90-92	112.70	116.52	3.37	3.44
B	C2An.3n			124.00	124.00	3.58	3.58
T	C3n.1n			129.74	129.74	4.18	4.18
B	C3n.1n			133.51	133.51	4.29	4.29
FO	*Nitzschia interfrigidaria*	32-1, 77-79	33-1, 75-77	133.79	138.37	?	?

Data sets are integrated from the work of numerous authors (see Table 2).

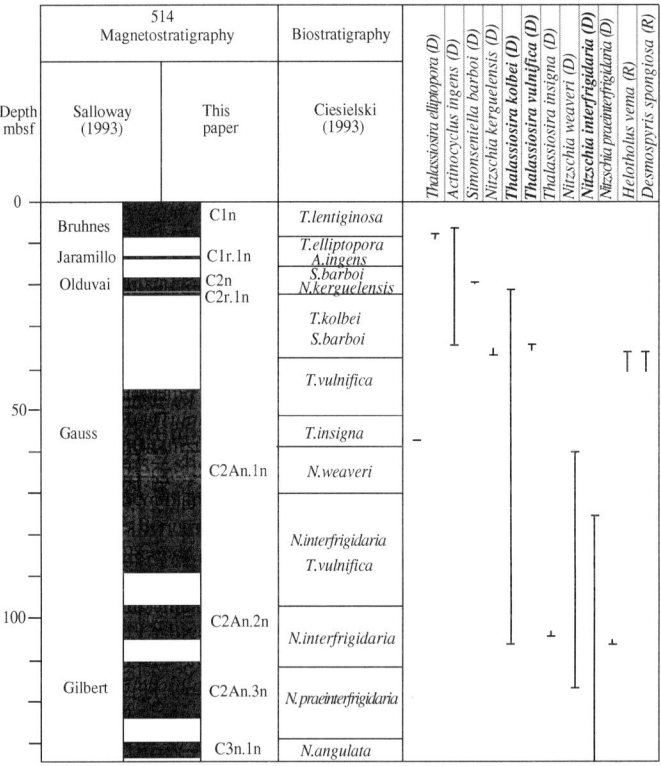

Fig. 7. Comparison of the DSDP Hole 514 polarity chronozones and diatom zonal assignments by previous workers with the chronozonal assignments of this study. The stratigraphical ranges of selected species are based on the original data sets (see text for references). D, diatoms; R, radiolaria; N, calcareous nannofossils. Emboldened events indicate primary biostratigraphical events recognized in this study. The polarity record shown represents the original interpretation calibrated to the GPTS of Cande & Kent (1995) and the interpretation from this study correlated with the GPTS of Cande & Kent (1995).

Table 7. *Composite dataset for DSDP Hole 514 containing the diatom, radiolarian and palaeomagnetic data calibrated to the GPTS of Cande & Kent (1995) using the biochronological interpretation based on this study*

	Event	Sample (top)	Sample (bottom)	Depth (t)	Depth (b)	Age (t)	Age (b)
LO	*Actinocyclus ingens*	3-1, 72-74	3-2, 70-72	6.34	7.82	0.59	0.72
LO	*Thalassiosira elliptopora*	3-2, 70-72	3-3, 49-51	7.82	9.11	0.72	0.84
B	C1n			8.43	8.43	0.78	0.78
T	C1r.1n			13.06	13.06	0.99	0.99
B	C1r.1n			13.55	13.55	1.07	1.07
T	C2n			18.52	18.52	1.77	1.77
LO	*Simonseniella barboi*	5-4, 54-56	6-1, 74-79	19.45	19.59	1.82	1.83
LO	*Thalassiosira kolbei˜*			21.12	22.26	1.92	2.15
B	C2n			21.60	21.60	1.95	1.95
T	C2r.1n			22.00	22.00	2.14	2.14
B	C2r.1n			22.50	22.50	2.15	2.15
FO	*Actinocyclus ingens*	9-1, 86-88	9-2, 70-72	32.74	34.22	2.35	2.37
LO	*Thalassiosira vulnifica*	9-2, 70-72	9-3, 63-65	34.22	35.65	2.37	2.40
FO	*Nitzschia kerguelensis*	9-3, 63-65	10-1, 33-35	35.65	36.75	2.40	2.42
LO	*Desmospyris spongiosa*			36.40	40.80	2.42	2.50
LO	*Helotholus vema*			36.40	40.80	2.42	2.50
T	C2An.1n			45.00	45.00	2.58	2.58
FO	*Cycladophora davisiana*			46.20	49.60	2.59	2.63
FO	*Hemidiscus cuneiformis*	35-1, 73-75	35-2, 73-75	47.15	48.63	2.60	2.62
LO	*Nitzschia weaveri*	15-2, 70-72	15-3, 70-72	60.62	62.12	2.74	2.76
LO	*Nitzschia interfrigidaria*	19-1, 72-74	19-2, 72-74	76.74	78.24	2.91	2.93
LO	*Nitzschia reinholdii*	19-1, 72-74	19-2, 72-74	76.74	78.24	2.91	2.93
FO	*Thalassiosira lentiginosa*			76.74	78.24	2.91	2.93
B	C2An.1n			89.00	89.00	3.04	3.04
T	C2An.2n			97.00	97.00	3.11	3.11
FO	*Thalassiosira insigna*	25-1, 95-97	26-1, 98-100	103.37	107.80	3.20	3.26
B	C2An.2n			105.00	105.00	3.22	3.22
FO	*Thalassiosira kolbei*			105.96	112.06	3.24	3.36
FO	*Nitzschia praeinterfrigidaria*			105.96	112.06	3.24	3.36
T	C2An.3n			110.57	110.57	3.33	3.33
FO	*Nitzschia weaveri*	27-2, 84-86	28-1, 90-92	112.70	116.52	3.37	3.44
B	C2An.3n			124.00	124.00	3.58	3.58
T	C3n.1n			129.74	129.74	4.18	4.18
B	C3n.1n			133.51	133.51	4.29	4.29
FO	*Nitzschia interfrigidaria*	32-1, 77-79	33-1, 75-77	133.79	138.37	?	?

Data sets are integrated from the work of numerous authors (see Table 2).

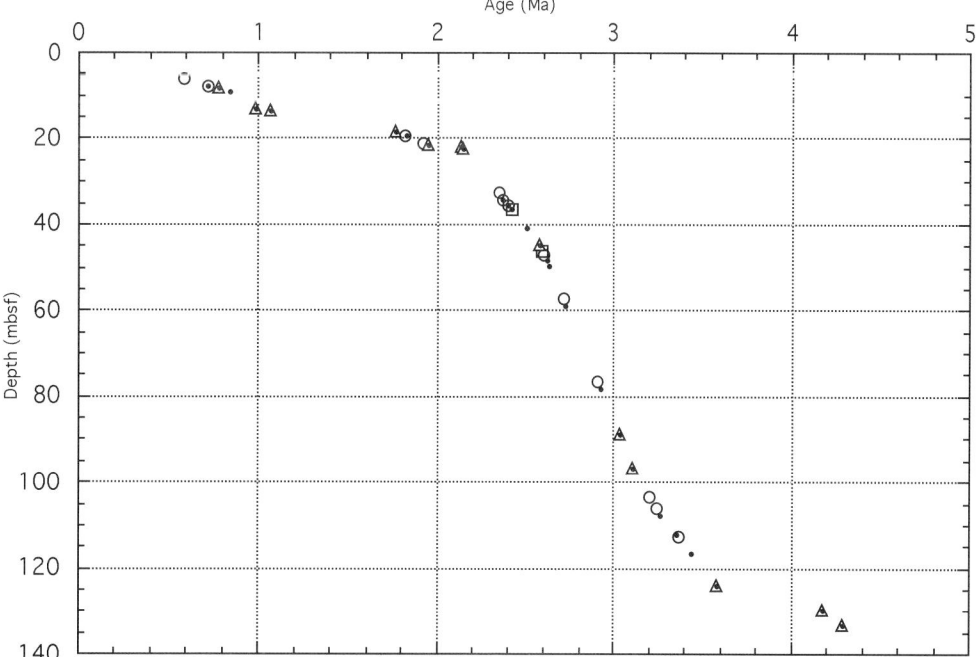

Fig. 8. A revised age–depth plot for DSDP Hole 514 using the biostratigraphical and polarity reversal events identified in this study and calibrated to the GPTS of Cande & Kent (1995). See Fig. 5 for legend of symbols.

Hole 689B

Thirty-three cores were recovered from a total penetration of 297.3 mbsf representing a recovery of 77% of the cored interval. Sediments recovered consist of Quaternary to upper Miocene diatom ooze, Miocene to Palaeocene calcareous ooze with a variable siliceous component, and Upper Cretaceous chalk. A good palaeomagnetic polarity reversal record was obtained by both shipboard and shorebase analysis (Barker et al. 1988; Spiess, 1990). Diatoms, calcareous nannofossils and radiolarians are present throughout the sequence. Moderately preserved, common to abundant diatoms are present throughout the Quaternary to lower Miocene sediments. Diatom abundance decreases with rare and poorly preserved organisms present in this interval (Gersonde & Burckle 1990). Calcareous nannofossils are rare in the Quaternary and upper Pliocene sediments, absent in the middle to lower Pliocene, common to abundant throughout the Miocene and abundant in the Oligocene. Preservation of the calcareous nannofossils varies and is moderate to good in the Miocene and moderate in the Oligocene (Wei & Wise 1990). Well-preserved radiolarians are abundant throughout the upper Pliocene to middle Miocene section. Radiolarian preservation is poor in the lower Miocene and upper Oligocene and good in the lower Oligocene (Barker et al. 1988).

Table 8 presents the integrated biostratigraphical and magnetostratigraphical events previously used (palaeomagnetics: Spiess 1990; diatoms: Gersonde & Burckle 1990; calcareous nannofossils: Wei & Wise 1990; radiolarians: Abelmann 1990; Lazarus 1990) and calibrated to the GPTS of Berggren et al. (1985a,b). An age–depth graph for Hole 113-689B using these data is shown in Fig. 9.

Figure 10 compares the polarity chronozones and diatom zones assigned by previous workers (Spiess 1990; Gersonde & Burckle 1990) with the chronozonal assignments in this study. The stratigraphical ranges of diatoms, radiolarians and calcareous nannofossils reflect the placement of the FO and LO of stratigraphically useful species as recognized by Gersonde & Burckle (1990), Abelmann (1990), Lazarus (1990), and Wei & Wise (1990), respectively. The revised chronozones illustrated are based on reassessing the data sets of these workers (Table 9). Comparison of the chronozonal assignments of Spiess (1990; Table 8) with those of this study indicates that only minimal revisions were completed. Modifications reflect in part the differences between the Berggren et al. (1985a,b) and Cande & Kent (1995) GPTS.

The integrated biostratigraphical and magnetostratigraphical data used in this study is presented in Table 9. We have limited the data sets to include the Quaternary to latest Eocene stratigraphical interval. The paucity of palaeomagnetic data available for the Eocene and Palaeocene of the Southern Ocean preclude reinterpretation of the data from Eocene or older sediments in this hole.

An age–depth graph for Hole 113-689B, using the stratigraphical markers determined in this study, is shown in Fig. 11. Four hiatuses are noted in the stratigraphical section. The first is placed in the upper 3 m of the section and is identified by the occurrence of older stratigraphical indicators such as the LO of *Desmospyris spongiosa*, the LO of *Helotholus vema* and the LO of *Thalassiosira kolbei* in the upper 0.6 m of the section. The youngest polarity reversal (top of Chron C2An.3n) occurs at a depth of 3.38 mbsf.

The second unconformity occurs between 15.17 and 18.09 mbsf. The upper boundary of this unconformity is constrained by a continuous string of polarity reversal (top of Chron C2An.3n through the top of C3n.4n) from 3.38 through 15.17 mbsf. This magnetic sequence is constrained by the FO of *Helotholus vema* (6.61 mbsf) and the FO of *Nitzschia barronii* (8.59–9.22 mbsf). The lower boundary is constrained by the LO of *Asteromphalus kennettii* at 18.38–19.30 mbsf and a continuous polarity reversal sequence from the top of Chron C4An at 18.09 mbsf to the base of C5ADn at 55.28 mbsf. This magnetic sequence is constrained by the occurrence of the following primary markers: FO *Denticulopsis praedimorpha* (47.45–48.09 mbsf), FO *Nitzschia denticuloides* (50.45–51.09 mbsf), FO *Denticulopsis hustedtii* (52.29–53.19 mbsf) and the LO of *Synedra jouseana* (54.69–55.07 mbsf). Also associated with this interval are the LO of *Cycladophora spongothorax* (20.24–22.16 mbsf) and the FO of *Acrosphaera labrata* (20.26–21.29 mbsf). This second unconformity spans approximately 5 million years.

The third unconformity is situated between 55.28 and 56.19 mbsf and represents about 2 million years. The upper boundary of this event is constrained by the LO of *Synedra jouseana* at and the lower boundary is constrained by a sequence of magnetics from the top of C5Cn.3n (56.53 mbsf) to the top of C6n (61.76 mbsf) and the FO of *Crucidenticula kanayae* (59.19–59.47 mbsf). The last occurrence of *Coscinodiscus rhombicus* (57.06–57.37 mbsf) is a potentially useful marker in this interval. This unconformity replaces two unconformities previously identified by Gersonde & Burckle (1990).

The oldest hiatus is placed between 61.76 and 63.37 mbsf and partitions the top of Chron C6An.1n (67.11 mbsf) from the base of C7n.1n (66.86 mbsf) to the top of C17n.1n (135.77 mbsf). The FO of *Isthmolithus recurvus* and the LO of *Reticulofenestra reticulata* at 131.7–132.73 mbsf occur within the lower part of this sequence. This oldest unconformity corresponds to an unconformity of similar age and depth identified by Spiess (1990) and Gersonde & Burckle (1990).

This study suggests that the stratigraphical section is more continuous than previously recognized. Gersonde & Burckle (1990) and Gersonde et al (1990) identified seven hiatuses in the Quaternary to Oligocene interval based on integrated bio- and magnetostratigraphy. Two of these breaks at 17.8 and 67.0 mbsf approximate hiatuses identified in this study. One hiatus at 58.8 mbsf reflects differences in the interpretation of chronozonal assignment of the polarity reversals. The others at 11.50, 43.80, and 64.7 mbsf reflect differences in biostratigraphical interpretation.

Spiess (1990) identified five hiatuses in the upper 140 m of Hole 113-689B based on pattern recognition of the polarity record. Two of these breaks placed at 17.65 and 66.86 mbsf coincide with those recorded by Gersonde & Burckle (1990), Gersonde et al. (1990) and this study. The breaks at 55.28 and 58.75 mbsf approximates breaks at 55.1 and 58.8 mbsf proposed by Gersonde and Burckle (1990) and Gersonde et al. (1990). The interpretation of the remaining hiatus at 59.65 mbsf reflects differences in the GPTS and chronozonal assignments used in each study.

Table 8. *Composite dataset for ODP Hole 689B containing the diatom, radiolarian, calcareous nannofossil and palaeomagnetic data calibrated to the GPTS of Cande & Kent (1995)*

	Event	Sample (top)	Sample (bottom)	Depth (t)	Depth (b)	Age (t)	Age (b)
	Unconformity			***	***	***	***
LO	*Actinocyclus ingens*	1H-1, 0	1H-1, 28-29	0.00	0.00	?	?
LO	*Antarctissa cylindrica*	1H-1, 0	1H-1, 25-26	0.00	0.26	?	?
LO	*Cycladophora pliocenica*	1H-1, 0	1H-1, 25-26	0.00	0.26	?	?
LO	*Eucyrtidium calvertense*	1H-1, 0	1H-1, 25-26	0.00	0.26	?	?
LO	*Desmospyris spongiosa*	1H-1, 0	1H-1, 25-26	0.00	0.26	?	?
LO	*Helotholus vema*	1H-1, 0	1H-1, 25-26	0.00	0.26	?	?

Table 8. (*continued*)

	Event	Sample (top)	Sample (bottom)	Depth (t)	Depth (b)	Age (t)	Age (b)
FO	*Cycladophora davisiana*	1H-1, 0	1H-1, 25-26	0.00	0.26	?	?
LO	*Nitzschia interfrigidaria*	1H-1, 0	1H-1, 28-29	0.00	0.29	?	?
LO	*Nitzschia barronii*	1H-1, 0	1H-1, 28-29	0.00	0.29	?	?
LO	*Thalassiosira inura*	1H-1, 28-29	1H-1, 56-58	0.29	0.58	?	?
LO	*Thalassiosira kolbei*	1H-1, 28-29	1H-1, 56-58	0.29	0.58	?	?
LO	*Nitzschia praeinterfrigidaria*	1H-2, 28-29	1H-2, 50-52	1.79	2.02	?	?
FO	*Nitzschia interfrigidaria*	1H-2, 28-29	1H-2, 50-52	1.79	2.02	?	?
T	C3n.1n			3.38	3.38	4.18	4.18
LO	*Prunopyle titan*	1H-3, 118-120	1H-4, 104-105	4.18	5.53	4.27	4.34
B	C3n.1n			4.40	4.40	4.29	4.29
LO	*Denticulopsis dimorpha*	2H-1, 31-33	2H-2, 115-116	5.63	6.45	4.34	4.38
FO	*Helotholus vema*	2H-2, 118-120	2H-3, 62-64	7.89	8.92	4.44	4.49
FO	*Nitzschia barronii*	2H-3, 28-29	2H-3, 90-92	8.59	9.22	4.47	4.55
T	C3n.2n			8.79	8.79	4.48	4.48
B	C3n.2n			9.63	9.63	4.62	4.62
FO	*Nitzschia reinholdii*	2H-4, 28-29	2H-4, 56-58	10.09	10.38	4.67	4.69
FO	*Thalassiosira kolbei*	2H-4, 56-58	2H-4, 115-116	10.38	10.98	4.69	4.75
T	C3n.3n			11.45	11.45	4.80	4.80
FO	*Thalassiosira inura*	2H-5, 27-28	2H-5, 55-57	11.58	11.87	4.84	4.89
B	C3n.3n			11.72	11.72	4.89	4.89
T	C3n.4n			15.17	15.17	4.98	4.98
	Unconformity			***	***	***	***
FO	*Thalassiosira oestrupii*	3H-2, 56-58	3H-2, 118-120	16.88	17.50	?	?
T	C3Bn			18.09	18.09	6.94	6.94
LO	*Asteromphalus kennettii*	3H-3, 56-58	3H-3, 148-150	18.38	19.30	7.02	7.68
B	C3Bn			18.67	18.67	7.09	7.09
T	C4n.1n			18.92	18.92	7.43	7.43
LO	*Actinocyclus ingens v. nodus*	3H-3, 148-150	3H-3, 56-58	19.30	19.88	7.68	8.07
FO	*Cosmiodiscus intersectus*	3H-5, 80-82	3H-5, 114-115	20.12	20.45	8.23	8.32
B	C4r.1n			20.17	20.17	8.26	8.26
LO	*Cycladophora spongothorax*	3H-4, 94-96	3H-5, 136-138	20.24	21.29	8.28	8.51
FO	*Acrosphaera labrata*	3H-4, 94-96	3H-5, 136-138	20.26	21.29	8.28	8.51
T	C4An			22.15	22.15	8.70	8.70
LO	*Reticulofenestra gelida*	5H-2, 30-32	5H-3, 30-32	22.60	24.30	8.88	9.66
B	C4Ar.1n			23.65	23.65	9.31	9.31
T	C4Ar.2n			23.81	23.81	9.58	9.58
B	C4Ar.2n			24.17	24.17	9.64	9.64
T	C5n.1n			24.67	24.67	9.74	9.74
LO	*Cycladophora humerus*	4H-2, 117-119	4H-3, 56-58	26.97	27.86	10.02	10.13
FO	*Acrosphaera australis*	4H-2, 117-119	4H-3, 56-58	26.97	27.86	10.02	10.13
FO	*Asteromphalus kennettii*	4H-3, 31-32	4H-3, 114-115	27.62	28.45	10.10	10.20
FO	*Eucyrtidium pseudoinflatum*	4H-4, 56-58	4H-4, 116-118	29.36	29.96	10.31	10.39
B	C5n.2n			34.55	34.55	10.95	10.95
B	C5r.2n			37.93	37.93	11.53	11.53
LO	*Actinomma golownini*	5H-5, 56-58	5H-5, 117-118	40.36	40.97	11.65	11.69
LO	*Nitzschia denticuloides*	5H-6, 28-29	5H-6, 114-115	41.59	42.45	11.72	11.76
FO	*Cycladophora spongothorax*	5H-CC	6H-1, 59	43.30	43.89	11.81	11.83
LO	*Dendrospyris megalocephalis*	5H-CC	6H-1, 59	43.30	43.89	11.81	11.83
FO	*Denticulopsis dimorpha*	6H-1, 27-28	6H-1, 114-115	43.58	44.45	11.82	11.86
LO	*Reticulofenestra hesslandii*	6H-1, 30-32	6H-2, 29-31	43.60	45.09	11.82	11.90
FO	*Dendrospyris megalocephalis*	6H-2, 58-60	6H-3, 91-93	45.38	47.21	11.91	12.41
T	5An.1n			45.93	45.93	11.94	11.94
LO	*Crucidenticula nicobarica*	6H-2, 114-115	6H-3, 28-29	45.95	46.59	11.95	12.15
B	5An.1n			46.41	46.41	12.08	12.08
LO	*Nitzschia grossepunctata*	6H-3, 28-29	6H-3, 114-115	46.59	47.45	12.15	12.51
T	5An.2n			46.68	46.68	12.18	12.18
B	5An.2n			47.18	47.18	12.40	12.40
LO	*Denticulopsis praedimorpha*	6H-3, 114-115	6H-4, 28-29	47.45	48.09	12.51	12.76
T	C5AAn			48.68	48.68	12.99	12.99
FO	*Actinomma golownini*	6H-4, 108-110	6H-5, 31-33	48.90	49.63	13.06	13.28
	C5AAn			49.18	49.18	13.14	13.14
T	C5ABn			49.68	49.68	13.30	13.30
FO	*Antarctissa deflandrei*	6H-5, 84-86	6H-5, 117-119	50.16	50.49	13.38	13.44
FO	*Nitzschia denticuloides*	6H-5, 114-115	6H-6, 28-29	50.45	51.09	13.43	13.54
B	C5ABn			50.91	50.91	13.51	13.51
T	C5ACn			51.93	51.93	13.70	13.70
FO	*Denticulopsis hustedtii*	6H-7, 28-29	7H-1, 28-29	52.29	53.19	13.99	14.26
LO	*Denticulopsis maccollumnii*	6H-7, 28-29	7H-1, 28-29	52.29	53.19	13.99	14.26

(*continued*)

Table 8. (continued)

	Event	Sample (top)	Sample (bottom)	Depth (t)	Depth (b)	Age (t)	Age (b)
B	C5ACn			52.41	52.41	14.08	14.08
FO	Cycladophora humerus	6H-7, 32-34	7H-1, 32-34	52.64	53.24	14.13	?
FO	Dendrospyris megalocephalis	6H-7, 32-34	7H-1, 32-34	52.64	53.24	14.13	?
T	C5ADn			52.85	52.85	14.18	14.18
LO	Synedra jouseana	7H-2, 28-29	7H-2, 65-67	54.69	55.07	?	?
LO	Thalassiosira spumellaroides	7H-2, 65-67	7H-2, 115-116	55.07	55.56		?
	Unconformity			***	***	***	***
FO	Actinocyclus ingens v. nodus	7H-3, 28-29	7H-3, 86-88	56.19	56.19	?	?
T	C5Bn.2n			56.53	56.53	15.03	15.03
LO	Coscinodiscus lewisianus	7H-3, 86-88	7H-3, 114-115	56.78	57.06	15.07	15.11
LO	Coscinodiscus rhombicus	7H-3, 115-116	7H-3, 145-147	57.06	57.37	15.11	15.16
FO	Eucyrtidium punctatum	7H-3, 145-147	7H-4, 148-150	57.37	58.90	15.16	?
B	C5Bn.2n			57.40	57.40	15.16	15.16
LO	Crucidenticula kanayae	7H4, 28-29	7H-4, 88-86	57.60	58.18	?	?
LO	Nitzschia maleinterpretaria	7H-4, 28-29	7H-4, 86-88	57.69	58.28	?	?
FO	Nitzschia grossepunctata	7H-4, 28-29	7H-4, 86-88	57.69	58.28	?	?
LO	Thalassiosira fraga	7H-4, 86-88	7H-4, 115-116	58.28	58.56	?	?
FO	Actinocyclus ingens	7H-4, 115-116	7H-5, 28-29	58.60	59.19	?	?
FO	Denticulopsis maccollumnii	7H-4, 115-116	7H-5, 28-29	58.60	59.19	?	?
	Unconformity			***	***	***	***
FO	Crucidenticula kanayae	7H-5, 28-29	7H-5, 55-57	59.19	59.47	?	?
B	C5Dn			60.40	60.40	17.62	17.62
FO	Cycladophora golli regipileus	7H-5, 89-91	7H-6, 117-119	61.21	61.49	18.01	18.15
FO	Crucidenticula nicobarica	7H-6, 115-116	7H-7, 28-29	61.56	62.19	18.18	18.49
FO	Cyrtocapsella tetrapera	7H-7, 21-23	7H-7, 56-58	61.23	61.87	18.02	18.31
T	C5En			61.76	61.76	18.28	18.28
FO	Cyrtocapsella longithorax	8H-1, 57-59	8H-2, 56-58	63.37	64.88	18.86	19.41
FO	Raphidodiscus marylandicus	8H-2, 114-115	8H-3, 28-29	65.15	65.79	19.51	19.74
FO	Thalassiosira fraga	8H-2, 114-115	8H-3, 28-29	65.15	65.79	19.51	19.74
LO	Chiasmolithus altus	8H-3, 30-32	8H-4, 30-32	65.85	67.32	19.76	20.29
LO	Reticulofenestra bisecta	8H-3, 30-32	8H-4, 30-32	65.85	67.32	19.76	20.29
LO	Rocella gelida	8H-3, 115-116	8H-3, 146-148	66.65	66.96	20.05	?
FO	Thalassiosira spumellaroides	8H-3, 144-146	8H-4, 28-29	66.96	67.29	?	24.91
B	C6n			66.86	66.86	20.13	20.13
	Unconformity			***	***	***	***
T	C7n.2n			67.11	67.11	24.84	24.84
LO	Rocella vigilans	8H-4, 28-29	8H-4, 57-59	67.29	67.59	24.91	25.05
B	7n.2n			67.88	67.88	25.18	25.18
T	C7An			68.61	68.61	25.50	25.50
B	C7An			68.88	68.88	25.65	25.65
T	C8n.1n			69.38	69.38	25.82	25.82
B	C8n.1n			70.11	70.11	25.95	25.95
T	C8n.2n			70.38	70.38	25.99	25.99
FO	Rocella gelida	8H-5, 114-115	8H-6, 28-29	71.15	71.79	26.07	26.13
FO	Rocella vigilans	8H-5, 114-115	8H-6, 28-29	71.15	71.79	26.07	26.13
B	C8n.2n			75.98	75.98	26.55	26.55
T	C9n			79.46	79.46	27.03	27.03
B	C9n			85.05	85.05	27.97	27.97
T	C10n.1n			89.82	89.82	28.28	28.28
B	C10n.1n			90.07	90.07	28.51	28.51
T	C10n.2n			91.01	91.01	28.58	28.58
B	C10n.2n			91.93	91.93	28.75	28.75
T	C11n.2n			100.23	100.23	29.77	29.77
B	C11n.2n			103.38	103.38	30.10	30.10
T	C12n			104.38	104.38	30.48	30.48
LO	Reticulofenestra umbilica	12H-4, 29-31	12H-5, 29-31	105.81	107.31	30.74	31.03
B	C12n			106.88	106.88	30.94	30.94
LO	Isthmolithus recurvus	13H-1, 130-132	13H-2, 130-132	111.02	113.42	31.83	32.35
T	C13n			116.71	116.71	33.06	33.06
B	C13n			119.70	119.70	33.55	33.55
LO	Reticulofenestra oamaruensis	13H-7, 28-30	14H-1, 130-132	119.90	121.50	33.58	33.78
FO	Reticulofenestra oamaruensis	14H-5, 130-132	14H-6, 130-132	127.50	127.70	34.55	34.59
T	C15n			128.33	128.33	34.66	34.66
FO	Isthmolithus recurvus	15H-2, 30-32	15H-2, 131-132	131.70	132.73	34.83	34.88
LO	Reticulofenestra reticulata	15H-2, 30-32	15H-2, 131-132	131.70	132.73	34.83	34.88
B	C15n			134.02	134.02	34.94	34.94

Data sets are integrated from the work of numerous authors (see Table 2).

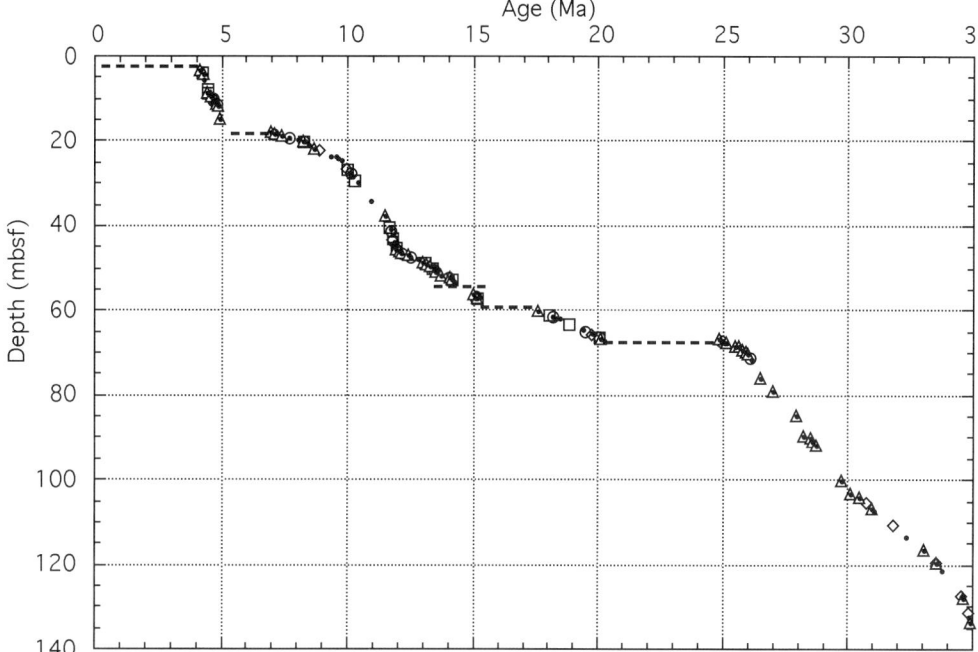

Fig. 9. An age–depth plot for DSDP Hole 689B utilizing the previously used biostratigraphical datums and polarity reversal interpretation calibrated to the GPTS of Cande & Kent (1995). See Fig. 5 for legend of symbols. Five stratigraphical breaks (dashed line) are based on Spiess (1990).

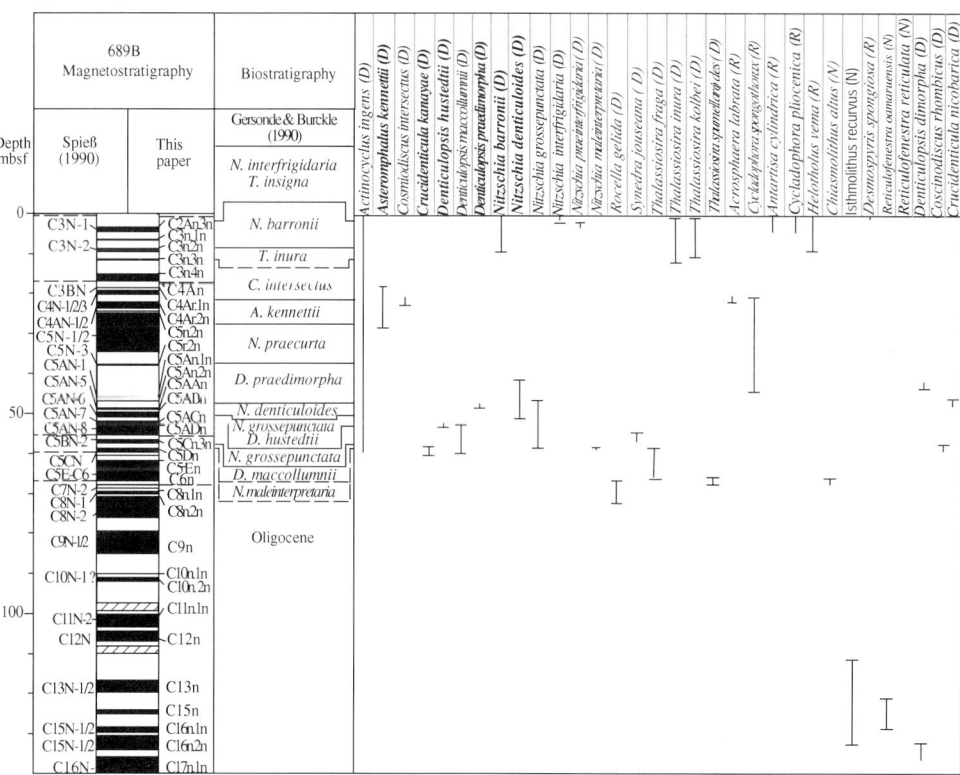

Fig. 10. Comparison of the ODP Hole 689B polarity chronozones and diatom zonal assignments by previous workers with the chronozonal assignments of this study. The stratigraphical ranges of selected stratigraphical indicators are based on the original data sets (see text for references). D, diatoms; R, radiolaria; N, calcareous nannofossils. Emboldened events indicate primary biostratigraphical events recognized in this study. The polarity record shown represents the original interpretation calibrated to the GPTS of Cande & Kent (1995) and the interpretation from this study correlated with the GPTS of Cande & Kent (1995). Diagonal lines indicate intervals of uncertainty.

Table 9. *Composite dataset for ODP Hole 689B containing the diatom, radiolarian, calcareous nannofossil and palaeomagnetic data calibrated to the GPTS of Cande & Kent (1995) using the biochronological interpretation based on this study*

	Event	Sample (top)	Sample (bottom)	Depth (t)	Depth (b)	Age (t)	Age (b)
	Unconformity			***	***	***	***
LO	Actinocyclus ingens	1H-1, 0	1H-1, 28-29	0.00	0.00	?	?
LO	Antarctissa cylindrica	1H-1, 0	1H-1, 25-26	0.00	0.26	?	?
LO	Cycladophora pliocenica	1H-1, 0	1H-1, 25-26	0.00	0.26	?	?
LO	Eucyrtidium calvertense	1H-1, 0	1H-1, 25-26	0.00	0.26	?	?
LO	Desmospyris spongiosa	1H-1, 0	1H-1, 25-26	0.00	0.26	?	?
LO	Helotholus vema	1H-1, 0	1H-1, 25-26	0.00	0.26	?	?
FO	Cycladophora davisiana	1H-1, 0	1H-1, 25-26	0.00	0.26	?	?
LO	Thalassiosira oestrupii	1H-1, 0	1H-1, 28-29	0.00	0.29	?	?
LO	Nitschia interfrigidaria	1H-1, 0	1H-1, 28-29	0.00	0.29	?	?
LO	Nitzschia barronii	1H-1, 0	1H-1, 28-29	0.00	0.29	?	?
LO	Thalassiosira inura	1H-1, 28-29	1H-1, 56-58	0.29	0.58	?	?
LO	Thalassiosira kolbei	1H-1, 28-29	1H-1, 56-58	0.29	0.58	?	?
LO	Nitzschia praeinterfrigidaria	1H-2, 28-29	1H-2, 50-52	1.79	2.02	?	?
FO	Nitschia interfrigidaria	1H-2, 28-29	1H-2, 50-52	1.79	2.02	?	?
T	C2An.3n			3.38	3.38	3.33	3.33
LO	Prunopyle titan	1H-3, 118-120	1H-4, 104-105	4.18	5.53	3.53	3.92
B	C2An.3n			4.40	4.40	3.58	3.58
T	C3n.1n			6.41	6.41	4.18	4.18
B	C3n.1n			6.61	6.61	4.29	4.29
LO	Denticulopsis dimorpha	2H-1, 31-33	2H-1, 115-116	5.63	6.45	3.95	4.02
FO	Helotholus vema	2H-2, 118-120	2H-3, 62-64	7.89	8.92	4.40	4.50
FO	Nitzschia barronii	2H-3, 28-29	2H-3, 90-92	8.59	9.22	4.46	4.55
T	C3n.2n			8.79	8.79	4.48	4.48
B	C3n.2n			9.63	9.63	4.62	4.62
FO	Nitzschia reinholdii	2H-4, 28-29	2H-4, 56-58	10.09	10.38	4.67	4.69
FO	Thalassiosira kolbei	2H-4, 56-58	2H-4, 115-116	10.38	10.98	4.68	4.75
T	C3n.3n			11.45	11.45	4.80	4.80
FO	Thalassiosira inura	2H-5, 27-28	2H-5, 55-57	11.58	11.87	4.84	4.89
B	C3n.3n			11.72	11.72	4.89	4.89
T	C3n.4n			15.17	15.17	4.98	4.98
FO	Thalassiosira oestrupii	3H-2, 56-58	3H-2, 118-120	16.88	17.50	?	?
	Unconformity			***	***	***	***
T	C4An			18.09	18.09	8.70	8.70
LO	Asteromphalus kennettii	3H-3 56-58	3H-3, 148-150	18.38	19.30	8.55	9.25
B	C4An			18.67	18.67	9.03	9.03
T	C4Ar.1n			18.92	18.92	9.23	9.23
LO	Actinocyclus ingens v. nodus	3H-3, 148-150	3H-3, 56-58	19.30	19.88	9.25	9.29
FO	Cosmiodiscus intersectus	3H-5, 80-82	3H-5, 114-115	20.12	20.45	9.31	9.35
B	C4Ar.1n			20.17	20.17	9.31	9.31
FO	Acrosphaera labrata	3H-4, 94-96	3H-5, 136-138	20.26	21.29	9.32	9.46
LO	Cycladophora spongothorax	3H-4, 94-96	3H-5, 136-138	20.24	22.16	9.32	9.58
T	C4Ar.2n			22.15	22.15	9.58	9.58
LO	Reticulofenestra gelida	5H-2, 30-32	5H-3, 30-32	22.60	24.67	9.60	9.92
B	C4Ar.2n			23.65	23.65	9.64	9.64
T	C5n.1n			23.81	23.81	9.74	9.74
B	C5n.1n			24.17	24.17	9.88	9.88
T	C5n.2n			24.67	24.67	9.92	9.92
LO	Cycladophora humerus	4H-2, 117-119	4H-3, 56-58	26.97	27.86	10.16	10.25
FO	Acrosphaera australis	4H-2, 117-119	4H-3, 56-58	26.97	27.86	10.16	10.25
FO	Asteromphalus kennettii	4H-3, 31-32	4H-3, 114-115	27.62	28.45	10.23	10.31
FO	Eucyrtidium pseudoinflatum	4H-4, 56-58	4H-4, 116-118	29.36	29.96	10.41	10.47
B	C5n.2n			34.55	34.55	10.95	10.95
B	C5r.2n			37.93	37.93	11.53	11.53
LO	Actinomma golownini	5H-5, 56-58	5H-5, 117-118	40.36	40.97	11.77	11.79
LO	Nitzschia denticuloides	5H-6, 28-29	5H-6, 114-115	41.59	42.45	11.80	11.83
FO	Cycladophora spongothorax	5H-CC	6H-1, 59	43.30	43.89	11.86	11.87
FO	Denticulopsis dimorpha	6H-1, 27-28	6H-1, 114-115	43.58	44.45	11.87	11.89
LO	Reticulofenestra hesslandii	6H-1, 30-32	6H-2, 29-31	43.60	45.09	11.87	11.91
FO	Dendrospyris megalocephalis	6H-2, 58-60	6H-3, 91-93	45.38	47.21	11.92	12.37
T	C5An.1n			45.93	45.93	11.94	11.94
LO	Crucidenticula nicobarica	6H-2, 114-115	6H-3, 28-29	45.95	46.59	11.95	12.14
B	C5An.1n			46.41	46.41	12.08	12.08
LO	Nitzschia grossepunctata	6H-3, 28-29	6H-3, 114-115	46.59	47.45	12.15	12.51
T	C5An.2n			46.68	46.68	12.18	12.18
B	C5An.2n			47.18	47.18	12.40	12.40
FO	Denticulopsis praedimorpha	6H-3, 114-115	6H-4, 28-29	47.45	48.09	12.51	12.76
T	C5AAn			48.68	48.68	12.99	12.99

Table 9. (continued)

	Event	Sample (top)	Sample (bottom)	Depth (t)	Depth (b)	Age (t)	Age (b)
FO	Actinomma golownini	6H-4, 108-110	6H-5, 31-33	48.90	49.63	13.06	13.28
B	C5AAn			49.18	49.18	13.14	13.14
T	C5ABn			49.68	49.68	13.30	13.30
FO	Antarctissa deflandrei	6H-5, 84-86	6H-5, 117-119	50.16	50.49	13.38	13.43
FO	Nitzschia denticuloides	6H-5, 114-115	6H-6, 28-29	50.45	51.09	13.43	13.54
B	C5ABn			50.91	50.91	13.51	13.51
T	C5ACn			51.93	53.19	13.70	13.70
FO	Denticulopsis hustedtii	6H-7, 28-29	7H-1, 28-29	52.29	53.19	13.99	14.24
LO	Denticulopsis maccollumnii	6H-7, 28-29	7H-1, 28-29	52.29	53.19	13.99	14.24
B	C5ACn			52.41	52.41	14.08	14.08
FO	Cycladophora humerus	6H-7, 32-34	7H-1, 32-34	52.64	53.24	14.13	14.25
FO	Desmospyris megalocephalis	6H-7, 32-34	7H-1, 32-34	52.64	53.24	14.13	14.25
T	C5ADn			52.85	52.85	14.18	14.18
LO	Thallassiosira spumellaroides	7H-2, 65-67	7H-2, 115-116	55.07	55.56	15.57	?
LO	Synedra jouseana	7H-2, 28-29	7H-2, 65-67	54.69	55.07	14.51	14.57
B	C5ADn			55.28	55.28	14.61	14.61
	Unconformity			***	***	***	***
FO	Actinocyclus ingens v. nodus	7H-3, 28-29	7H-3, 86-88	56.19	56.19	?	?
T	C5Cn.3n			56.53	56.53	16.56	16.56
LO	Coscinodiscus lewisianus	7H-3, 86-88	7H-3, 114-115	56.78	57.06	16.61	16.66
LO	Coscinodiscus rhombicus	7H-3, 115-116	7H-3, 145-147	57.06	57.37	16.66	16.72
FO	Eucyrtidium punctatum	7H-3, 145-147	7H-4, 148-150	57.37	58.90	16.72	17.33
B	C5Cn.3n			57.40	57.40	16.73	16.73
LO	Crucidenticula kanayae	7H4, 28-29	7H-4, 88-86	57.60	58.18	16.81	17.05
LO	Nitzschia maleinterpretaria	7H-4, 28-29	7H-4, 86-88	57.69	58.28	16.85	17.09
FO	Nitzschia grossepunctata	7H-4, 28-29	7H-4, 86-88	57.69	58.28	16.85	17.09
LO	Thalassiosira fraga	7H-4, 86-88	7H-4, 114-115	58.28	58.56	17.09	17.20
FO	Actinocyclus ingens	7H-4, 115-116	7H-5, 28-29	58.60	59.19	17.22	17.45
FO	Denticulopsis maccollumnii	7H-4, 115-116	7H-5, 28-29	58.60	59.19	17.22	17.45
T	C5Dn			58.75	58.75	17.28	17.28
FO	Crucidenticula kanayae	7H-5, 28-29	7H-5, 55-57	59.19	59.47	17.45	17.55
B	C5Dn			59.65	59.65	17.62	17.62
T	C5En			60.03	60.03	18.28	18.28
B	C5En			60.40	60.40	18.78	18.78
FO	Cycladophora golli regipileus	7H-5, 89-91	7H-6, 117-119	61.21	61.49	18.94	19.00
FO	Crucidenticula nicobarica	7H-6, 115-116	7H-7, 28-29	61.56	62.19	19.01	19.14
FO	Cyrtocapsella tetrapera	7H-7, 21-23	7H-7, 56-58	61.23	61.87	18.94	?
T	C6n			61.76	61.76	19.05	19.05
	Unconformity	***	***	***	***	***	***
FO	Cyrtocapsella longithorax	8H-1, 57-59	8H-2, 56-58	63.37	64.88	?	?
FO	Raphidodiscus marylandicus	8H-2, 114-115	8H-3, 28-29	65.15	65.79	?	?
FO	Thalassiosira fraga	8H-2, 114-115	8H-3, 28-29	65.15	65.79	?	?
LO	Chiasmolithus altus	8H-3, 30-32	8H-4, 30-32	65.85	67.32	?	?
LO	Reticulofenestra bisecta	8H-3, 30-32	8H-4, 30-32	65.85	67.32	?	24.93
LO	Rocella gelida	8H-3, 115-116	8H-3, 146-148	66.65	66.96	?	24.84
B	C7n.1n			66.86	66.86	24.78	24.78
T	C7n.2n			67.11	67.11	24.84	24.84
FO	Thalassiosira spumellaroides	8H-3, 144-146	8H-4, 28-29	66.96	67.29	24.77	24.92
LO	Rocella vigilans	8H-4, 28-29	8H-4, 57-59	67.29	67.59	24.91	25.05
B	7n.2n			67.88	67.88	25.18	25.18
T	C7An			68.61	68.61	25.50	25.50
B	C7An			68.88	68.88	25.65	25.65
T	C8n.1n			69.38	69.38	25.82	25.82
B	C8n.1n			70.11	70.11	25.95	25.95
T	C8n.2n			70.38	70.38	25.99	25.99
FO	Rocella vigilans	8H-5, 114-115	8H-6, 28-29	71.15	71.79	26.07	26.13
FO	Rocella gelida	8H-5, 114-115	8H-6, 28-29	71.15	71.79	26.07	26.13
B	C8n.2n			75.98	75.98	26.55	26.55
T	C9n			79.46	79.46	27.03	27.03
B	C9n			85.05	85.05	27.97	27.97
T	C10n.1n			89.82	89.82	28.28	28.28
B	C10n.1n			90.07	90.07	28.51	28.51
T	C10n.2n			91.01	91.01	28.58	28.58
B	C10n.2n			91.93	91.93	28.75	28.75
T	C11n.1n			100.23	100.23	29.40	29.40
B	C11n.2n			103.38	103.38	29.66	29.66
T	C12n			104.38	104.38	30.48	30.48
LO	Reticulofenestra umbilica	12H-4, 29-31	12H-5, 29-31	105.81	107.31	30.74	31.03

(continued)

Table 9. (*continued*)

	Event	Sample (top)	Sample (bottom)	Depth (t)	Depth (b)	Age (t)	Age (b)
B	C12n			106.88	106.88	30.94	30.94
LO	*Isthmolithus recurvus*	13H-1, 130-132	13H-2, 130-132	111.02	113.42	31.83	32.35
T	C13n			116.71	116.71	33.06	33.06
B	C13n			119.70	119.70	33.55	33.55
LO	*Reticulofenestra oamaruensis*	13H-7, 28-30	14H-1, 130-132	119.90	121.50	33.60	34.01
T	C15n			124.09	124.09	34.66	34.66
B	C15n			125.07	125.07	34.94	34.94
FO	*Reticulofenestra oamaruensis*	14H-5, 130-132	14H-6, 130-132	127.50	127.70	35.24	35.26
T	C16n.1n			128.33	128.33	35.34	35.34
B	C16n.1n			129.70	129.70	35.53	35.53
T	C16n.2n			130.05	130.05	35.69	35.69
FO	*Isthmolithus recurvus*	15H-2, 30-32	15H-2, 131-132	131.70	132.73	35.82	36.01
LO	*Reticulofenestra reticulata*	15H-2, 30-32	15H-2, 131-132	131.70	132.73	35.82	36.01
B	C16n.2n			134.02	134.02	36.34	36.34
T	C17n.1n			135.77	135.77	36.20	36.20

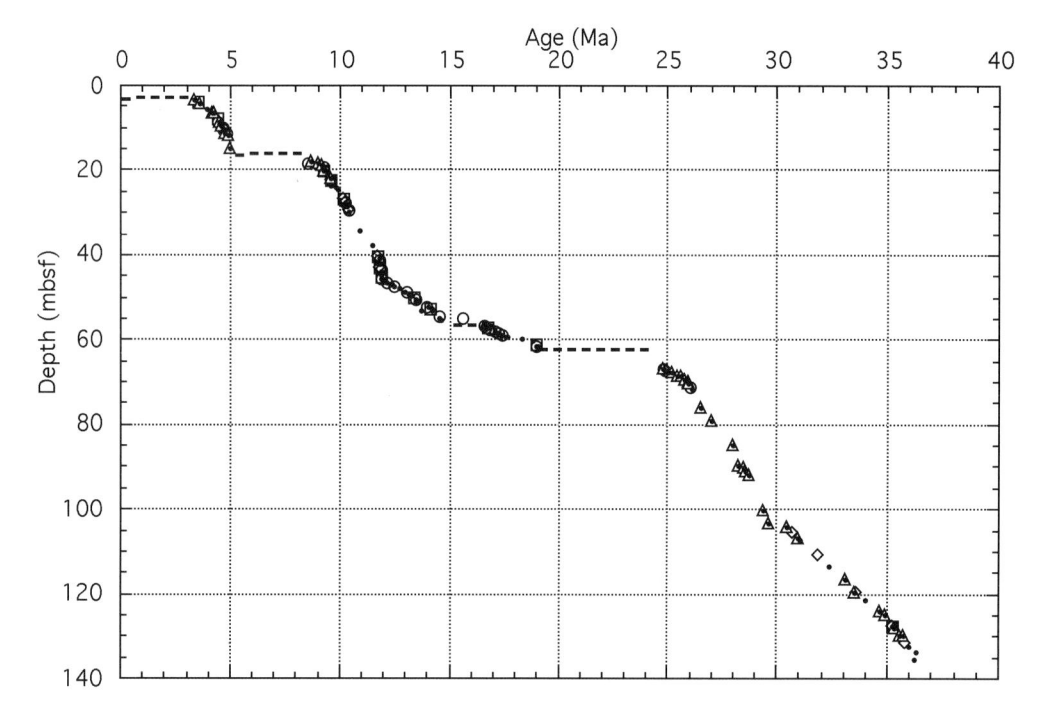

Fig. 11. A revised age–depth plot for DSDP Hole 689B utilizing the biostratigraphical and polarity reversal events identified in this study and calibrated to the GPTS of Cande & Kent (1995). See Fig. 5 for legend of symbols. Four stratigraphical breaks (dashed line) are based on the results of this study.

Hole 690B

Twenty-five cores were recovered from a total penetration of 213.4 mbsf representing a recovery of 100% of the cored interval. Sediments recovered consist of Quaternary calcareous ooze, Pliocene to upper Miocene diatom ooze, and Miocene to upper Palaeocene calcareous ooze. A siliceous component varies throughout the Miocene to upper Eocene section (Barker *et al.* 1988). A good palaeomagnetic polarity reversal record was obtained by both shipboard and shorebase analysis (Barker *et al.* 1988; Spiess 1990). Diatoms, calcareous nannofossils and radiolarians are present throughout the sequence. Diatoms are present through the Pleistocene to Oligocene sequence. Abundance and preservation varies with common, moderately preserved diatoms present in the Pleistocene to upper Miocene sediments; few, moderately preserved diatoms recorded in the middle and lower Miocene; and rare, poorly preserved diatoms observed in the Oligocene (Gersonde & Burckle 1990). Calcareous nannofossils are rare in the Pleistocene, absent in the Pliocene to upper Miocene, and abundant throughout the remaining sequence. Preservation of calcareous nannofossils is moderate in the Miocene to Oligocene interval (Wei & Wise 1990). Radiolarians are abundant and well preserved in the Pleistocene to lower Miocene and abundant and poorly preserved in the Oligocene. Radiolarians are absent from the remaining sequence (Barker *et al.* 1988). The previously used biostratigraphical and magnetostratigraphical data (palaeomagnetics: Spiess 1990; diatoms: Gersonde & Burckle 1990; calcareous nannofossils: Wei & Wise 1990; radiolarians: Abelmann 1990; Lazarus 1990) integrated and calibrated to the GPTS of Berggren *et al.* (1985a, b) is presented in Table 10. An age–depth graph for Hole 113-690B using this data is shown in Fig. 12.

Figure 13 compares the polarity chronozones and diatom zones assigned by previous workers (Spiess 1990; Gersonde & Burckle 1990, respectively) with the chronozonal assignments in this study. The stratigraphical ranges of diatoms, radiolarians and calcareous nannofossils reflect the placement of the FO and LO of stratigraphically useful species as recognized by Gersonde & Burckle (1990), Abelmann (1990), Lazarus (1990), and Wei & Wise (1990), respectively. The revised chronozones illustrated are based on reassessing the data sets of these workers (Table 11). Comparison of the chronozonal assignments of Spiess (1990; Table 10) with those of this study (Table 11) indicates that significant reinterpretation was completed for the upper 50 m of the section. Within the upper 50 m, differences in chronozonal assignments reflect in part differences in the GPTS used in each study.

An age–depth graph for Hole 113-690B using the stratigraphical markers determined in this study is shown in Fig. 14. Four unconformities are identified in this stratigraphical sequence at approximately 18.3, 42.0, 51.5 and 90 mbsf. The youngest unconformity is constrained above by the LO of *Thallassiosira kolbei* (1.84–2.39 mbsf), the FO of *Nitzschia interfrigidaria* (8.39–9.25 mbsf), the FO of *Helotholus vema* (9.84–11.7 mbsf) and the FO of *Nitzschia barronii* at (11.39–11.98 mbsf), The interval above the unconformity contains the base of C2r.1n (2 mbsf) to the base of C3n.2n (11.78 mbsf). The lower boundary of this unconformity is constrained by the FO of *Denticulopsis praedimorpha* (36.75–37.39 mbsf), the FO of *Nitzschia denticulodes* (38.89–39.75 mbsf), the FO of *Denticulopsis hustedtii* (40.39–41.08 mbsf), the LO of *Synedra jouseana* (41.32–42 mbsf) and a continuous polarity sequence from the top of C4Ar.1n (18.57 mbsf) to the base of C5ADn (41.68).

The second stratigraphical break is placed between 41.68 and 42.16 mbsf. The upper boundary of this unconformity is constrained by the LO of *Denticulopsis hustedtii* and the LO of *Synedra jouseana*. The lower boundary is constrained by a polarity sequence from the top of C5Cn.3n (42.16 mbsf) to the base of C6n (51.28 mbsf) and the FO of *Crucidenticula kanayae* (44.08–44.31) mbsf.

A third unconformity is recognized at a depth of 51.5 mbsf. The upper boundary is constrained by the base of Chron C6n. The lower boundary is constrained by a sequence of magnetic events from the top of C8n.2n (53.25 mbsf) to the base of C12n (84.01 mbsf). No primary biostratigraphical markers occur within the sequence.

The oldest hiatus occurs at approximately 90 mbsf. The upper boundary of this break is constrained by the the base of C12n (84.01 mbsf). The lower boundary is constrained by the base of C16n.1n (91.95 mbsf), the top of C16n.2n (92.21 mbsf) and the FO of *Isthmolithus recurvus* (92.41- 93.42 mbsf) and LO of *Reticulofenestra reticulata* at 92.41- 93.42 mbsf.

This study suggests that the stratigraphical section is more continuous than previously recognized. Gersonde & Burckle (1990) and Gersonde *et al.* (1990) identified 11 hiatuses at 2.0, 7.2, 14.9, 18.8, 21.0, 31.2, 39.0, 43.7, 47.1, 48.7 and 51.2 mbsf. These breaks are based primarily on an integrated bio and magnetostratigraphy. Gersonde *et al.* (1990) identify the hiatuses at approximately 17.5 and 51.2 mbsf as the most important stratigraphical breaks. The unconformities at approximately 18.45 and 51.5 mbsf recognized in this study occur at approximately the same depths in the hole to similar unconformities identified by Gersonde & Burckle (1990).

Spiess (1990) identified nine hiatuses within an interval of approximately 37.8 my. Breaks at 2.0, 7.23, 18.83, 21.07, 31.2, 43.66, 51.28 mbsf coincide with Neogene hiatuses recognized by Gersonde & Burckle (1990) and Gersonde *et al.* (1990). Earlier hiatuses at 72.18 and 91.58 mbsf occur within the upper Eocene to Oligocene sequence, which is not considered by Gersonde & Burckle (1990) and Gersonde *et al* (1990). The hiatus at approximately 90 mbsf is coincides with the hiatus at 91.58 mbsf recorded by Spiess 1990.

Table 10. *Composite dataset for ODP Hole 690B containing the diatom, radiolarian, calcareous nannofossil and palaeomagnetic data calibrated to the GPTS of Cande and Kent (1995)*

	Event	Sample (top)	Sample (bottom)	Depth (t)	Depth (b)	Age (t)	Age (b)
FO	*Nitzschia kerguelensis*	1H-1, 0	1H-1, 73-75	0.00	0.75	?	?
LO	*Antarctissa cylindrica*	1H-1, 0	2H-2, 0	0.00	2.10	?	?
LO	*Cycladophora pliocenica*	1H-1, 0	2H-2, 0	0.00	2.10	?	?
LO	*Eucyrtidium calvertense*	1H-1, 0	2H-2, 0	0.00	2.10	?	?
LO	*Desmospyris spongiosa*	1H-1, 0	2H-2, 0	0.00	2.10	?	?
LO	*Helotholus vema*	1H-1, 0	2H-2, 0	0.00	2.10	?	?
FO	*Cycladophora davisiana*	1H-1, 0	2H-2, 0	0.00	2.10	?	?
LO	*Actinocyclus ingens*	1H-1, 73-75	1H-2, 32-34	0.75	1.84	?	?
LO	*Thalassiosira kolbei*	1H-2, 32-34	2H-1, 28-29	1.84	2.39	?	?
LO	*Nitzschia interfrigidaria*	1H-2, 32-34	2H-1, 28-29	1.84	2.39	?	?
LO	*Thalassiosira inura*	1H-2, 32-34	2H-1, 28-29	1.84	2.39	?	?
LO	*Nitzschia barronii*	1H-2, 32-34	2H-1, 28-29	1.84	2.39	?	?

(*continued*)

Table 10. (*continued*)

	Event	Sample (top)	Sample (bottom)	Depth (t)	Depth (b)	Age (t)	Age (b)
	Unconformity			***	***	***	***
T	? Reunion			3.45	3.45	?	?
B	? Reunion			3.70	3.70	?	?
T	C2An.1n			4.83	4.83	2.58	2.58
T	*Simonseniella barboi*	2H-3, 114-115	2H-4, 28-29	6.25	6.89	2.93	3.09
B	2An.1n			6.71	6.71	3.04	3.04
LO	*Nitzschia praeinterfrigidaria*	2H-4, 28-29	2H-4, 114-115	6.89	7.75	3.09	?
T	C2An.2n			6.98	6.98	3.11	3.11
	Unconformity			***	***	***	***
FO	*Thalassiosira lentiginosa*	2H-4, 114-115	2H-5, 28-29	7.75	8.39	?	?
FO	*Nitzschia interfrigidaria*	2H-5, 28-29	2H-5, 114-115	8.39	9.25	?	?
LO	*Prunopyle titan*	2H-6, 24	2H-CC	9.84	11.70	?	?
FO	*Helotholus vema*	2H-6, 24	2H-CC	9.84	11.70	?	?
T	C3n.1n			10.23	10.23	4.18	4.18
B	C3n.1n			10.48	10.48	4.29	4.29
T	C3n.2n			11.21	11.21	4.48	4.48
FO	*Nitzschia barronii*	2H-7, 27-29	3H-1, 27-28	11.39	11.98	4.52	4.63
B	C3n.2n			11.78	11.78	4.62	4.62
LO	*Denticulopsis dimorpha*	3H-1, 27-28	3H-1, 115-116	11.98	12.86	4.63	4.68
FO	*Thalassiosira inura*	3H-2, 114-115	3H-3, 73-75	14.36	15.21	4.76	4.81
FO	*Nitzschia praeinterfrigidaria*	3H-3, 73-75	3H-3, 125-127	15.45	15.97	4.82	4.85
LO	*Cycladophora spongothorax*	3H-4, 26	3H-6, 26	16.46	19.46	4.88	?
LO	*Cycladophora humerus*	3H-CC	4H-2, 26-28	16.76	18.26	4.89	4.98
T	C3n.4n			18.32	18.32	4.98	4.98
FO	*Nitzschia praeinterfrigidaria*	3H-3, 73-75	3H-3, 125-127	18.45	18.87	?	?
LO	*Asteromphalus kennettii*	3H-5, 73-75	3H-5, 114-115	18.45	18.87	?	?
	Unconformity			***	***	***	***
B	unidentified anomaly			18.57	18.57	5.23	5.23
FO	*Cosmiodiscus intersectus*	3H-5, 114-115	3H-6, 23-25	18.87	19.95	?	?
B	C4An			20.07	20.07	9.03	9.03
	Unconformity			***	***	***	***
LO	*Cycladophora humerus*	3H-CC	4H-2, 26-28	21.40	23.16	?	10.00
B	C5n.1n			22.53	22.53	9.88	9.88
T	C5n.2n			22.76	22.76	9.92	9.92
FO	*Eucyrtidium pseudoinflatum*	4H-2, 26	4H-4, 26	23.16	26.16	10.00	10.58
LO	*Actinomma golownini*	4H-2, 26	4H-4, 26	23.16	26.16	10.00	10.58
FO	*Asteromphalus kennettii*	4H-1, 114-115	4H-3, 26-27	24.06	24.67	10.17	10.29
LO	*Denticulopsis praedimorpha*	4H-5, 26-27	4H-5, 114-115	27.67	28.55	10.88	?
B	C5n.2n			28.03	28.03	10.95	10.95
	Unconformity			***	***	***	***
LO	*Nitzschia denticuloides*	5H-1, 28-29	5H-1, 115-117	31.39	32.27	?	?
T	C5An.1n			31.72	31.72	11.94	11.94
B	C5An.1n			32.45	32.45	12.08	12.08
FO	*Cycladophora spongothorax*	5H-2, 26	5H-4, 26	32.86	35.86	12.12	12.93
LO	*Reticulofenestra gelida*	5H-1, 30-32	5H-2, 30-32	32.90	34.40	12.12	12.51
LO	*Reticulofenestra hesslandii*	5H-1, 30-32	5H-2, 30-32	32.90	34.40	12.12	12.51
T	C5An.2n			33.47	33.47	12.18	12.18
B	C5An.2n			33.95	33.95	12.40	12.40
LO	*Crucidenticula nicobarica*	5H-3, 27-29	5H-3, 115-117	34.39	35.25	12.51	12.72
FO	*Denticulopsis dimorpha*	5H-3, 115-117	5H-4, 27-29	35.25	35.89	12.72	12.88
T	C5AAn			36.35	36.35	12.99	12.99
FO	*Denticulopsis praedimorpha*	5H-4, 115-117	5H-5, 27-29	36.75	37.39	13.09	13.25
B	C5AAn			36.95	36.95	13.14	13.14
T	C5ABn			37.20	37.20	13.30	13.30
FO	*Actinomma golownini*	5H-5, 49-51	5H-6, 49-51	37.61	39.11	13.41	13.82
	C5ABn			37.97	37.97	13.51	13.51
FO	*Nitzschia denticuloides*	5H-6, 27-29	5H-6, 115-117	38.89	39.75	13.68	14.00
LO	*Nitzschia grossepunctata*	5H-6, 27-29	5H-6, 115-117	38.89	39.75	13.68	14.00
T	C5ACn			38.97	38.97	13.70	13.70
FO	*Cycladophora humerus*	5H-6, 49-51	6H-1, 51-52	39.11	41.32	13.75	14.51
FO	*Antarctissa deflandrei*	5H-6, 49-51	6H-1, 51-52	39.11	41.32	13.75	14.51
B	C5ACn			39.95	39.95	14.08	14.08
T	C5ADn			40.20	40.20	14.18	14.18
LO	*Denticulopsis maccollumnii*	5H-7, 27-29	6H-1, 27-29	40.39	41.08	14.24	14.44
FO	*Denticulopsis hustedtii*	5H-7, 27-29	6H-1, 27-29	40.39	41.08	14.24	14.44
LO	*Nitzschia maleinterpretaria*	6H-1, 50-52	6H-1, 115-117	41.32	42.00	14.51	14.74
LO	*Synedra jouseana*	6H-1, 50-52	6H-1, 115-117	41.32	42.00	14.51	14.74
FO	*Actinocyclus ingens v. nodus*	6H-1, 50-52	6H-1, 115-117	41.32	42.00	14.51	14.74
B	C5ADn			41.68	41.68	14.61	14.61

Table 10. (continued)

	Event	Sample (top)	Sample (bottom)	Depth (t)	Depth (b)	Age (t)	Age (b)
T	C5Bn.1n			42.16	42.16	14.80	14.80
B	C5Bn.1n			42.41	42.41	14.89	14.89
LO	Crucidenticula kanayae	6H-2, 49-51	6H-2,115-117	42.81	43.45	?	?
FO	Nitzschia grossepunctata	6H-2, 115-117	6H-3, 27-29	43.45	44.08	?	?
FO	Actinocyclus ingens	6H-2, 115-117	6H-3, 27-29	43.45	44.08	?	?
FO	Nitzschia maleinterpretaria	6H-2, 115-117	6H-3, 27-29	43.45	44.08	?	?
	Unconformity			***	***	***	***
FO	Denticulopsis maccollumnii	6H-3, 27-29	6H-3, 49-51	44.08	44.31	?	?
FO	Crucidenticula kanayae	6H-3, 27-29	6H-3, 49-51	44.08	44.31	?	?
B	In C5Dn			44.68	44.68	?	?
LO	Thalassiosira fraga	6H-3, 115-117	6H-4, 27-29	44.95	45.58	?	?
FO	Cycladophora golli regipileus	6H-5, 49-51	6H-6, 49-51	45.41	47.31	?	?
B	C5Dn			45.68	45.68	17.62	17.62
B	In C5En			46.68	46.68	?	?
FO	Cyrtocapsella longithorax	6H-6, 49-51	6H-7, 49-51	48.81	50.31	?	?
FO	Crucidenticula nicobarica	6H-5, 27-29	6H-5, 114-115	47.09	47.95	?	?
LO	Thalassiosira spumellaroides	6H-6, 49-51	6H-6, 115-116	48.59	49.46	?	?
FO	Cyrtocapsella tetrapera	6H-6, 49-51	7H-1, 49-51	48.81	50.89	?	?
FO	Thalassiosira fraga	6H-7, 27-29	7H-1, 27-29	50.09	50.69	?	?
LO	Rocella gelida	6H-7, 27-29	7H-1, 27-29	50.09	50.69	?	?
LO	Rocella vigilans	7H-1, 27-29	7H-1, 49-51	50.68	50.91	?	?
FO	Thalassiosira aspinosa	7H-1, 27-29	7H-1, 49-51	50.68	50.91	?	?
FO	Raphidodiscus marylandicus	7H-1, 27-29	7H-1, 49-51	50.68	50.91	?	?
LO	Lisitzina ornata	7H-1, 49-51	7H-1, 114-115	50.91	51.55	?	?
FO	Thalassiosira spumellaroides	7H-1, 49-51	7H-1, 114-115	50.91	51.55	?	?
	Unconformity			***	***	***	***
LO	Chiasmolithus altus			51.72	52.20	?	?
T	C8n.1n			53.25	53.25	25.82	25.82
FO	Rocella gelida	7H-3, 114-115	7H-4, 27-29	54.55	55.19	25.93	26.01
FO	Rocella vigilans	7H-3, 114-115	7H-4, 27-29	54.55	55.19	25.93	26.01
B	C8n.1n			54.75	54.75	25.95	25.95
T	C8n.2n			55.00	55.00	25.99	25.99
FO	Lisitzina ornata	7H-6, 27-29	7H-6, 114-115	58.19	59.05	26.34	26.43
B	C8n.2n			60.11	60.11	26.55	26.55
T	C9n			60.99	60.99	27.03	27.03
B	C9n			68.48	68.48	27.97	27.97
LO	Reticulofenestra bisecta	9H-2, 28-30	9H-3, 28-30	71.58	73.10	?	?
	Unconformity			***	***	***	***
B	C11n.2n			73.93	73.93	30.10	30.10
T	C12n			80.76	80.76	30.48	30.48
FO	Cyclicargolithus abisectus	9H-6, 28-30	9H-7, 28-30	82.68	84.18	30.75	?
B	C12n			84.01	84.01	30.94	30.94
LO	Reticulofenestra umbilica	10H-4, 26-28	10H-5, 26-28	84.18	85.68	?	?
LO	Isthmolithus recurvus	11H-1, 29-31	11H-1, 130-132	89.39	90.09	?	?
	Unconformity			***	***	***	***
B	C16n.2n			91.95	91.95	36.34	36.34
FO	Isthmolithus recurvus	11H-3, 29-31	11H-3, 130-132	92.41	93.42	36.36	36.39
LO	Reticulofenestra reticulata	11H-3, 29-31	11H-3, 130-132	92.41	93.42	36.36	36.39
	In C16r			95.7	95.7	?	?
T	17n.1n			96.59	96.59	36.62	36.62
B	17n.1n			99.04	99.04	37.47	37.47

Data sets are integrated from the work of numerous authors (see Table 2).

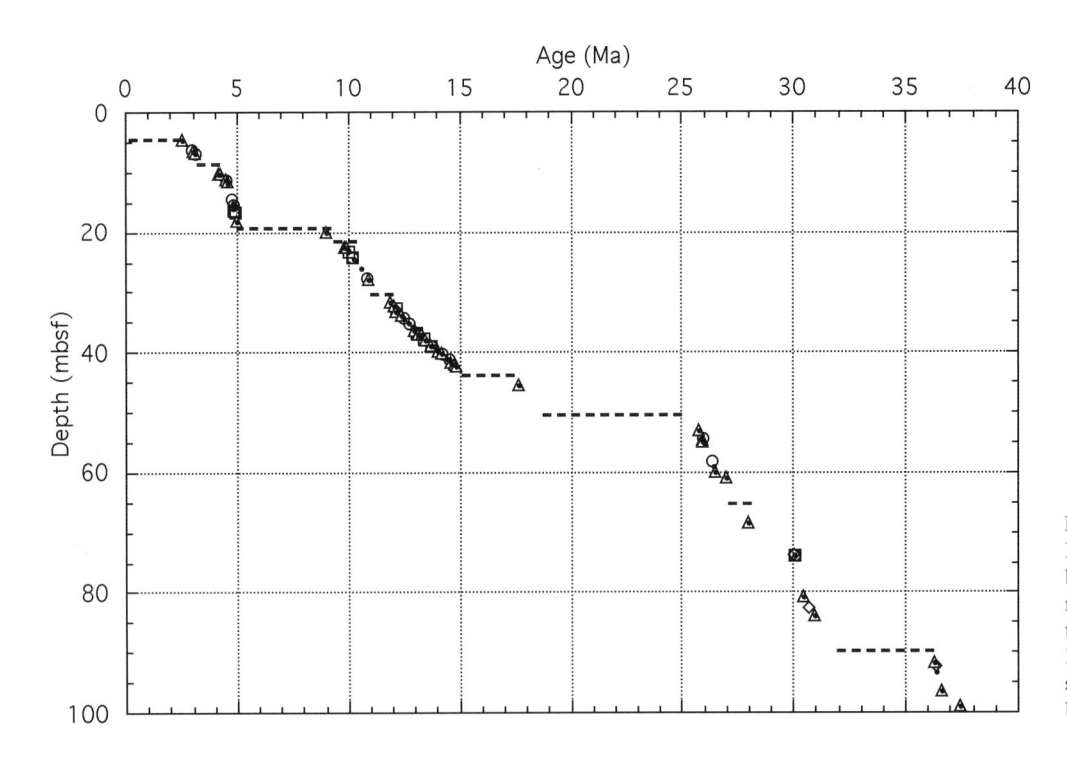

Fig. 12. An age–depth plot for DSDP Hole 690B utilizing the previously used biostratigraphical datums and polarity reversal interpretation calibrated to the GPTS of Cande & Kent (1995). See Fig. 5 for legend of symbols. Nine stratigraphical breaks (dashed line) are based on Spiess (1990).

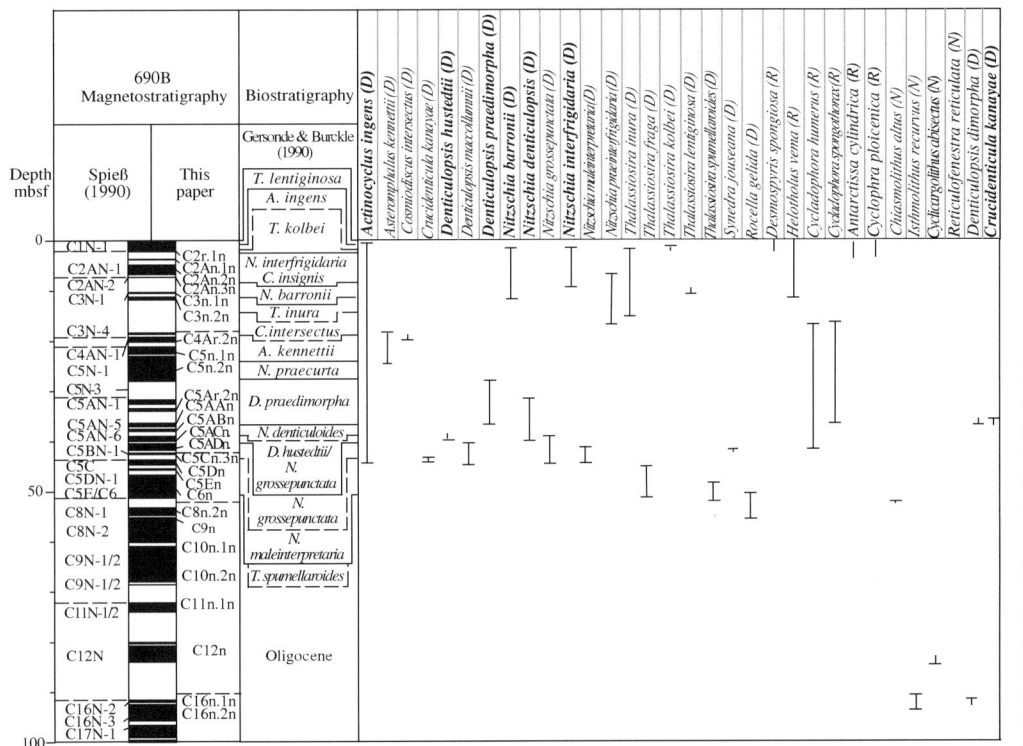

Fig. 13. Comparison of the ODP Hole 690B polarity chronozones and diatom zonal assignments by previous workers with the chronozonal assignments of this study. The stratigraphical ranges of selected stratigraphical indicators are based on the original data sets (see text for references). D, diatoms; R, radiolaria; N, calcareous nannofossils. Emboldened events indicate primary biostratigraphical events recognized in this study. The polarity record shown represents the original interpretation calibrated to the GPTS of Cande & Kent (1995) and the interpretation from this study correlated with the GPTS of Cande & Kent (1995).

Table 11. *Composite dataset for ODP Hole 690B containing the diatom, radiolarian, calcareous nannofossil and palaeomagnetic data calibrated to the GPTS of Cande & Kent (1995) using the biochronological interpretation based on this study*

	Event	Sample (top)	Sample (bottom)	Depth (t)	Depth (b)	Age (t)	Age (b)
FO	*Nitzschia kerguelensis*	1H-1, 0	1H-1, 73-75	0.00	0.75	0.00	0.81
LO	*Antarctissa cylindrica*	1H-1, 0	2H-2, 0	0.00	2.10	0.00	2.26
LO	*Cycladophora pliocenica*	1H-1, 0	2H-2, 0	0.00	2.10	0.00	2.26
LO	*Eucyrtidium calvertense*	1H-1, 0	2H-2, 0	0.00	2.10	0.00	2.26
LO	*Desmospyris spongiosa*	1H-1, 0	2H-2, 0	0.00	2.10	0.00	2.26
LO	*Helotholus vema*	1H-1, 0	2H-2, 0	0.00	2.10	0.00	2.26
FO	*Cycladophora davisiana*	1H-1, 0	2H-2, 0	0.00	2.10	0.00	2.26
LO	*Actinocyclus ingens*	1H-1, 73-75	1H-2, 32-34	0.75	1.84	0.81	1.98
LO	*Thalassiosira kolbei*	1H-2, 32-34	2H-1, 28-29	1.84	2.39	1.98	2.27
LO	*Nitzschia interfrigidaria*	1H-2, 32-34	2H-1, 28-29	1.84	2.39	1.98	2.27
LO	*Thalassiosira inura*	1H-2, 32-34	2H-1, 28-29	1.84	2.39	1.98	2.27
LO	*Nitzschia barronii*	1H-2, 32-34	2H-1, 28-29	1.84	2.39	1.98	2.27
B	C2r.1n			2.00	2.00	2.15	2.15
T	C2An.1n			3.45	3.45	2.58	2.58
B	C2An.1n			3.70	3.70	3.04	3.04
T	C2An.2n			4.83	4.83	3.11	3.11
LO	*Simonseniella barboi*	2H-3, 114-115	2H-4, 28-29	6.25	6.89	3.19	3.29
B	C2An.2n			6.71	6.71	3.22	3.22
LO	*Nitzschia praeinterfrigidaria*	2H-4, 28-29	2H-4, 114-115	6.89	7.75	3.29	3.68
T	C2An.3n			6.98	6.98	3.33	3.33
B	C2An.3n			7.23	7.23	3.58	3.58
FO	*Thalassiosira lentiginosa*	2H-4, 114-115	2H-5, 28-29	7.75	8.39	3.68	3.81
FO	*Nitzschia interfrigidaria*	2H-5, 28-29	2H-5, 114-115	8.39	9.25	3.81	3.98
LO	*Prunopyle titan*	2H-6, 24	2H-CC	9.84	11.70	4.10	4.60
FO	*Helotholus vema*	2H-6, 24	2H-CC	9.84	11.70	4.10	4.60
T	C3n.1n			10.23	10.23	4.18	4.18
B	C3n.1n			10.48	10.48	4.29	4.29
T	C3n.2n			11.21	11.21	4.48	4.48
FO	*Nitzschia barronii*	2H-7, 27-29	3H-1, 27-28	11.39	11.98	4.52	?
B	C3n.2n			11.78	11.78	4.62	4.62
LO	*Denticulopsis dimorpha*	3H-1, 27-28	3H-1, 115-116	11.98	12.86	4.63	4.68
FO	*Thalassiosira inura*	3H-2, 114-115	3H-3, 73-75	14.36	15.21	?	?
FO	*Nitzschia praeinterfrigidaria*	3H-3, 73-75	3H-3, 125-127	15.45	15.97	?	?
LO	*Cycladophora spongothorax*	3H-4, 26	3H-6, 26	16.46	19.46	?	?
LO	*Cycladophora humerus*	3H-CC	4H-2, 26-28	16.70	18.26	?	?
	Unconformity			***	***	***	***
T	C4Ar.1n			18.32	18.32	9.23	9.23
LO	*Asteromphalus kennettii*	3H-5, 73-75	3H-5, 114-115	18.45	18.87	9.27	9.31
B	C4Ar.1n			18.57	18.57	9.31	9.31
T	C4Ar.2n			18.83	18.83	9.58	9.58
FO	*Cosmiodiscus intersectus*	3H-5, 114-115	3H-6, 23-25	18.87	19.95	9.58	9.63
B	C4Ar.2n			20.07	20.07	9.64	9.64
B	C5n.1n			22.53	22.53	9.88	9.88
T	C5n.2n			22.76	22.76	9.92	9.92
FO	*Eucyrtidium pseudoinflatum*	4H-2, 26	4H-4, 26	23.16	26.16	10.00	10.58
LO	*Actinomma golownini*	4H-2, 26	4H-4, 26	23.16	26.16	10.00	10.58
FO	*Asteromphalus kennettii*	4H-1, 114-115	4H-3, 26-27	24.06	24.67	10.17	10.29
LO	*Cycladophora humerus*	3H-CC	4H-2, 26-28	24.30	26.08	10.22	10.57
LO	*Denticulopsis praedimorpha*	4H-5, 26-27	4H-5, 114-115	27.67	28.55	10.88	11.02
B	C5n.2n			28.03	28.03	10.95	10.95
T	C5r.2n			31.72	31.72	11.48	11.48
LO	*Nitzschia denticuloides*	5H-1, 28-29	5H-1, 115-117	31.39	32.27	11.46	11.52
B	C5r.2n			32.45	32.45	11.53	11.53
FO	*Cycladophora spongothorax*	5H-2, 26	5H-4, 26	32.86	35.86	12.03	12.96
LO	*Reticulofenestra gelida*	5H-1, 30-32	5H-2, 30-32	32.90	34.40	12.08	12.85
LO	*Reticulofenestra hesslandii*	5H-1, 30-32	5H-2, 30-32	32.90	34.40	12.08	12.85
T	C5Ar.2n			33.47	33.47	12.78	12.78
B	C5Ar.2n			33.95	33.95	12.82	12.82
LO	*Crucidenticula nicobarica*	5H-3, 27-29	5H-3, 115-117	34.39	35.25	12.85	12.91
FO	*Denticulopsis dimorpha*	5H-3, 115-117	5H-4, 27-29	35.25	35.89	12.91	12.96
T	C5AAn			36.35	36.35	12.99	12.99
FO	*Denticulopsis praedimorpha*	5H-4, 115-117	5H-5, 27-29	36.75	37.39	13.09	13.35
B	C5AAn			36.95	36.95	13.14	13.14
T	C5ABn			37.20	37.20	13.30	13.30
FO	*Actinomma golownini*	5H-5, 49-51	5H-6, 49-51	37.61	39.11	13.41	13.75
B	C5ABn			37.97	37.97	13.51	13.51
FO	*Nitzschia denticuloides*	5H-6, 27-29	5H-6, 115-117	38.89	39.75	13.68	14.02

(*continued*)

Table 11. (*continued*)

	Event	Sample (top)	Sample (bottom)	Depth (t)	Depth (b)	Age (t)	Age (b)
LO	*Nitzschia grossepunctata*	5H-6, 27-29	5H-6, 115-117	38.89	39.75	13.68	14.02
T	C5ACn			38.97	38.97	13.70	13.70
FO	*Cycladophora humerus*	5H-6, 49-51	6H-1, 51-52	39.11	41.32	13.75	14.51
FO	*Antarctissa deflandrei*	5H-6, 49-51	6H-1, 51-52	39.11	41.32	13.75	14.51
B	C5ACn			39.95	39.95	14.08	14.08
T	C5ADn			40.20	40.20	14.18	14.18
LO	*Denticulopsis maccollumnii*	5H-7, 27-29	6H-1, 27-29	40.39	41.08	14.24	14.44
FO	*Denticulopsis hustedtii*	5H-7, 27-29	6H-1, 27-29	40.39	41.08	14.24	14.44
LO	*Nitzschia maleinterpretaria*	6H-1, 50-52	6H-1, 115-117	41.32	42.00	14.51	?
LO	*Synedra jouseana*	6H-1, 50-52	6H-1, 115-117	41.32	42.00	14.51	?
FO	*Actinocyclus ingens v. nodus*	6H-1, 50-52	6H-1, 115-117	41.32	42.00	14.51	?
B	C5ADn			41.68	41.68	14.61	14.61
	Unconformity			***	***	***	***
T	C5Cn.3n			42.16	42.16	16.56	16.56
B	C5Cn.3n			42.41	42.41	16.73	16.73
LO	*Crucidenticula kanayae*	6H-2, 49-51	6H-2, 115-117	42.81	43.45	16.91	17.19
FO	*Nitzschia grossepunctata*	6H-2, 115-117	6H-3, 27-29	43.45	44.08	17.19	17.42
FO	*Actinocyclus ingens*	6H-2, 115-117	6H-3, 27-29	43.45	44.08	17.19	17.42
FO	*Nitzschia maleinterpretaria*	6H-2, 115-117	6H-3, 27-29	43.45	44.08	17.19	17.42
T	C5Dn			43.66	43.66	17.28	17.28
FO	*Denticulopsis maccollumnii*	6H-3, 27-29	6H-3, 49-51	44.08	44.31	17.42	17.50
FO	*Crucidenticula kanayae*	6H-3, 27-29	6H-3, 49-51	44.08	44.31	17.42	17.50
B	C5Dn			44.68	44.68	17.62	17.62
LO	*Thalassiosira fraga*	6H-3, 115-117	6H-4, 27-29	44.95	45.58	17.86	18.59
T	C5En			45.41	45.41	18.28	18.28
FO	*Cycladophora golli regipileus*	6H-5, 49-51	6H-6, 49-51	45.41	47.31	18.28	19.20
B	C5En			45.68	45.68	18.78	18.78
T	C6n			46.68	46.68	19.05	19.05
FO	*Cyrtocapsella longithorax*	6H-6, 49-51	6H-7, 49-51	47.31	48.81	19.20	19.55
FO	*Crucidenticula nicobarica*	6H-5, 27-29	6H-5, 115-117	48.58	49.45	19.50	19.70
LO	*Thalassiosira spumellaroides*	6H-6, 27-29	6H-6, 49-51	48.58	48.81	19.50	19.55
FO	*Cyrtocapsella tetrapera*	6H-6, 49-51	7H-1, 49-51	48.81	50.89	19.55	20.04
FO	*Thalassiosira fraga*	6H-7, 27-29	7H-1, 27-29	50.09	50.69	19.85	19.99
LO	*Rocella gelida*	6H-7, 27-29	7H-1, 27-29	50.09	50.69	19.85	19.99
LO	*Rocella vigilans*	7H-1, 27-29	7H-1, 49-51	50.68	50.91	19.99	20.04
FO	*Thalassiosira aspinosa*	7H-1, 27-29	7H-1, 49-51	50.68	50.91	19.99	20.04
FO	*Raphidodiscus marylandicus*	7H-1, 27-29	7H-1, 49-51	50.68	50.91	19.99	20.04
LO	*Lisitzina ornata*	7H-1, 49-51	7H-1, 114-115	50.91	51.55	20.04	?
FO	*Thalassiosira spumellaroides*	7H-1, 49-51	7H-1, 114-115	50.91	51.55	20.04	?
B	C6n			51.28	51.28	20.13	20.13
	Unconformity			***	***	***	***
LO	*Chiasmolithus altus*			51.72	52.20	?	?
T	C8n.2n			53.25	53.25	25.99	25.99
FO	*Rocella gelida*	7H-3, 114-115	7H-4, 27-29	54.55	55.19	26.48	27.06
FO	*Rocella vigilans*	7H-3, 114-115	7H-4, 27-29	54.55	55.19	26.48	27.06
B	C8n.2n			54.75	54.75	26.55	26.55
T	C9n			55.00	55.00	27.03	27.03
FO	*Lisitzina ornata*	7H-6, 27-29	7H-6, 114-115	58.19	59.05	27.62	27.78
B	C9n			60.11	60.11	27.97	27.97
T	C10n.1n			60.99	60.99	28.28	28.28
B	C10n.1n			67.98	67.98	28.51	28.51
T	C10n.2n			68.23	68.23	28.58	28.58
B	C10n.2n			68.48	68.48	28.75	28.75
LO	*Reticulofenestra bisecta*	9H-2, 28-30	9H-3, 28-30	71.58	73.10	29.29	29.54
T	C11n.1n			72.18	72.18	29.40	29.40
B	C11n.1n			73.93	73.93	29.66	29.66
FO	*Cyclicargolithus abisectus*	9H-6, 28-30	9H-7, 28-30	82.68	84.18	30.77	?
T	C12n			80.76	80.76	30.48	30.48
B	C12n			84.01	84.01	30.94	30.94
LO	*Reticulofenestra umbilica*	10H-4, 26-28	10H-5, 26-28	84.18	85.68	?	?
LO	*Isthmolithus recurvus*	11H-1, 29-31	11H-1, 130-132	89.39	90.09	?	?
	Unconformity			***	***	***	***
B	C16n.1n			91.95	91.95	35.53	35.53
T	C16n.2n			92.21	92.21	35.69	35.69
FO	*Isthmolithus recurvus*	11H-3, 29-31	11H-1, 130-132	92.41	93.42	35.73	35.92
LO	*Reticulofenestra reticulata*	11H-3, 29-31	11H-1, 130-132	92.41	93.42	35.73	35.92
B	C16n.2n			95.70	95.70	36.34	36.34

Data sets are integrated from the work of numerous authors (see Table 2).

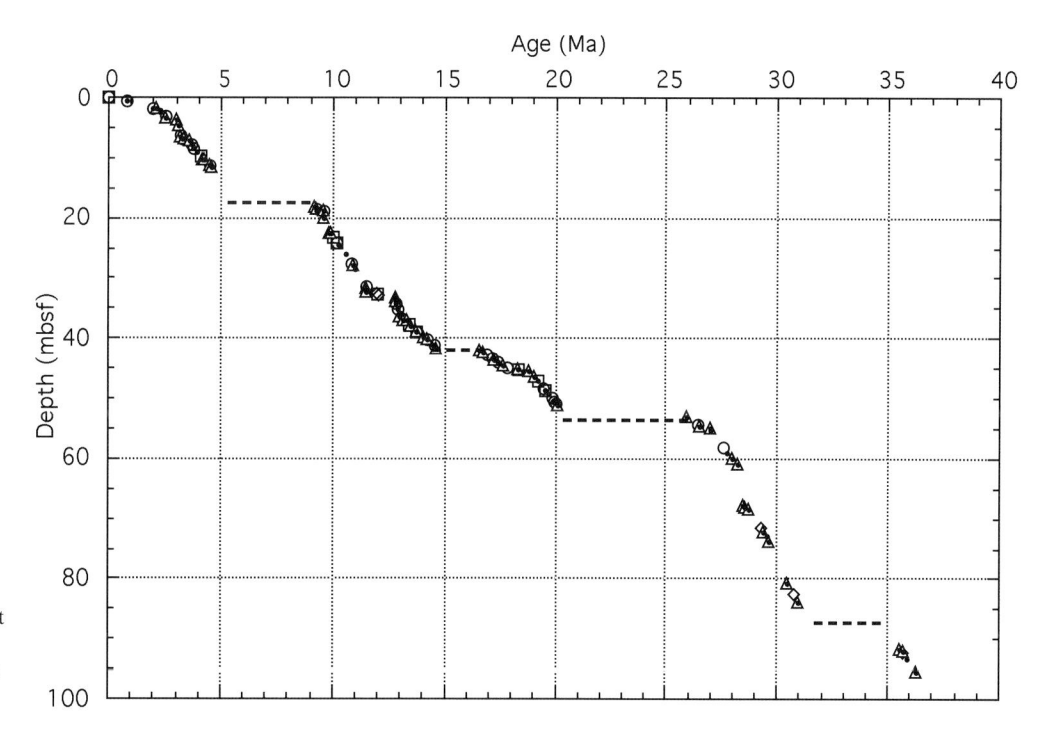

Fig. 14. A revised age–depth plot for DSDP Hole 690B utilizing the biostratigraphical and polarity reversal events identified in this study and calibrated to the GPTS of Cande & Kent (1995). See Fig. 5 for legend of symbols. Four stratigraphical breaks (dashed line) are based on the results of this study.

Hole 695A

Thirty-six cores were recovered from a total penetration of 341.1 mbsf representing a recovery of 75% of the cored interval. Sediments recovered consist of Quaternary to lower Pliocene diatom ooze with a variable terrigenous content and lower Pliocene to upper Miocene muds with diatoms (Barker *et al.* 1988). A good palaeomagnetic polarity reversal record was obtained using shipboard data (Barker *et al.* 1988). Diatoms and radiolarians are present throughout the sequence. Rare to abundant and poor to well preserved diatoms are recorded throughout the section (Gersonde & Burckle 1990). Calcareous nannofossils are absent from this site. Well-preserved radiolarians are common to abundant throughout the section with fewer radiolarians observed in the lowermost portion of the sequence (Barker *et al.* 1988). The previously used biostratigraphical and magnetostratigraphical data (palaeomagnetics: Barker *et al.* 1988; diatoms: Gersonde & Burckle 1990; calcareous nannofossils: none; radiolarians: none) integrated and calibrated to the GPTS of Berggren *et al.* (1985a, b) is presented in Table 12. An age–depth graph for Hole 113-695A using this data is shown in Fig. 15.

Figure 16 compares the polarity chronozones and diatom assigned by previous workers (Barker *et al.* 1988; Gersonde & Burckle 1990) with the chronozonal assignments in this study. The stratigraphical ranges reflect the placement of FO and LO of stratigraphically useful species as recognized by Gersonde & Burckle (1990). The revised chronozones illustrated are based on reassessing the data sets of these workers (Table 13). Comparison of the chronozonal assignments recorded in Barker *et al.* (1988; Table 12) with those of this study (Table 13) indicates that only minimal reinterpretations occur resulting from differences in the specific GPTS used in each study.

An age–depth graph for Hole 113-695A using the stratigraphical markers determined in this study is shown in Fig. 17. Stratigraphical control is based predominantly on a continuous sequence of polarity reversals extending from the base of C2n at 5.5 mbsf to the base of C3n.2n at 191 mbsf. This sequence of reversals is correlated with the GPTS using the FO of *Nitzschia praeinterfrigidaria* (186.32–195.19 mbsf) and the FO of *Nitzschia barronii* (127.9–13.6 mbsf). The first occurrence of *Nitzschia barronii* at 4.02–4.18 Ma is approximately 0.25 million years younger than at other sites in this study. The results of this study are similar to those presented by Gersonde & Burckle (1990) and Barker *et al.* (1988).

Table 12. *Composite dataset for ODP Hole 695A containing the diatom, radiolarian, calcareous nannofossil and palaeomagnetic data calibrated to the GPTS of Cande and Kent (1995)*

	Event	Sample (top)	Sample (bottom)	Depth (t)	Depth (b)	Age (t)	Age (b)
LO	*Antarctissa cylindrica*	1H-1, 0	1H-CC	0.00	6.40	0.00	2.27
LO	*Cycladophora pliocenica*	1H-1, 0	1H-CC	0.00	6.40	0.00	2.27
LO	*Actinocyclus ingens*	1H1, 55-56	1H-2, 30-31	0.56	1.81	0.20	0.64
LO	*Thalassiosira vulnifica*	1H4, 30-31	1H-4, 100-101	1.81	5.51	0.64	1.95
FO	*Nitzschia kerguelensis*	1H3, 30-31	1H4, 30-31	3.31	4.80	1.17	1.70
B	C2n			5.50	5.50	1.95	1.95
LO	*Thalassiosira insigna*	1H-CC	2H-6, 70-71	6.40	11.11	2.11	2.77
LO	*Desmospyris spongiosa*	1H-CC	2H-CC	6.40	12.40	2.11	2.89
LO	*Helotholus vema*	1H-CC	2H-CC	6.40	12.40	2.11	2.89
FO	*Cycladophora davisiana*	1H-CC	2H-CC	6.40	12.40	2.11	2.89
T	C2An.1n			9.00	9.00	2.58	2.58
LO	*Nitzschia praeinterfrigidaria*	2H-6, 70-71	2H-CC	11.11	12.40	2.77	2.92
FO	*Thalassiosira vulnifica*	2H-CC	3H-4, 22-22	12.40	17.12	2.89	3.11
LO	*Nitzschia interfrigidaria*	2H-CC	3H-4, 22-22	12.40	17.12	2.89	3.11
B	C2An.1n			14.00	14.00	3.04	3.04
T	C2An.2n			17.00	17.00	3.11	3.11
LO	*Nitzschia barronii*	3H-4, 21-22	3H-5, 75-76	17.12	19.16	3.11	3.12
LO	*Nitzschia weaveri*	3H-4, 21-22	3H-5, 75-76	17.12	19.60	3.11	3.12
FO	*Thalassiosira insigna*	3H-CC	4H-CC	22.00	31.60	3.13	3.16
LO	*Thalassiosira inura*	5HCC	6H-CC	41.20	50.80	3.20	3.30
B	C2An.2n			48.00	48.00	3.22	3.22
FO	*Nitzschia interfrigidaria*	6H-CC	7H-CC	50.80	60.40	3.30	3.71
T	C2An.3n			52.00	52.00	3.33	3.33
B	C2An.3n			53.00	53.00	3.58	3.58
T	C3n.1n			87.00	87.00	4.18	4.18
B	C3n.1n			103.50	103.50	4.29	4.29
T	C3n.2n			137.00	137.00	4.48	4.48
FO	*Nitzschia barronii*	15H-CC	16H-CC	127.90	137.60	4.42	4.48
FO	*Nitzschia weaveri*	18X-CC	19X-CC	147.30	157.00	4.54	4.60
LO	*Denticulopsis hustedtii*	19X-CC	20X-CC	157.00	166.80	4.60	4.68
LO	*Prunopyle titan*	19X-CC	20X-CC	157.00	166.80	4.60	4.68
B	C3n.2n			160.00	160.00	4.62	4.62
T	C3n.3n			176.50	176.50	4.80	4.80
LO	*Cosmiodiscus intersectus*	21X-CC	22X-CC	176.50	186.20	4.80	4.89
FO	*Nitzschia praeinterfrigidaria*	22X-CC	23X-CC	186.20	195.90	4.83	4.86
B	C3n.3n			205.60	205.60	4.89	4.89

Data sets are integrated from the work of numerous authors (see Table 2).

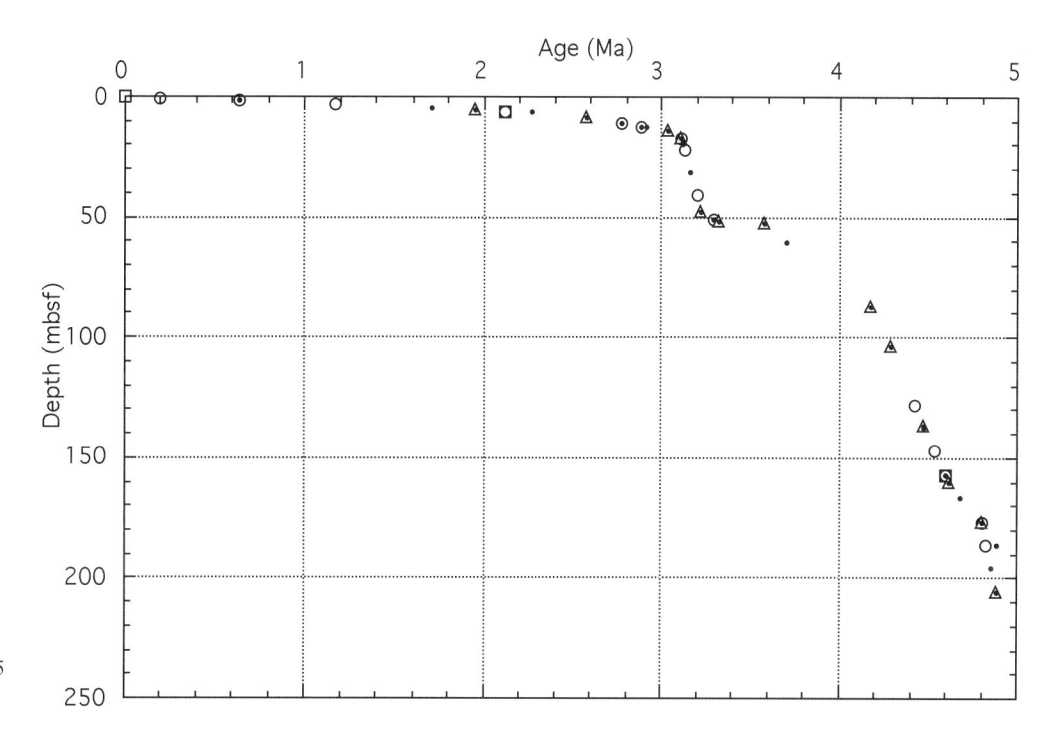

Fig. 15. An age–depth plot for DSDP Hole 695A utilizing the previously used biostratigraphical datums and polarity reversal interpretation calibrated to the GPTS of Cande & Kent (1995). See Fig. 5 for legend of symbols.

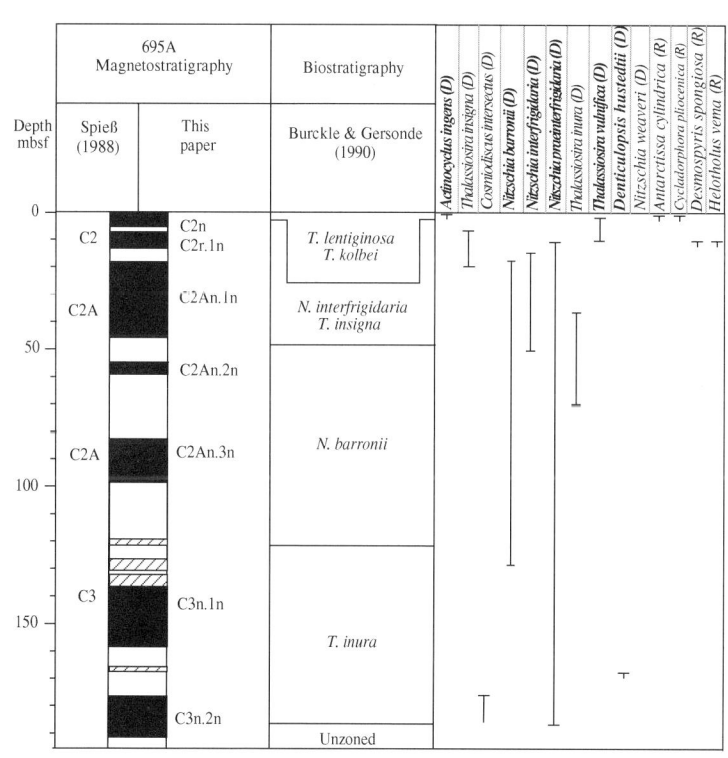

Fig. 16. Comparison of the ODP Hole 695A polarity chronozones and diatom zonal assignments by previous workers with the chronozonal assignments of this study. The stratigraphical ranges of selected stratigraphical indicators are based on the original data sets (see text for references). D, diatoms; R, radiolaria; N, calcareous nannofossils. Emboldened events indicate primary biostratigraphical events recognized in this study. The polarity record shown represents the original interpretation calibrated to the GPTS of Cande & Kent (1995) and the interpretation from this study correlated with the GPTS of Cande & Kent (1995).

Table 13. *Composite dataset for ODP Hole 695A containing the diatom, radiolarian, calcareous nannofossil and palaeomagnetic data calibrated to the GPTS of Cande & Kent (1995) using the biochronological interpretation based on this study*

	Event	Sample (top)	Sample (bottom)	Depth (t)	Depth (b)	Age (t)	Age (b)
LO	*Antarctissa cylindrica*	1H-1, 0	1H-CC	0.00	6.40	0.00	2.27
LO	*Cycladophora pliocenica*	1H-1, 0	1H-CC	0.00	6.40	0.00	2.27
LO	*Actinocyclus ingens*	1H1, 55-56	1H-2, 30-31	0.56	1.81	0.20	0.64
LO	*Thalassiosira vulnifica*	1H4, 30-31	1H-4, 100-101	1.81	5.51	0.64	1.95
FO	*Nitzschia kerguelensis*	1H3, 30-31	1H4, 30-31	3.31	4.80	1.17	1.70
B	C2n			5.50	5.50	1.95	1.95
LO	*Thalassiosira insigna*	1H-CC	2H-6, 70-71	6.40	11.11	2.00	2.14
LO	*Desmospyris spongiosa*	1H-CC	2H-CC	6.40	12.40	2.00	2.15
LO	*Helotholus vema*	1H-CC	2H-CC	6.40	12.40	2.00	2.15
LO	*Cycladophora davisiana*	1H-CC	2H-CC	6.40	12.40	2.00	2.15
T	C2r.1n			9.00	9.00	2.14	2.14
LO	*Nitzschia praeinterfrigidaria*	2H-6, 70-71	2H-CC	11.11	12.40	2.14	2.15
FO	*Thalassiosira vulnifica*	2H-CC	3H-4, 22-22	12.40	17.12	2.15	2.61
LO	*Nitzschia interfrigidaria*	2H-CC	3H-4, 22-22	12.40	17.12	2.15	2.61
B	C2r.1n			14.00	14.00	2.15	2.15
T	C2An.1n			17.00	17.00	2.58	2.58
LO	*Nitzschia barronii*	3H-4, 21-22	3H-5, 75-76	17.12	19.16	2.58	2.61
LO	*Nitzchia weaveri*	3H-4, 21-22	3H-5, 75-76	17.12	19.60	2.58	2.61
FO	*Thalassiosira insigna*	3H-CC	4H-CC	22.00	31.60	2.65	2.80
LO	*Thalassiosira inura*	5HCC	6H-CC	41.20	50.80	2.94	3.08
B	C2An.1n			48.00	48.00	3.04	3.04
FO	*Nitzschia interfrigidaria*	6H-CC	7H-CC	50.80	60.40	3.09	3.24
T	C2An.2n			52.00	52.00	3.11	3.11
B	C2An.2n			53.00	53.00	3.22	3.22
T	C2An.3n			87.00	87.00	3.33	3.33
B	C2An.3n			103.50	103.50	3.58	3.58
FO	*Nitzschia barronii*	15H-CC	16H-CC	127.90	137.60	4.02	4.18
T	C3n.1n			137.00	137.00	4.18	4.18
FO	*Nitzchia weaveri*	18X-CC	19X-CC	147.30	157.00	4.23	4.27
LO	*Denticulopsis hustedtii*	19X-CC	20X-CC	157.00	166.80	4.28	4.37
LO	*Prunopyle titan*	19X-CC	20X-CC	157.00	166.80	4.28	4.37
B	C3n.1n			160.00	160.00	4.29	4.29
T	C3n.2n			176.50	176.50	4.48	4.48
LO	*Cosmiodiscus intersectus*	21X-CC	22X-CC	176.50	186.20	4.48	?
FO	*Nitzschia praeinterfrigidaria*	22X-CC	23X-CC	186.20	195.90	4.57	?
B	C3n.2n			191.00	191.00	4.62	4.62

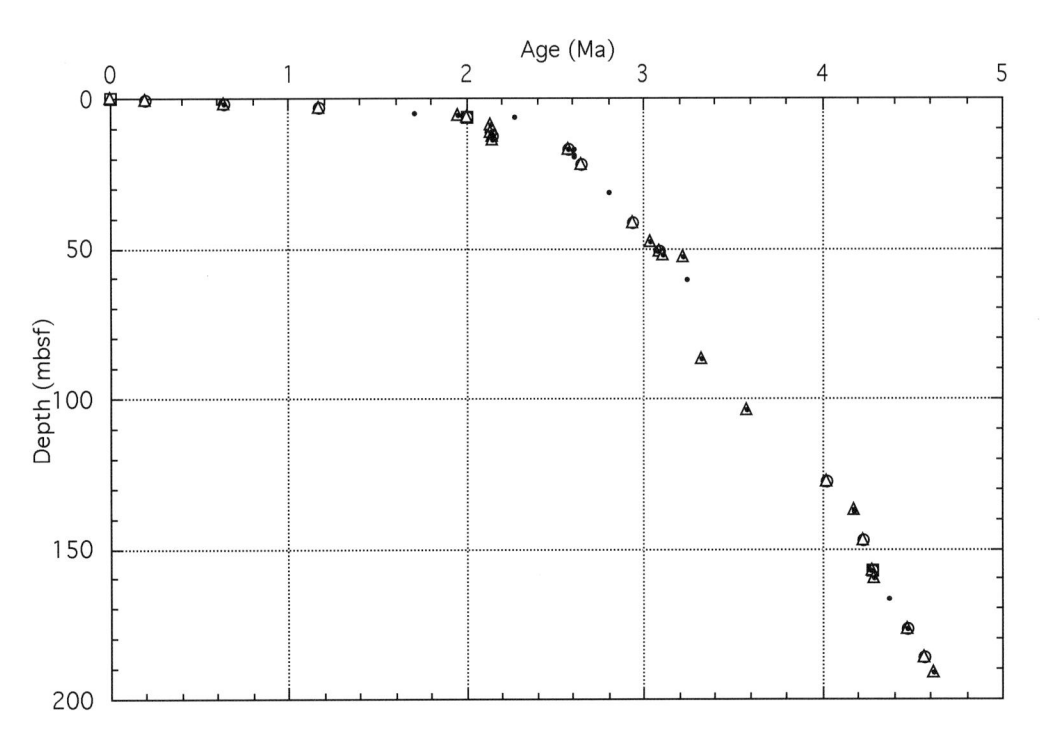

Data sets are integrated from the work of numerous authors (see Table 2).

Fig. 17. A revised age–depth plot for DSDP Hole 695A utilizing the biostratigraphical and polarity reversal events identified in this study and calibrated to the GPTS of Cande & Kent (1995). See Fig. 5 for legend of symbols.

Hole 696A

Twelve cores were recovered from a total penetration of 103 mbsf representing a recovery of 55% of the cored interval. Sediments recovered consist of Quaternary to lower Pliocene muds with a variable siliceous component (Barker, Kennett et al., 1988). A good palaeomagnetic polarity reversal record was obtained using shipboard data (Barker et al. 1988). Diatoms and radiolarians are present throughout the sequence. Rare to common and moderately preserved diatoms are observed in the sequence. Radiolarians are common to abundant and well preserved throughout the section (Barker et al. 1988). Calcareous nannofossils are not recorded (Barker et al. 1988). The integrated biostratigraphical and magnetostratigraphical data previously used (palaeomagnetics: Barker et al. 1988; diatoms: Gersonde & Burckle 1990; calcareous nannofossils: none; radiolarians: none) and calibrated to the GPTS of Berggren et al. (1985a, b) is shown in Table 14. An age–depth graph for Hole 113-690B using this data is shown in Fig. 18.

Figure 19 compares the polarity chronozones and diatom zones assigned by previous workers (Barker et al. 1988; Gersonde & Burckle 1990) with the chronozonal assignments in this study. The stratigraphical ranges reflect the placement of FO and LO of stratigraphically useful diatom species as recognized by Gersonde & Burckle (1990). The revised chronozones and diatom zones illustrated are based on reassessing the data sets of these workers (Table 15). Comparison of the chronozonal assignments stated in Barker et al. (1988; Table 14) with those in this study (Table 15) indicates that only minimal reinterpretations were completed and reflect differences in the GPTS used in each study.

Figure 20 shows an age–depth graph for Hole 113-696A using the stratigraphical markers determined in this study. The section is interpreted to be continuous based on both palaeomagnetic and biostratigraphical control. Palaeomagnetic control consists of a continuous polarity reversal sequence representing the base of C3n.1n at 6.0 mbsf to the base of C3Bn at 74.0 mbsf. Biostratigraphical markers used to calibrate this polarity sequence to the GPTS include the FO of *Nitzschia barronii* (11.21–12 mbsf), and the FO of *Nitzschia praeinterfrigidaria* (22.4–31.2 mbsf). The last occurrence of *Thallassiosira vulnifica* has an age of 1.64–1.95 Ma which is a minimum of 0.40 million years younger than its age at other sites. No primary biostratigraphical control points were recognized for the lower 30 m of the section.

Table 14. *Composite dataset for ODP Hole 696A containing the diatom, radiolarian, calcareous nannofossil and palaeomagnetic data calibrated to the GPTS of Cande and Kent (1995)*

	Event	Sample (top)	Sample (bottom)	Depth (t)	Depth (b)	Age (t)	Age (b)
LO	Thalassiosira inura	1H-2	1H-2, 55-56	0.00	3.56	?	?
LO	Actinocyclus ingens	1H-2	1H-2, 55-56	0.00	3.56	?	?
LO	Denticulopsis hustedtii	1H-2	1H-2, 55-56	0.00	3.56	?	?
LO	Nitzschia barronii	1H-1, 0	1H-2, 55-56	0.00	3.56	?	?
B	C3n.1n			6.00	6.00	4.29	4.29
T	C3n.2n			9.00	9.00	4.48	4.48
LO	Thalassiosira vulnifica	2H-6, 120-121	2H-CC	11.21	12.00	4.51	4.52
FO	Thalassiosira vulnifica	2H-6, 120-121	2H-CC	11.21	12.00	4.51	4.52
LO	Thalassiosira insigna	2H-6, 120-121	2H-CC	11.21	12.00	4.51	4.52
FO	Thalassiosira insigna	2H-6, 120-121	2H-CC	11.21	12.00	4.51	4.52
FO	Nitzschia barronii	2H-6, 120-121	2H-CC	11.21	12.00	4.51	4.52
LO	Thalassiosira complicata	2H-CC	3H-1, 50-51	12.00	12.51	4.52	4.52
LO	Nitzschia praeinterfrigidaria	3H-1, 50-51	3H-CC	12.51	21.60	4.52	4.64
FO	Nitzschia praeinterfrigidaria	4H-1, 80-87	4H-CC	22.40	31.20	4.65	4.76
T	C3n.3n			34.00	34.00	4.80	4.80
T	C3n.4n			45.00	45.00	4.98	4.98
FO	Nitzschia kerguelensis	6H-4, 120-121	6H-CC	46.51	50.40	5.13	5.65
B	C3n.4n			47.50	47.50	5.23	5.23
T	C3An.1n			52.00	52.00	5.89	5.89
B	C3An.1n			63.50	63.50	6.14	6.14
FO	Rouxia heteropolara	8H-4, 20-21	8H-CC	64.71	69.60	6.18	6.77
FO	Thalassiosira inura	8H-4, 20-21	8H-CC	64.71	69.60	6.18	6.77
T	C3An.2n			67.00	67.00	6.27	6.27
B	C3An.2n			68.00	68.00	6.57	6.57
T	C3Bn			71.00	71.00	6.94	6.94
B	C3Bn			74.00	74.00	7.09	7.09
FO	Thalassiosira oestrupii	9H-4, 120-121	9H-CC	75.31	79.30	?	?
FO	Nitzschia praecurta	25R-CC	26R-CC	289.00	298.70	?	?

Data sets are integrated from the work of numerous authors (see Table 2).

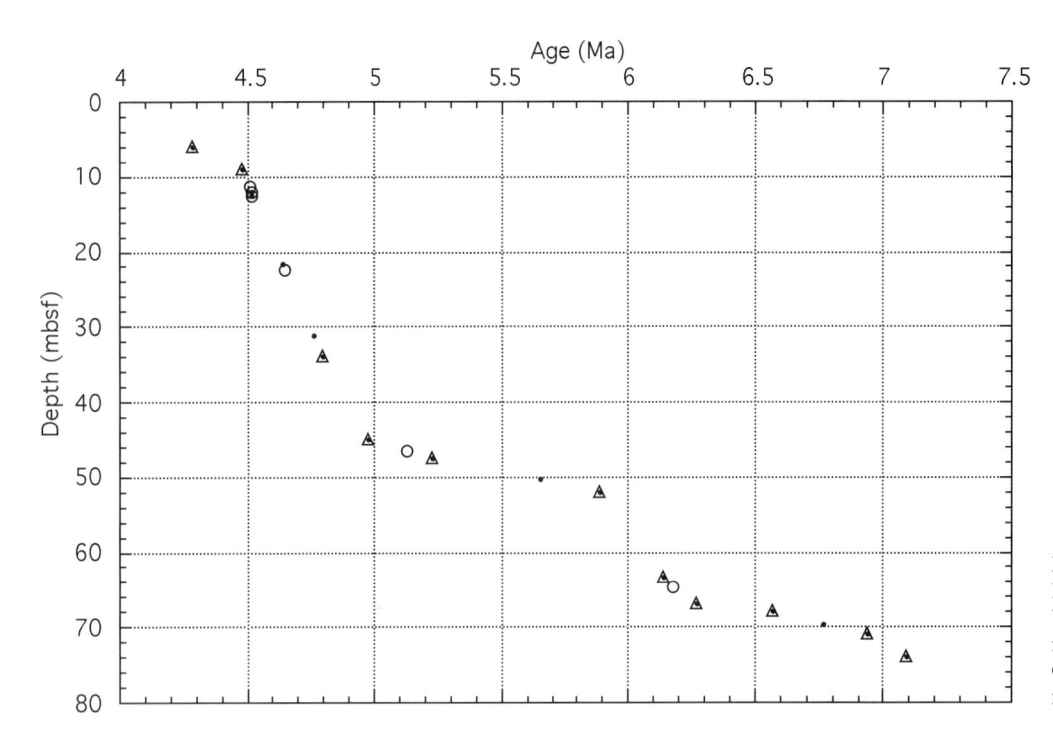

Fig. 18. An age–depth plot for DSDP Hole 696A utilizing the previously used biostratigraphical datums and polarity reversal interpretation calibrated to the GPTS of Cande & Kent (1995). See Fig. 5 for legend of symbols.

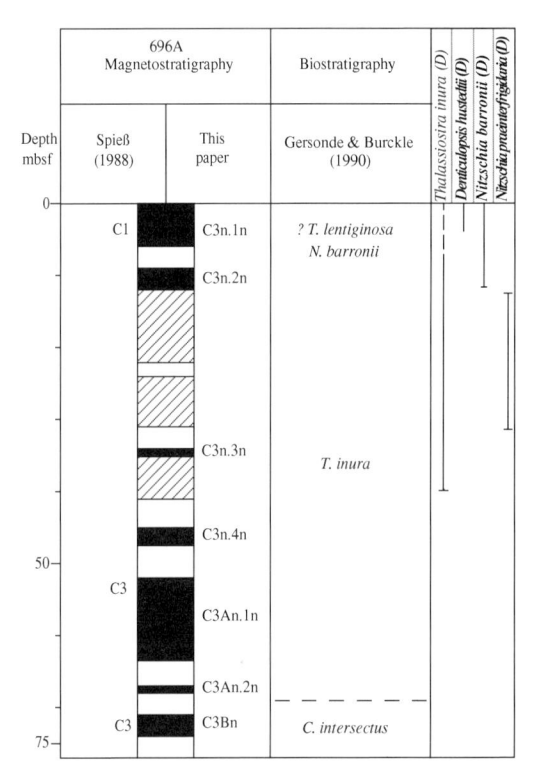

Fig. 19. Comparison of the ODP Hole 696A polarity chronozones and diatom zonal assignments by previous workers with the chronozonal assignments of this study. The stratigraphical ranges of selected stratigraphical indicators are based on the original data sets (see text for references). D, diatoms; R, radiolaria; N, calcareous nannofossils. Emboldened events indicate primary biostratigraphical events recognized in this study. The polarity record shown represents the original interpretation calibrated to the GPTS of Cande & Kent (1995) and the interpretation from this study correlated with the GPTS of Cande & Kent (1995). Diagonal lines indicate intervals of uncertainty.

Table 15. *Composite dataset for ODP Hole 696A containing the diatom, radiolarian, calcareous nannofossil and palaeomagnetic data calibrated to the GPTS of Cande & Kent (1995) using the biochronological interpretation based on this study*

	Event	Sample (top)	Sample (bottom)	Depth (t)	Depth (b)	Age (t)	Age (b)
LO	*Thalassiosira inura*	1H-2	1H-2, 55-56	0.00	3.56	?	?
LO	*Actinocyclus ingens*	1H-2	1H-2, 55-56	0.00	3.56	?	?
LO	*Denticulopsis hustedtii*	1H-2	1H-2, 55-56	0.00	3.56	?	?
LO	*Nitzschia barronii*	1H-1, 0	1H-2, 55-56	0.00	3.56	?	?
B	C3n.1n			6.00	6.00	4.29	4.29
T	C3n.2n			9.00	9.00	4.48	4.48
LO	*Thalassiosira vulnifica*	2H-6, 120-121	2H-CC	11.21	12.00	4.51	4.52
FO	*Thalassiosira vulnifica*	2H-6, 120-121	2H-CC	11.21	12.00	4.51	4.52
LO	*Thalassiosira insigna*	2H-6, 120-121	2H-CC	11.21	12.00	4.51	4.52
FO	*Thalassiosira insigna*	2H-6, 120-121	2H-CC	11.21	12.00	4.51	4.52
FO	*Nitzschia barronii*	2H-6, 120-121	2H-CC	11.21	12.00	4.51	4.52
LO	*Thalassiosira complicata*	2H-CC	3H-1, 50-51	12.00	12.51	4.52	4.52
LO	*Nitzschia praeinterfrigidaria*	3H-1, 50-51	3H-CC	12.51	21.60	4.52	4.64
FO	*Nitzschia praeinterfrigidaria*	4H-1, 80-87	4H-CC	22.40	31.20	4.65	4.76
T	C3n.3n			34.00	34.00	4.80	4.80
T	C3n.4n			45.00	45.00	4.98	4.98
FO	*Nitzschia kerguelensis*	6H-4, 120-121	6H-CC	46.51	50.40	5.13	5.65
B	C3n.4n			47.50	47.50	5.23	5.23
T	C3An.1n			52.00	52.00	5.89	5.89
B	C3An.1n			63.50	63.50	6.14	6.14
FO	*Rouxia heteropolara*	8H-4, 20-21	8H-CC	64.71	69.60	6.18	6.77
FO	*Thalassiosira inura*	8H-4, 20-21	8H-CC	64.71	69.60	6.18	6.77
T	C3An.2n			67.00	67.00	6.27	6.27
B	C3An.2n			68.00	68.00	6.57	6.57
T	C3Bn			71.00	71.00	6.94	6.94
B	C3Bn			74.00	74.00	7.09	7.09
FO	*Thalassiosira oestrupii*	9H-4, 120-121	9H-CC	75.31	79.30	?	?
FO	*Nitzschia praecurta*	25R-CC	26R-CC	289.00	298.70	?	?

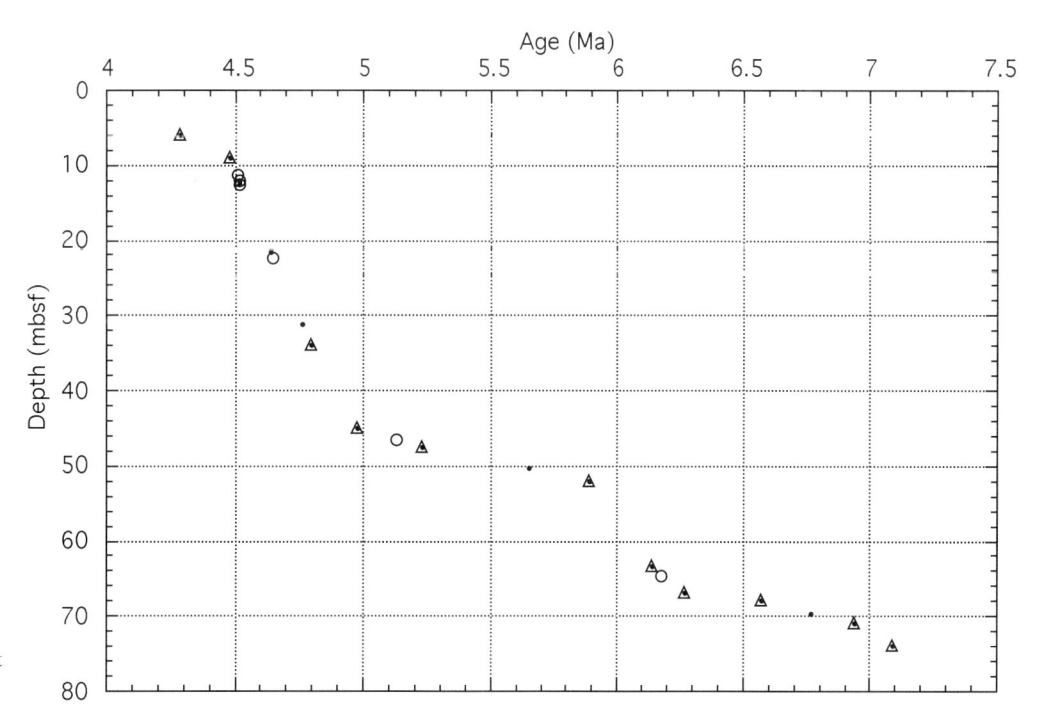

Data sets are integrated from the work of numerous authors (see Table 2).

Fig. 20. A revised age–depth plot for DSDP Hole 696A utilizing the biostratigraphical and polarity reversal events identified in this study and calibrated to the GPTS of Cande & Kent (1995). See Fig. 5 for legend of symbols.

Hole 697B

Thirty-two cores were recovered from a total penetration of 322.9 mbsf representing a recovery of 62% of the cored interval. Sediments recovered consist of Quaternary to lower Pliocene mud with a siliceous component. A good palaeomagnetic polarity reversal record was obtained through shipboard analysis (Barker et al. 1988). Variable core recovery occasionally limits the interpretation of the magnetostratigraphy. Diatoms and radiolarians are present throughout the section. Diatom abundance and preservation varies, ranging from rare to common and exhibiting poor to moderate preservation. Radiolarians are common in the Pliocene with moderate to poor preservation (Barker et al. 1988). The integrated biostratigraphical and magnetostratigraphical data previously used (palaeomagnetics: Barker et al. 1988; diatoms: Gersonde & Burckle 1990; calcareous nannofossils: none; radiolarians: Barker et al. 1988) and calibrated to the GPTS of Berggren et al. (1985a, b) are presented in Table 16. An age–depth graph for Hole 113-690B using this data is shown in Figure 21.

Figure 22 compares the polarity chronozones and diatom zones assigned by previous workers (Barker et al. 1988; Gersonde & Burckle 1990) with the chronozonal assignments in this study. The stratigraphical ranges of diatoms and radiolarians reflect the placement of the FO and LO of stratigraphically useful species as recognized by Gersonde & Burckle (1990), Abelmann (1990) and Lazarus (1990), respectively. The revised chronozones and diatom zones illustrated are based on reassessing the data sets of these workers (Table 17). Comparison of the chronozonal assignments of Barker et al. (1988; Table 16) with those of this study (Table 17) indicates that only minimal reinterpretations were completed in this hole reflecting differences in the GPTS used in each study.

An age–depth graph for Hole 113-697B using the stratigraphical markers determined in this study is shown in Fig. 23. The section is interpreted to be continuous based on both palaeomagnetic and biostratigraphical control. Palaeomagnetic control consists of a continuous sequence of polarity reversals extending from the base of C1n at 37.0 mbsf to the base of C3n.2n at 200 mbsf. Biostratigraphical markers used to calibrate this polarity sequence include the LO of *Thalassiosira vulnifica* (104.7–114.3 mbsf) the FO of *Nitzschia interfrigidaria* (186.6–196.2 mbsf), and FO of *Nitzschia barronii* (207.56–215 mbsf) and the FO of *Denticulopsis hustedtii* (293–302.6 mbsf).

Table 16. *Composite dataset for ODP Hole 697B containing the diatom, radiolarian, calcareous nannofossil and palaeomagnetic data calibrated to the GPTS of Cande & Kent (1995)*

	Event	Sample (top)	Sample (bottom)	Depth (t)	Depth (b)	Age (t)	Age (b)
B	C1n			37.00	37.00	0.78	0.78
LO	*Actinocyclus ingens*	2H-CC	3H-CC	37.10	46.80	0.79	1.10
T	C1r.1n			40.05	40.05	0.99	0.99
B	C1r.1n			46.00	46.00	1.07	1.07
LO	*Nitzschia barronii*	3H-CC	4H-3, 145-146	46.80	51.26	1.10	1.25
LO	*Thalassiosira kolbei*	5H-CC	6H-CC	66.20	76.00	1.74	1.95
T	C2n			67.00	67.00	1.77	1.77
B	C2n			76.00	76.00	1.95	1.95
LO	*Thalassiosira vulnifica*	7H-CC	8H-CC	85.70	95.40	2.17	2.40
LO	*Desmospyris spongiosa*	8H-CC	9H-CC	95.40	104.70	2.39	2.82
LO	*Helotholus vema*	8H-CC	9H-CC	95.40	104.70	2.39	2.82
T	C2An.1n			103.50	103.50	2.58	2.58
FO	*Thalassiosira vulnifica*	9H-CC	10H-CC	104.70	114.30	2.61	2.82
LO	*Denticulopsis hustedtii*	9H-CC	10H-CC	104.70	114.30	2.61	2.82
FO	*Thalassiosira kolbei*	9H-CC	10H-CC	104.70	114.30	2.61	2.82
LO	*Thalassiosira inura*	10H-CC	11H-CC	114.30	119.80	2.82	2.94
LO	*Thalassiosira insigna*	11H-CC	12X-CC	119.80	128.60	2.94	3.08
B	C2An.1n			124.00	124.00	3.04	3.04
FO	*Cycladophora davisiana*	12X-CC	13X-CC	128.60	138.20	3.08	3.34
LO	*Nitzschia interfrigidaria*	12X-CC	13X-3, 31-32	128.60	131.92	3.08	3.11
T	C2An.2n			132.50	132.50	3.11	3.11
FO	*Actinocyclus ingens*	13X-CC	14X-CC	138.20	147.90	3.21	3.34
B	C2An.2n			139.00	139.00	3.22	3.22
T	C2An.3n			147.00	147.00	3.33	3.33
FO	*Thalassiosira lentiginosa*	15X-2, 48-49	15X-CC	151.39	157.50	3.40	3.49
LO	*Thalassiosira complicata*	16X-2, 10-11	16X-2, 71-72	159.11	159.72	3.52	3.53
LO	*Nitzschia praeinterfrigidaria*	16X-2, 71-72	16X-CC	157.72	167.20	3.50	3.65
B	C2An.3n			163.00	163.00	3.58	3.58
LO	*Simonseniella barboi*	17X-CC	18X-CC	176.90	186.60	3.86	4.05
FO	*Nitzschia interfrigidaria*	18X-CC	19X-CC	186.60	196.20	4.05	4.35
T	C3n.1n			193.00	193.00	4.18	4.18
FO	*Nitzschia barronii*	21X-2, 15-16	21X-CC	207.56	215.50	?	?
FO	*Thalassiosira oestrupii*	21X-2, 15-16	21X-CC	207.56	215.50	?	?
FO	*Rouxia heteropolara*	21X-CC	22X-CC	215.50	225.20	?	?
FO	*Cosmiodiscus intersectus*	24X-CC	25X-CC	244.50	254.20	?	?
FO	*Nitzschia praecurta*	25X-CC	26X-CC	254.20	263.20	?	?
FO	*Thalassiosira inura*	28X-3, 114-115	28X-CC	277.75	283.30	?	?
FO	*Denticulopsis hustedtii*	29X-CC	30X-CC	293.00	302.60	?	?

Data sets are integrated from the work of numerous authors (see Table 2)

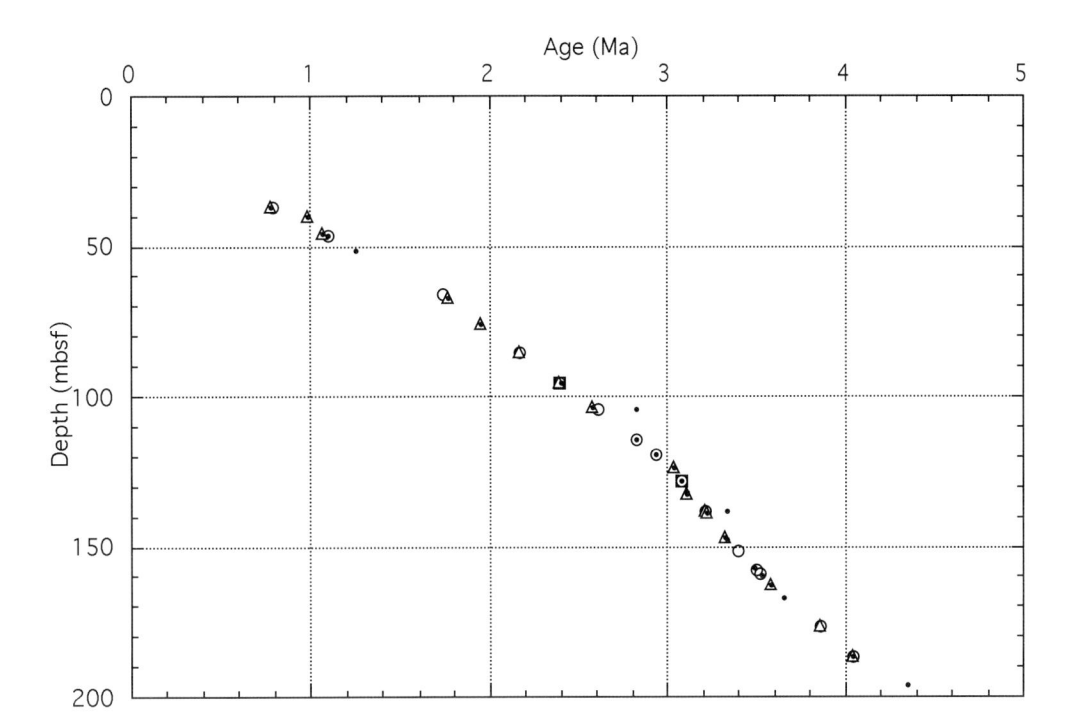

Fig. 21. An age–depth plot for DSDP Hole 697B utilizing the previously used biostratigraphical datums and polarity reversal interpretation calibrated to the GPTS of Cande & Kent (1995). See Fig. 5 for legend of symbols.

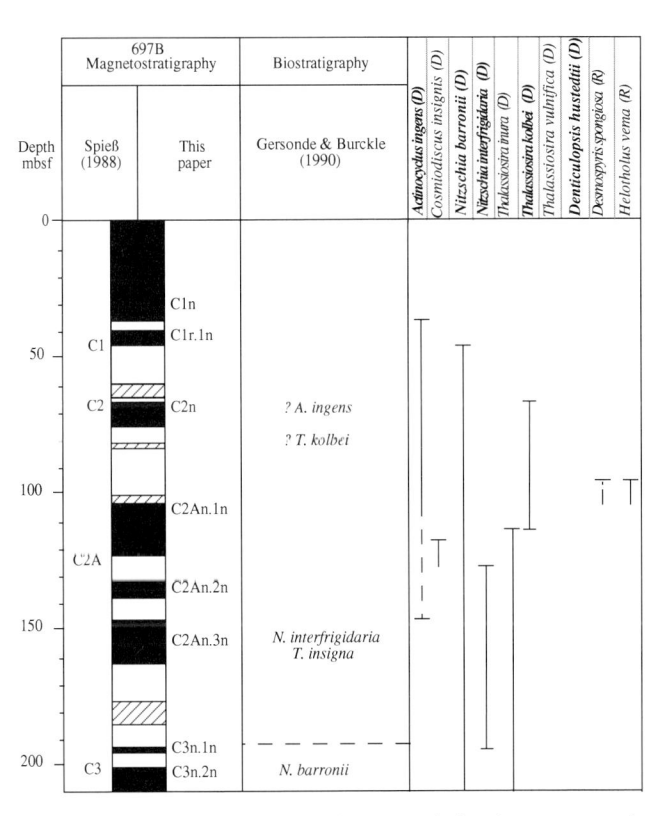

Fig. 22. Comparison of the ODP Hole 697B polarity chronozones and diatom zonal assignments by previous workers with the chronozonal assignments of this study. The stratigraphical ranges of selected stratigraphical indicators are based on the original data sets (see text for references). D, diatoms; R, radiolaria; N, calcareous nannofossils. Emboldened events indicate primary biostratigraphical events recognized in this study. The polarity record shown represents the original interpretation calibrated to the GPTS of Cande & Kent (1995) and the interpretation from this study correlated with the GPTS of Cande & Kent (1995). Diagonal lines indicate intervals of uncertainty.

Table 17. *Composite dataset for ODP Hole 697B containing the diatom, radiolarian, calcareous nannofossil and palaeomagnetic data calibrated to the GPTS of Cande & Kent (1995) using the biochronological interpretation based on this study*

	Event	Sample (top)	Sample (bottom)	Depth (t)	Depth (b)	Age (t)	Age (b)
B	C1n			37.00	37.00	0.78	0.78
LO	*Actinocyclus ingens*	2H-CC	3H-CC	37.10	46.80	0.79	1.10
T	C1r.1n			40.05	40.05	0.99	0.99
B	C1r.1n			46.00	46.00	1.07	1.07
LO	*Nitzschia barronii*	3H-CC	4H-3, 145-146	46.80	51.26	1.10	1.25
LO	*Thalassiosira kolbei*	5H-CC	6H-CC	66.20	76.00	1.74	1.95
T	C2n			67.00	67.00	1.77	1.77
B	C2n			76.00	76.00	1.95	1.95
LO	*Thalassiosira vulnifica*	7H-CC	8H-CC	85.70	95.40	2.17	2.39
LO	*Desmospyris spongiosa*	8H-CC	9H-CC	95.40	104.70	2.39	2.82
LO	*Helotholus vema*	8H-CC	9H-CC	95.40	104.70	2.39	2.82
T	C2An.1n			103.50	103.50	2.58	2.58
FO	*Thalassiosira vulnifica*	9H-CC	10H-CC	104.70	114.30	2.61	2.82
LO	*Denticulopsis hustedtii*	9H-CC	10H-CC	104.70	114.30	2.61	2.82
FO	*Thalassiosira kolbei*	9H-CC	10H-CC	104.70	114.30	2.61	2.82
LO	*Thalassiosira inura*	10H-CC	11H-CC	114.30	119.80	2.82	2.94
LO	*Thalassiosira insigna*	11H-CC	12X-CC	119.80	128.60	2.94	3.08
B	C2An.1n			124.00	124.00	3.04	3.04
FO	*Cycladophora davisiana*	12X-CC	13X-CC	128.60	138.20	3.08	3.34
LO	*Nitzschia interfrigidaria*	12X-CC	13X-3, 31-32	128.60	131.92	3.08	3.11
T	C2An.2n			132.50	132.50	3.11	3.11
FO	*Actinocyclus ingens*	13X-CC	14X-CC	138.20	147.90	3.21	3.34
B	C2An.2n			139.00	139.00	3.22	3.22
T	C2An.3n			147.00	147.00	3.33	3.33
FO	*Thalassiosira lentiginosa*	15X-2, 48-49	15X-CC	151.39	157.50	3.40	3.50
LO	*Thalassiosira complicata*	16X-2, 10-11	16X-2, 71-72	159.11	159.72	3.52	3.53
LO	*Nitzschia praeinterfrigidaria*	16X-2, 71-72	16X-CC	157.72	167.20	3.50	3.65
B	C2An.3n			163.00	163.00	3.58	3.58
LO	*Simonseniella barboi*	17X-CC	18X-CC	176.90	186.60	3.86	4.05
FO	*Nitzschia interfrigidaria*	18X-CC	19X-CC	186.60	196.20	4.05	4.35
T	C3n.1n			193.00	193.00	4.18	4.18
B	C3n.1n			194.50	194.50	4.29	4.29
T	C3n.2n			200.00	200.00	4.48	4.48
FO	*Nitzschia barronii*	21X-2, 15-16	21X-CC	207.56	215.50	?	?
FO	*Thalassiosira oestrupii*	21X-2, 15-16	21X-CC	207.56	215.50	?	?
FO	*Rouxia heteropolara*	21X-CC	22X-CC	215.50	225.20	?	?
FO	*Cosmiodiscus intersectus*	24X-CC	25X-CC	244.50	254.20	?	?
FO	*Nitzschia praecurta*	25X-CC	26X-CC	254.20	263.20	?	?
FO	*Thalassiosira inura*	28X-3, 114-115	28X-CC	277.75	283.30	?	?
FO	*Denticulopsis hustedtii*	29X-CC	30X-CC	293.00	302.60	?	?

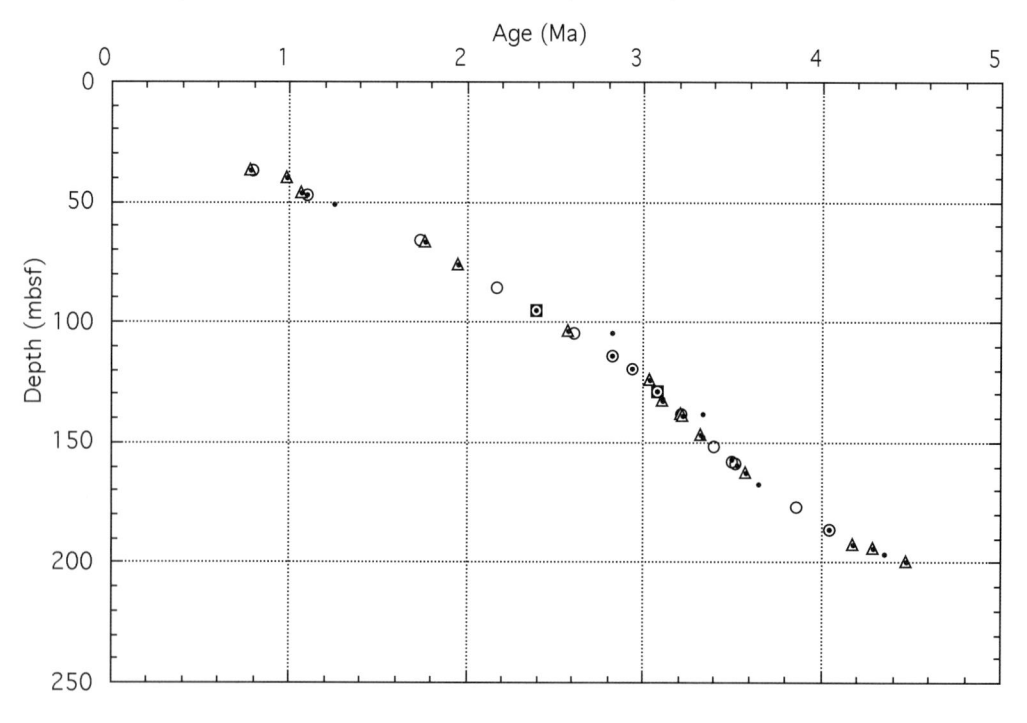

Data sets are integrated from the work of numerous authors (see Table 2).

Fig. 23. A revised age–depth plot for DSDP Hole 697B utilizing the biostratigraphical and polarity reversal events identified in this study and calibrated to the GPTS of Cande & Kent (1995). See Fig. 5 for legend of symbols.

Hole 699A

Fifty-six cores were recovered from a total penetration of 518.1 mbsf representing a recovery of 68% of the cored interval. Sediments recovered consist of Quaternary to lower Oligocene siliceous ooze with calcareous nannofossils and clay, lower Oligocene to middle/upper Eocene calcareous ooze with diatoms and radiolarians, and middle/upper Eocene to upper Palaeocene chalk (Ciesielski et al. 1988). A good palaeomagnetic polarity reversal record was obtained (Hailwood & Clement 1991a). Diatoms and radiolarians are present throughout the sequence. Diatoms are present in the Quaternary to upper Eocene sediments and typically exhibit moderate to good preservation. Diatom preservation is poor in the upper lower Oligocene interval (Ciesielski et al. 1988). Radiolarians are generally abundant and well preserved in the Quaternary to upper Eocene sediments. Radiolarians were not observed in sediments older than the late Eocene (Ciesielski et al. 1988). Calcareous nannofossils are absent from the Quaternary to lower Miocene sediments and abundant with poor preservation exhibited for the remaining sequence (Crux 1991). The integrated biostratigraphical and magnetostratigraphical data previously used (palaeomagnetics: Hailwood & Clement 1991a; diatoms: Ciesielski et al. 1988; Ciesielski 1991; calcareous nannofossils: Crux 1991; Wei 1991; radiolarians: Ciesielski et al. 1988) and calibrated to the GPTS of Berggren et al. (1985a, b) are presented in Table 18. An age–depth graph for Hole 113-699A using this data is shown in Fig. 24.

Figure 25 compares the polarity chronozones and diatom zones assigned by previous workers (Hailwood & Clement 1991a; Ciesielski 1991) with the chronozonal assignments in this study. In addition, this figure illustrates the stratigraphical ranges of diatoms, radiolarians and calcareous nannofossils. The stratigraphical ranges reflect the placement of FO and LO of stratigraphically useful species as recognized by Ciesielski (1991), Ciesielski et al. (1988), Crux (1991) and Wei (1991), respectively. The revised chronozones and diatom zones illustrated are based on reassessing the data sets of these workers (Table 19). Comparison of the chronozonal assignments of Hailwood & Clement (1991a; Table 18) with those of this study (Table 19) indicates that minimal reinterpretations were completed.

An age–depth graph for Hole 114-699A using the stratigraphical markers determined in this study is presented in Fig. 26. Two hiatuses are placed in this section. The younger occurring between 62.39 and 62.56 mbsf and the older occurring between 70.29 and 87.57 mbsf. The upper boundary of the younger hiatus is constrained by a continuous sequence of polarity reversals from the base of C1n at 10.59 mbsf to the base of C3n.4n at 62.39 mbsf. This polarity sequence is calibrated to the GPTS with the LO of *Thalassiosira kolbei* (21.70–23.70 mbsf), the LO of *Thallassiosira vulnifica* (27.60–28.40 mbsf), the FO of *Nitzchia interfrigidaria* (31.00–32.00 mbsf) and the FO of *Nitzschia praeinterfrigidaria* (39.03–40.53 mbsf). The lower boundary of this hiatus is constrained by the base of C5ACn at 68.84 mbsf and the FO of *Nitzschia denticuloides* (67.5–69.20 mbsf). The upper boundary of the older hiatus is constrained by the base of C5ACn and the top of C5ADn at 70.29 mbsf. The lower boundary is constrained by a sequence of polarity reversals from C6Cn.1n (87.57 mbsf) to the top of C11n.1n (208.09 mbsf) calibrated to the GPTS. The continuity of this magnetic sequence is punctuated by uncertain intervals. There are also no primary biostratigraphical markers within this interval. The LO of *Reticulofenestra oamaruensis* (283.10–284.90 mbsf) provides a potentially useful event within this sequence.

This study suggests that the stratigraphy of the section is more continuous than previously recognized. Ciesielski et al. (1988) identified a possible hiatus at 31.0–32.2 mbsf and three unconformities at 39.03–43.07 mbsf, 54.52–62.15 mbsf and 67.2–69.20 mbsf. Recognition of these stratigraphical breaks are based on both palaeomagnetic and biostratigraphical markers. Hiatuses identified at 54.52–62.15 mbsf and 67.2–69.20 mbsf approximate those identified in this study. The potential hiatus at 31.0–32.2 mbsf and the hiatus at 39.03–40.37 mbsf are not recognized in this study and reflect differences in the polarity interpretations.

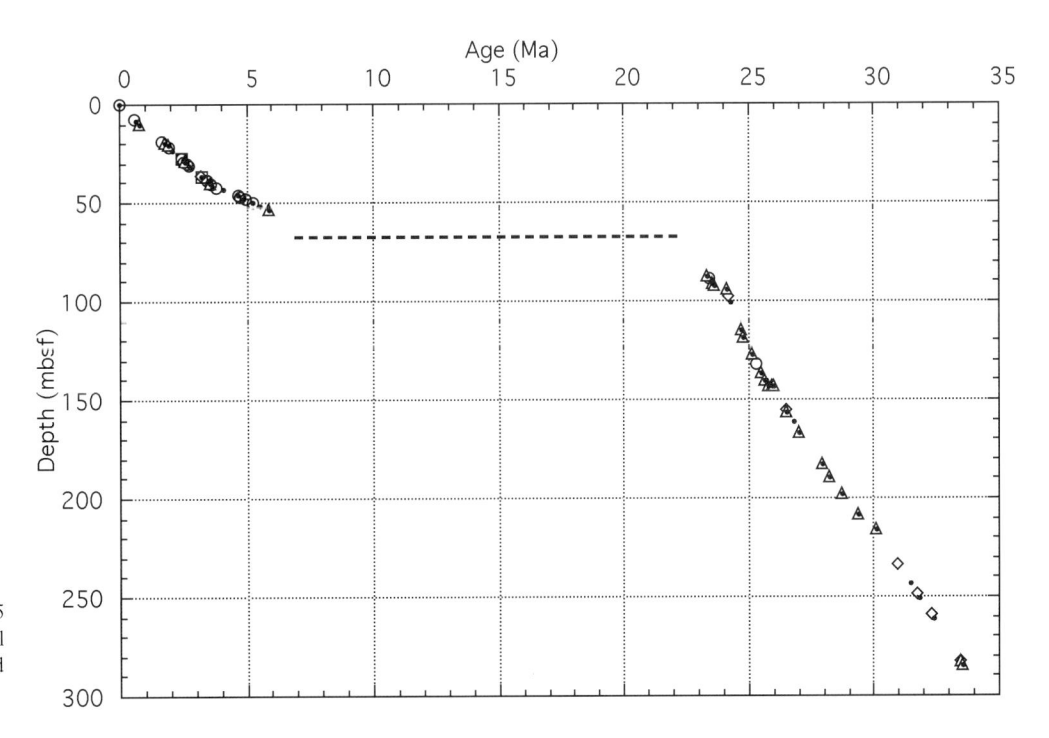

Fig. 24. An age–depth plot for DSDP Hole 699A utilizing the previously used biostratigraphical datums and polarity reversal interpretation calibrated to the GPTS of Cande & Kent (1995). See Fig. 5 for legend of symbols. One stratigraphical break (dashed line) is based on Hailwood & Clement (1991a).

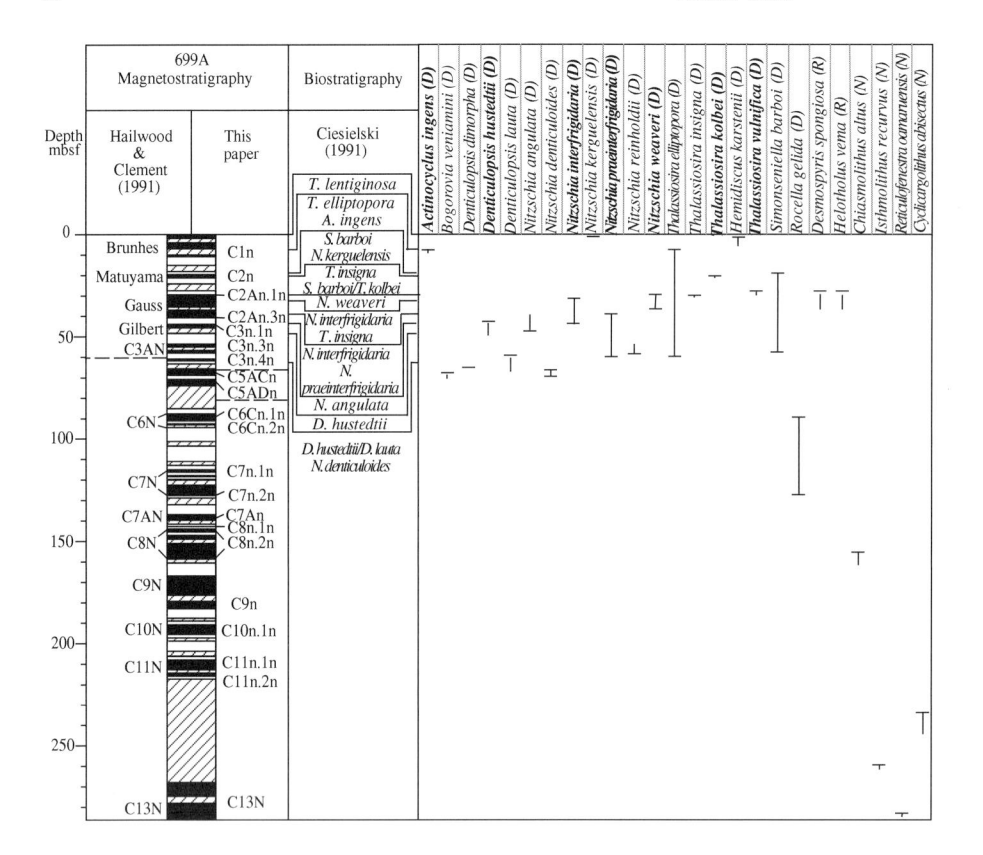

Fig. 25. Comparison of the ODP Hole 699A polarity chronozones and diatom zonal assignments by previous workers with the chronozonal assignments of this study. The stratigraphical ranges of selected stratigraphical markers are based on the original data sets (see text for references). D, diatoms; R, radiolaria; N, calcareous nannofossils. Emboldened events indicate primary biostratigraphical events recognized in this study. The polarity record shown represents the original interpretation calibrated to the GPTS of Cande & Kent (1995) and the interpretation from this study correlated with the GPTS of Cande & Kent (1995).

Table 18. *Composite dataset for ODP Hole 699A containing the diatom, radiolarian, calcareous nannofossil and palaeomagnetic data calibrated to the GPTS of Cande and Kent (1995)*

	Event	Sample (top)	Sample (bottom)	Depth (t)	Depth (b)	Age (t)	Age (b)
LO	Hemidiscus karstenii	1H-1, 0	1H-1, 40-42	0.00	0.42	0.00	0.03
LO	Actinocyclus ingens	1H-5, 140-142	1H-6, 140-142	7.42	8.92	0.55	0.66
LO	Thalassiosira elliptopora	1H-5, 140-142	1H-6, 140-142	7.42	8.92	0.55	0.66
B	C1n			10.59	10.59	0.78	0.78
LO	Simonseniella barboi	3H-1, 67-70	3H-2, 67-69	18.80	20.90	1.68	1.92
T	C2n			19.59	19.59	1.77	1.77
B	C2n			21.19	21.19	1.95	1.95
LO	Thalassiosira kolbei	3H-3, 60	3H-4, 110	21.70	23.70	1.93	2.10
LO	Thalassiosira vulnifica	3H-CC	4H-1, 124	27.60	28.40	2.43	2.49
LO	Desmospyris spongiosa	3H-CC	4H-CC	27.60	37.10	2.44	3.26
LO	Helotholus vema	3H-CC	4H-CC	27.60	37.10	2.43	3.26
LO	Nitzschia reinholdii	4H-1, 122-126	4H-2, 37-39	28.82	29.49	2.53	2.64
T	C2An.1n			29.45	29.45	2.58	2.58
LO	Thalassiosira insigna	4H-2, 143	4H-3, 40	30.53	31.00	2.68	2.72
LO	Nitzschia interfrigidaria	4H-3, 30	4H-4, 9	31.00	32.20	2.72	2.82
LO	Nitzschia weaveri	4H-3, 40	4H-4, 10	31.00	32.20	2.72	2.83
LO	Prunopyle titan	4H-CC	5H-CC	37.10	46.60	3.26	4.64
FO	Nitzschia weaveri	5H-2, 41	5H-3, 27	39.03	40.00	3.43	3.52
LO	Nitzschia praeinterfrigidaria	5H-2, 41	5H-3, 27	39.03	40.53	3.43	3.57
FO	Nitzschia weaveri	5H-3, 41-43	5H-4, 41-43	40.53	42.03	3.57	3.70
B	C2An.3n			40.69	40.69	3.58	3.58
FO	Nitzschia interfrigidaria	5H-4, 43	5H-5, 50	42.03	43.60	3.82	4.10
FO	Nitzschia angulata	5H-CC	6H-1, 3	46.60	46.63	4.64	4.64
LO	Denticulopsis hustedtii	6H-1, 42	6H-2, 6	47.02	48.16	4.71	4.92
LO	Hemidiscus cuneiformis	6H-2, 40-42	6H-3, 40-42	48.52	50.02	4.98	5.25
FO	Nitzschia praeinterfrigidaria	6H-3, 40-42	6H-4, 40-42	50.02	51.52	5.25	5.52
T	C3An.1n			53.61	53.61	5.89	5.89
FO	Thalassiosira lentiginosa	5H-5, 41-43	6H-1, 40-42	56.49	57.95	?	?
FO	Thalassiosira elliptopora	7H-1, 39-43	7H-2, 44-47	56.53	58.60	?	?
LO	Denticulopsis lauta	7H-3, 5	7H-5, 5	59.15	62.15	?	?
	Unconformity			***	***	***	***
FO	Nitzschia reinholdii	7H-5, 42-46	8H-1, 40-42	62.56	66.02	?	?
FO	Thalassiosira torokina	7H-5, 42-46	8H-1, 40-42	62.56	66.02	?	?
FO	Hemidiscus cuneiformis	7H-5, 42-46	8H-1, 40-42	62.56	66.02	?	?
FO	Hemidiscus karstenii	7H-5, 42-46	8H-1, 40-42	62.56	66.02	?	?
LO	Denticulopsis dimorpha	7H-CC	8H-1, 40	65.00	66.00	?	?
FO	Simonseniella barboi	8H-1, 40-42	8H-2, 40-42	66.02	67.52	?	?

Table 18. (*continued*)

	Event	Sample (top)	Sample (bottom)	Depth (t)	Depth (b)	Age (t)	Age (b)
LO	*Denticulopsis lauta*	8H-1, 40-42	8H-2, 40-42	66.02	67.52	?	?
LO	*Nitzschia denticuloides*	8H-1, 55	8H-1, 142	66.15	67.02	?	?
FO	*Nitzschia denticuloides*	8H-2, 42	8H-3, 60	67.52	69.20	?	?
LO	*Bogorovia veniamini*	8H-2, 42	8H-3, 60	67.52	69.20	?	?
T	C6Cn.1n			87.57	87.57	23.35	23.35
LO	*Rocella gelida*	10H-3, 143	10H-5, 4	89.02	90.64	23.43	23.51
B	C6Cn.1n			91.17	91.17	23.54	23.54
T	C6Cn.2n			92.49	92.49	23.68	23.68
B	C6Cn.3n			94.47	94.47	24.12	24.12
LO	*Reticulofenestra bisecta*	11H-3, 75	11H-5, 59	97.85	100.69	24.22	24.30
T	C7n.1n			115.35	115.35	24.73	24.73
B	C7n.1n			118.50	118.50	24.78	24.78
B	C7n.2n			127.69	127.69	25.18	25.18
FO	*Rocella gelida*	14H-CC	15H-CC	132.10	141.60	25.33	25.71
T	C7An			137.09	137.09	25.50	25.50
B	C7An			140.38	140.38	25.65	25.65
T	C8n.1n			143.84	143.84	25.82	25.82
B	C8n.1n			143.04	143.04	25.95	25.95
T	C8n.2n			143.84	143.84	25.99	25.99
LO	*Chiasmolithus altus*	17H-4, 30	18H-1, 110	155.90	161.70	26.54	26.80
B	C8n.2n			156.14	156.14	26.55	26.55
T	C9n			166.95	166.95	27.03	27.03
B	C9n			182.69	182.69	27.97	27.97
T	10n.1n			189.34	189.34	28.28	28.28
B	10n.2n			198.35	198.35	28.75	28.75
T	C11n.1n			208.09	208.09	29.40	29.40
B	C11n.2n			216.05	216.05	30.10	30.10
FO	*Cyclicargolithus abisectus*	26X-1, 4	27X-1, 56	233.64	243.66	30.98	31.49
LO	*Reticulofenestra umbilica*	27X-4, 116	27X-6, 62	248.76	251.22	31.75	31.87
LO	*Isthmolithus recurvus*	29X-1, 75	29X-2, 75	259.85	261.35	32.31	32.38
LO	*Reticulofenestra oamaruensis*	31X-4, 50	31X-5, 80	283.10	284.90	33.48	?
B	C13n			284.55	284.55	33.55	33.55

Data sets are integrated from the work of numerous authors (see Table 2).

Table 19. *Composite dataset for ODP Hole 699A containing the diatom, radiolarian, calcareous nannofossil and palaeomagnetic data calibrated to the GPTS of Cande & Kent (1995) using the biochronological interpretation based on this study*

	Event	Sample (top)	Sample (bottom)	Depth (t)	Depth (b)	Age (t)	Age (b)
LO	*Hemidiscus karstenii*	1H-1, 0	1H-1, 40-42	0.00	0.42	0.00	0.03
LO	*Actinocyclus ingens*	1H-5, 140-142	1H-6, 140-142	7.42	8.92	0.54	0.55
LO	*Thalassiosira elliptopora*	1H-5, 140-142	1H-6, 140-142	7.42	8.92	0.55	0.66
B	C1n			10.59	10.59	0.78	0.78
LO	*Simonseniella barboi*	3H-1, 67-70	3H-2, 67-69	18.80	20.90	1.68	1.92
T	C2u			19.59	19.59	1.77	1.77
B	C2u			21.19	21.19	1.95	1.95
LO	*Thalassiosira kolbei*	3H-3, 60	3H-4, 110	21.70	23.70	1.99	2.14
LO	*Thalassiosira vulnifica*	3H-CC	4H-1, 124	27.60	28.40	2.43	2.49
LO	*Desmospyris spongiosa*	3H-CC	4H-CC	27.60	37.10	2.43	3.26
LO	*Helotholus vema*	3H-CC	4H-CC	27.60	37.10	2.44	3.26
LO	*Nitzschia reinholdii*	4H-1, 122-126	4H-2, 37-39	28.82	29.49	2.53	2.64
T	C2An.1n			29.45	29.45	2.58	2.58
LO	*Thalassiosira insigna*	4H-2, 143	4H-3, 40	30.53	31.00	2.68	2.72
LO	*Nitzschia interfrigidaria*	4H-3, 30	4H-4, 9	31.00	32.00	2.72	2.81
LO	*Nitzschia weaveri*	4H-3, 40	4H-4, 10	31.00	32.20	2.72	2.82
LO	*Prunopyle titan*	4H-CC	5H-CC	37.10	46.60	3.26	4.64
FO	*Nitzschia weaveri*	5H-2, 41	5H-3, 27	39.03	40.37	3.43	3.55
LO	*Nitzschia praeinterfrigidaria*	5H-2, 41	5H-3, 27	39.03	40.53	3.43	3.56
FO	*Nitzschia weaveri*	5H-3, 41-43	5H-4, 41-43	40.53	42.03	3.56	3.70
B	C2An.3n			40.69	40.69	3.58	3.58
FO	*Nitzschia interfrigidaria*	5H-4, 43	5H-5, 50	42.03	43.60	3.82	4.11
T	C3n.1n			43.99	43.99	4.18	4.18
FO	*Nitzschia angulata*	5H-CC	6H-1, 3	46.60	46.63	4.35	4.37
LO	*Denticulopsis hustedtii*	6H-1, 42	6H-2, 6	47.02	48.16	4.38	4.49
LO	*Hemidiscus cuneiformis*	6H-2, 40-42	6H-3, 40-42	48.52	50.02	4.47	4.63
FO	*Nitzschia praeinterfrigidaria*	6H-3, 40-42	6H-4, 40-42	50.02	51.52	4.57	4.74

(*continued*)

Table 19. (continued)

	Event	Sample (top)	Sample (bottom)	Depth (t)	Depth (b)	Age (t)	Age (b)
T	C3n.3n			53.61	53.61	4.80	4.80
FO	Nitzschia reinholdii	6H-6, 43	7H-2, 44	54.53	58.04	4.81	4.89
FO	Thalassiosira lentiginosa	5H-5, 41-43	6H-1, 40-42	56.49	57.95	4.86	4.89
FO	Thalassiosira elliptopora	7H-1, 39-43	7H-2, 44-47	56.53	58.60	4.86	4.91
B	C3n.3n			58.00	58.00	4.89	4.89
LO	Denticulopsis lauta	7H-3, 5	7H-5, 5	59.15	62.15	4.90	5.19
T	C3n.4n			60.86	60.86	4.98	4.98
B	C3n.4n			62.39	62.39	5.23	5.23
	Unconformity			***	***	***	***
FO	Nitzschia reinholdii	7H-5, 42-46	8H-1, 40-42	62.56	66.02	?	?
FO	Thalassiosira torokina	7H-5, 42-46	8H-1, 40-42	62.56	66.02	?	?
FO	Hemidiscus cuneiformis	7H-5, 42-46	8H-1, 40-42	62.56	66.02	?	?
FO	Hemidiscus karstenii	7H-5, 42-46	8H-1, 40-42	62.56	66.02	?	?
LO	Denticulopsis dimorpha	7H-CC	8H-1, 40	65.00	66.00	?	?
FO	Simonseniella barboi	8H-1, 40-42	8H-2, 40-42	66.02	67.52	?	?
LO	Denticulopsis lauta	8H-1, 40-42	8H-2, 40-42	66.02	67.52	?	?
LO	Nitzschia denticuloides	8H-1, 55	8H-1, 142	66.15	67.02	?	?
FO	Nitzschia denticuloides	8H-2, 42	8H-3, 60	67.52	69.20	?	?
LO	Bogorovia veniamini	8H-2, 42	8H-3, 60	67.52	69.20	?	?
B	C5ACn			68.84	68.84	14.08	14.08
T	C5ADn			70.29	70.29	14.18	14.18
	Unconformity			***	***	***	***
T	C6Cn.1n			87.57	87.57	23.35	23.35
LO	Rocella gelida	10H-3, 143	10H-5, 4	89.02	90.64	23.43	23.51
B	C6Cn.1n			91.17	91.17	23.54	23.54
T	C6Cn.2n			92.49	92.49	23.68	23.68
B	C6Cn.2n			94.47	94.47	23.80	23.80
LO	Reticulofenestra bisecta	11H-3, 75	11H-5, 59	97.85	100.69	23.83	23.86
T	C6Cn.3n			115.35	115.35	24.00	24.00
B	C6Cn.3n			116.22	116.22	24.12	24.12
T	C7n.1n			118.09	118.09	24.73	24.73
B	C7n.1n			118.50	118.50	24.78	24.78
T	C7n.2n			119.22	119.22	24.84	24.84
B	C7n.2n			127.69	127.69	25.18	25.18
FO	Rocella gelida	14H-CC	15H-CC	132.10	141.60	25.33	25.76
T	C7An			137.09	137.09	25.50	25.50
T	C8n.1n			142.74	142.74	25.82	25.82
B	C8n.1n			143.04	143.04	25.95	25.95
T	C8n.2n			143.84	143.84	25.99	25.99
LO	Chiasmolithus altus	17H-4, 30	18H-1, 110	155.90	161.70	26.53	26.80
B	C8n.2n			156.14	156.14	26.55	26.55
T	C9n			166.95	166.95	27.03	27.03
B	C9n			182.69	182.69	27.97	27.97
T	C10n.1n			189.34	189.34	28.28	28.28
B	C10n.1n			198.35	198.35	28.51	28.51
T	C11n.1n			208.09	208.09	29.40	29.40
B	C11n.2n			216.05	216.05	30.10	30.10
LO	Cyclicargolithus abisectus	26X-1, 4	27X-1, 56	233.64	243.66	30.99	31.49
LO	Reticulofenestra umbilica	27X-4, 116	27X-6, 62	248.76	251.22	31.75	31.87
LO	Isthmolithus recurvus	29X-1, 75	29X-2, 75	259.85	261.35	32.31	32.38
LO	Reticulofenestra oamaruensis	31X-4, 50	31X-5, 80	283.10	284.90	33.48	?
B	C13n			284.55	284.55	33.55	33.55

Data sets are integrated from the work of numerous authors (see Table 2).

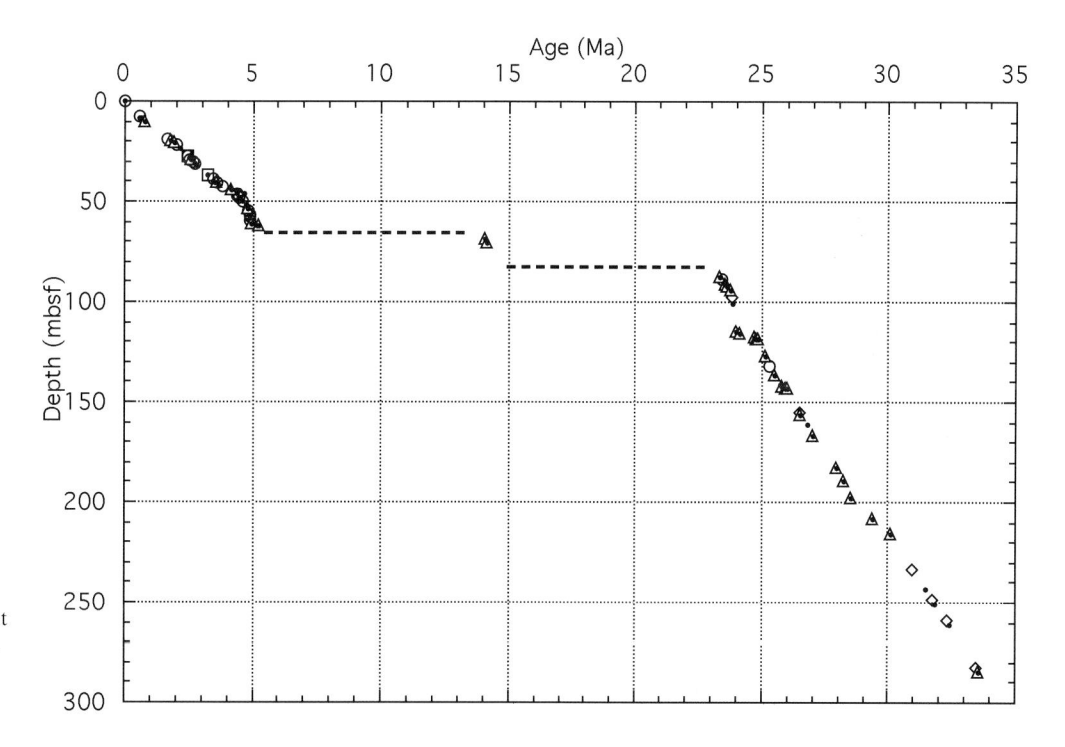

Fig. 26. A revised age–depth plot for DSDP Hole 699A utilizing the biostratigraphical and polarity reversal events identified in this study and calibrated to the GPTS of Cande & Kent (1995). See Fig. 5 for legend of symbols. Two stratigraphical breaks (dashed line) are based on the results of this study.

Hole 701A

Eight cores were recovered from a total penetration of 74.8 mbsf representing a recovery of 93% of the cored interval. Sediments recovered consist of Quaternary and upper Pliocene diatom ooze with ash and mud. A good palaeomagnetic polarity reversal record was obtained by both shipboard and shorebase analysis (Ciesielski *et al.* 1988; Clement & Hailwood 1991). Common to abundant diatoms and radiolarians occur throughout the sequence. Calcareous nannofossils are present in the sediments, but consist of long ranging species which are probably reworked (Ciesielski *et al.* 1988). The integrated biostratigraphical and magnetostratigraphical data previously used (palaeomagnetics: Clement & Hailwood 1991; diatoms: Ciesielski *et al.* 1988; Ciesielski 1991; calcareous nannofossils: none; radiolarians: Ciesielski *et al.* 1988) and calibrated to the GPTS of Berggren *et al.* (1985*a, b*) are shown in Table 20. An age–depth graph for Hole 114-701A using this data is shown in Fig. 27.

Figure 28 compares the polarity chronozones and diatom zones assigned by previous workers (Clement & Hailwood 1991; Ciesielski *et al.* 1988) with the assignments in this study. The stratigraphical ranges for both diatoms and radiolarians reflect the placement of the FO and LO of stratigraphically useful species as recognized by Ciesielski *et al.* (1988). The revised chronozones and diatom zones illustrated are based on reassessing the data sets of these workers (Table 21). Comparison of the chronozonal assignments of Clement & Hailwood (1991; Table 20) with those of this study (Table 21) indicates that only minimal reinterpretations were completed reflecting differences in the chronozonal nomenclature and GPTS used in each study.

Figure 29 shows an age–depth graph for Hole 114-701A using the stratigraphical markers determined in this study. Seven biostratigraphical and palaeomagnetic control points are recognized in this sequence and provide evidence suggesting a continuous stratigraphical record. The results of this study are similar to those presented by Ciesielski *et al.* (1988).

Table 20. *Composite dataset for ODP Hole 701A containing the diatom, radiolarian, calcareous nannofossil and palaeomagnetic data calibrated to the GPTS of Cande and Kent (1995)*

	Event	Sample (top)	Sample (bottom)	Depth (t)	Depth (b)	Age (t)	Age (b)
LO	*Hemidiscus karstenii*	IH-4, 58	IH-4, 116	5.08	5.66	0.23	0.26
B	C1n			17.02	17.02	0.78	0.78
LO	*Actinocyclus ingens*	2H, CC	3H-1, 75	17.80	18.55	0.81	0.83
LO	*Simonseniella barboi*	4H, CC	5H-2, 140	36.80	39.70	1.48	1.58
LO	*Thalassiosira kolbei*	5H-5, 114	6H-4, 104	43.94	51.84	1.73	1.98
T	C2n			44.93	44.93	1.77	1.77
B	C2n			50.95	50.95	1.95	1.95
LO	*Thalassiosira vulnifica*	6H, CC	7H-3, 77	55.80	59.57	2.12	2.26
LO	*Helotholus vema*			55.80	65.30	2.12	2.46
LO	*Desmospyris spongiosa*			55.80	65.30	2.12	2.46
FO	*Cycladophora davisiana*			55.80	65.30	2.12	2.46
LO	*Nitzschia interfrigidaria*	8H-1, 19	8H-3, 60	65.49	68.90	2.47	?
LO	*Thalassiosira insigna*	8H-1, 19	8H-3, 60	65.49	68.90	2.47	?
T	C2An.1n			68.73	68.73	2.58	2.58
LO	*Nitzschia weaveri*	8H-3, 60	8H-4, 46	68.90	70.26	?	?

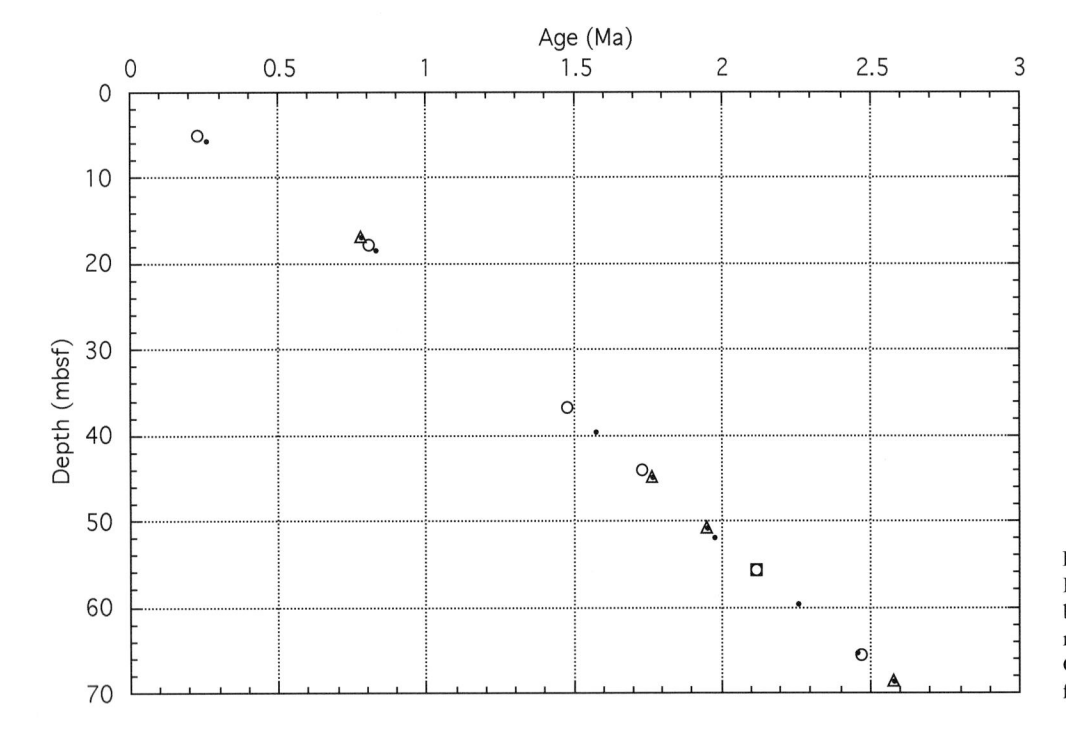

Data sets are integrated from the work of numerous authors (see Table 2).

Fig. 27. An age–depth plot for DSDP Hole 701A utilizing the previously used biostratigraphical datums and polarity reversal interpretation calibrated to the GPTS of Cande & Kent (1995). See Fig. 5 for legend of symbols.

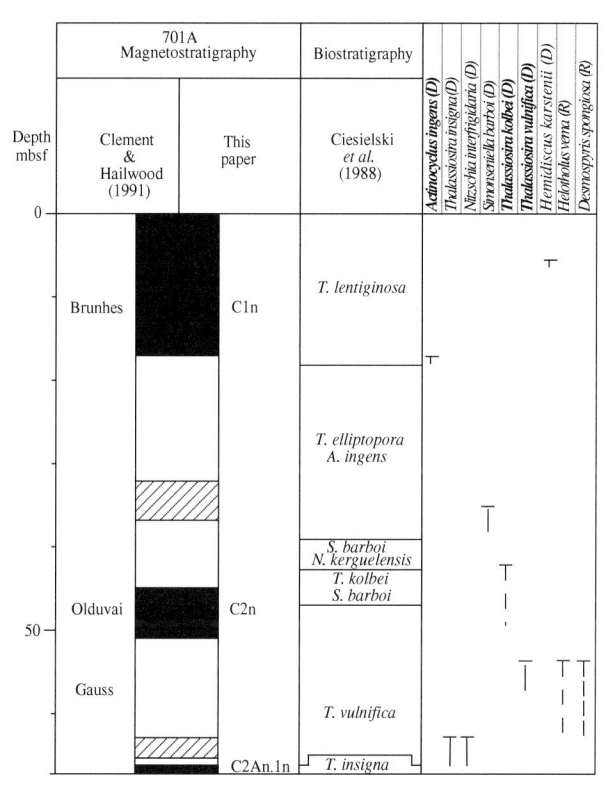

Fig. 28. Comparison of the ODP Hole 701A polarity chronozones and diatom zonal assignments by previous workers with the chronozonal assignments of this study. The stratigraphical ranges of selected stratigraphical markers are based on the original data sets (see text for references). D, diatoms; R, radiolaria; N, calcareous nannofossils. Emboldened events indicate primary biostratigraphical events recognized in this study. The polarity record shown represents the original interpretation calibrated to the GPTS of Cande & Kent (1995) and the interpretation from this study correlated with the GPTS of Cande & Kent (1995). Diagonal lines indicate intervals of uncertainty.

Table 21. *Composite dataset for ODP Hole 701A containing the diatom, radiolarian, calcareous nannofossil and palaeomagnetic data calibrated to the GPTS of Cande & Kent (1995) using the biochronological interpretation based on this study*

	Event	Sample (top)	Sample (bottom)	Depth (t)	Depth (b)	Age (t)	Age (b)
LO	*Hemidiscus karstenii*	1H-4, 58	1H-4, 116	5.08	5.66	0.23	0.26
B	C1n			17.02	17.02	0.78	0.78
LO	*Actinocyclus ingens*	2H-CC	3H-1, 75	17.80	18.55	0.81	0.83
LO	*Simonseniella barboi*	4H-CC	5H-2, 140	36.80	39.70	1.48	1.58
LO	*Thalassiosira kolbei*	5H-5, 114	6H-4, 104	43.94	51.84	1.73	1.98
T	C2n			44.93	44.93	1.77	1.77
B	C2n			50.95	50.95	1.95	1.95
LO	*Thalassiosira vulnifica*	6H-CC	7H-3, 77	55.80	59.57	2.12	2.26
LO	*Helotholus vema*			55.80	65.30	2.12	2.46
LO	*Desmospyris spongiosa*			55.80	65.30	2.12	2.46
FO	*Cycladophora davisiana*			55.80	65.30	2.12	2.46
LO	*Nitzschia interfrigidaria*	8H-1, 19	8H-3, 60	65.49	68.90	2.47	?
LO	*Thalassiosira insigna*	8H-1, 19	8H-3, 60	65.49	68.90	2.47	?
T	C2An.1n			68.73	68.73	2.58	2.58
LO	*Nitzschia weaveri*	8H-3, 60	8H-4, 46	68.90	70.26	?	?

Data sets are integrated from the work of numerous authors (see Table 2).

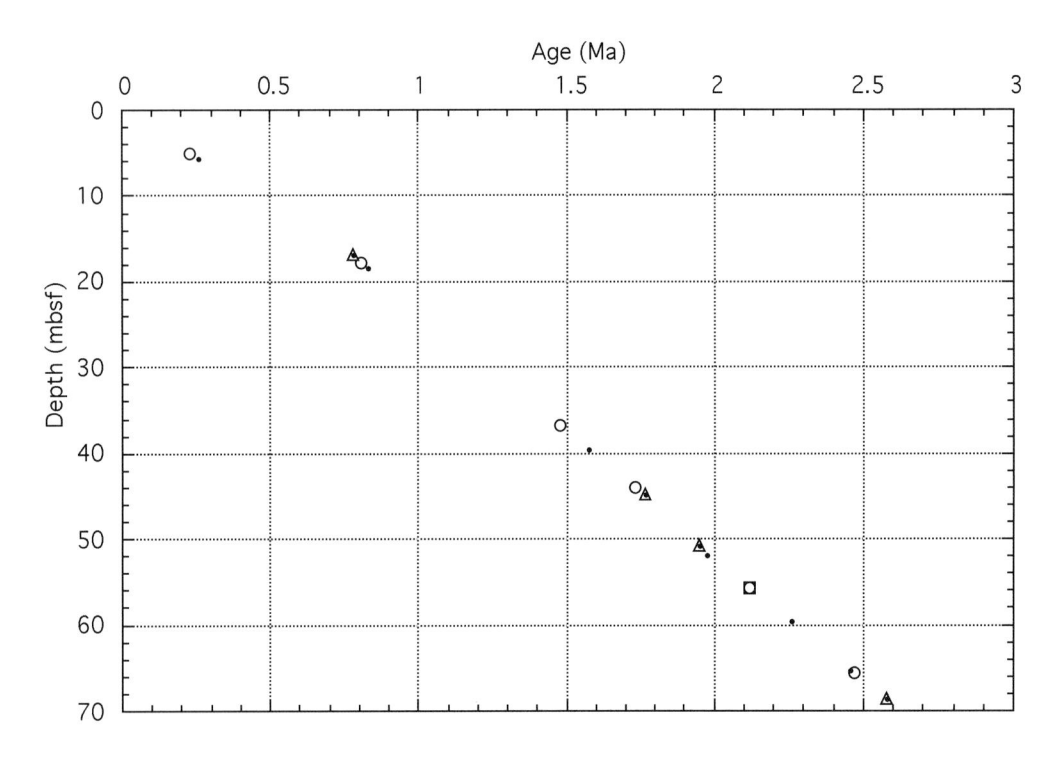

Fig. 29. A revised age–depth plot for DSDP Hole 701A utilizing the biostratigraphical and polarity reversal events identified in this study and calibrated to the GPTS of Cande & Kent (1995). See Fig. 5 for legend of symbols.

Hole 704B

Seventy-two cores were recovered from a total penetration of 671.7 mbsf representing a recovery of 74% of the cored interval. Sediments recovered consist of Quaternary to upper Pliocene calcareous siliceous ooze and upper Pliocene to upper Miocene siliceous calcareous ooze, upper Miocene to middle Miocene calcareous nannofossil ooze, and middle Miocene to lower Oligocene chalk (Ciesielski *et al.* 1988). A good palaeomagnetic polarity reversal record was obtained by both shipboard and shorebase analysis (Ciesielski *et al.* 1988; Hailwood & Clement 1991*b*). Diatoms, calcareous nannofossils and radiolarians are present throughout the sequence. Diatoms are abundant and well preserved throughout the sequence with a decrease in abundance and preservation in the upper and lower Miocene. Calcareous nannofossils are present through the section. Radiolarians are generally abundant and well preserved through the section. The integrated biostratigraphical and magnetostratigraphical data previously used (palaeomagnetics: Hailwood & Clement 1991*b*; diatoms: Ciesielski *et al.* 1988; Ciesielski 1991; calcareous nannofossils: none; radiolarians: none) and calibrated to the GPTS of Berggren *et al.* (1985*a,b*) are shown in Table 22. An age–depth graph for Hole 114-704B using this data is shown in Fig. 30.

Figure 31 compares the polarity chronozones and diatom zones assigned by previous workers. Hailwood & Clement (1991*b*), Ciesielski *et al.* (1988) and Ciesielski (1991) with the assignments in this study. In addition, this figure illustrates the stratigraphical ranges of diatoms and calcareous nannofossils. The stratigraphical ranges reflect the placement of FO and LO of stratigraphically useful species as recognized by Ciesielski *et al.* (1988) and Ciesielski (1991). The revised chronozones and diatom zones illustrated are based on reassessing the data sets of these workers (Table 23). Comparison of the chronozonal assignments of Hailwood & Clement (1991*b*; Table 22) with those of this study (Table 23) indicate that the minimal reinterpretations reflect in part differences in the GPTS used in each study.

Figure 32 shows an age–depth graph for Hole 114-704B using the stratigraphical markers determined in this study. The LO of *Thallassiosira kolbei* (111.20–113.50 mbsf) and the last occurrence of *Thallassiosira vulnifica* (149.70–156.70 mbsf) have ages approximately 0.50 million years older than the ages of these events at other sites. These differences reflect the paucity of the polarity reversal record in the upper 200 mbsf. One hiatus is placed in the section at 452.15–453.60 mbsf. The upper boundary of this break is constrained by the First occurrence of *Nitzschia denticuloides* (439.00–442.49 mbsf), the FO of *Denticulopsis hustedtii* (445.51–451.67 mbsf) and the base of C5Bn.2n (443.00 mbsf). The lower boundary is constrained by the base of C5Cn.3n at 453.60 mbsf. Ciesielski *et al.* (1988) recognize two hiatuses within the sequence between 287 and 308 mbsf and between 425 and 432 mbsf. Hailwood & Clement (1991*b*) recognize one hiatus at approximately 440 mbsf. With the current data set, we interpret a hiatus at approximately 452 mbsf which is contemporaneous with that identified by Hailwood & Clement (1991*b*).

Table 22. *Composite dataset for ODP Hole 704B containing the diatom, radiolarian, calcareous nannofossil and palaeomagnetic data calibrated to the GPTS of Cande & Kent (1995)*

	Event	Sample (top)	Sample (bottom)	Depth (t)	Depth (b)	Age (t)	Age (b)
LO	*Hemidiscus karstenii*	1H-1, 0	1H2, 150	0.00	3.00	0.00	0.07
LO	*Pseudoemiliana lacunosa*	1H-1, 0	1H-CC	0.00	6.70	0.00	0.15
LO	*Actinocyclus ingens*	3H-CC	4H-2, 95	25.70	28.15	0.58	0.64
B	C1n			34.51	34.51	0.78	0.78
B	C1r.1n			44.27	44.27	1.07	1.07
LO	*Simonseniella barboi*	11H-CC	12H-CC	101.70	111.20	2.12	2.30
LO	*Thalassiosira kolbei*	12H-CC	13H-2, 125	111.20	113.50	2.30	2.34
LO	*Helicosphaera sellii*	13H-CC	14H-CC	120.70	130.20	2.47	2.65
LO	*Calcidiscus macintyrei*	16H-CC	17H-CC	147.20	156.70	2.96	3.13
LO	*Discoaster browerii*	16H-CC	17H-CC	147.20	156.70	2.96	3.13
LO	*Thalassiosira vulnifica*	17X-2, 100	17X-CC	149.70	156.70	3.00	3.13
FO	*Thalassiosira vulnifica*	19X-1, 21	19X-1, 75	166.41	166.95	3.31	3.32
FO	*Nitzschia weaveri*	19X-CC	20X-CC	175.70	185.20	3.48	3.66
LO	*Reticulofenestra pseudoumbilica*	21X-CC	22X-CC	194.70	204.20	3.83	4.01
FO	*Nitzschia angulata*	23X-2, 70	23X-CC	206.40	213.70	4.05	4.18
T	C3n.1n			213.70	213.70	4.18	4.18
LO	*Hemidiscus cuneiformis*	23X-3, CC	24X-1, 80-82	213.70	214.52	4.18	4.25
LO	*Simonseniella barboi*	23X-3, CC	24X-1, 80-82	213.70	214.52	4.18	4.25
LO	*Thalassiosira miocenica*	23X-3, CC	24X-1, 80-82	213.70	214.52	4.18	4.25
LO	*Denticulopsis hustedtii*	23X-CC	24X-CC	213.70	223.20	4.18	5.42
B	C3n.1n			215.00	215.00	4.29	4.29
LO	*Thalassiosira insigna*	24X-2, 80-82	24X-3, 80-82	216.02	217.52	4.42	4.57
LO	*Nitzschia reinholdii*	24X-2, 80-82	24X-3, 80-82	216.02	217.52	4.42	4.57
T	C3n.2n			216.50	216.50	4.48	4.48
LO	*Cosmiodiscus intersectus*	24X-3, 80-82	24X-4, 80-82	217.52	219.02	4.52	4.57
LO	*Nitzschia fossilis*	24X-4, 80-82	24X-5, 80-82	219.02	220.52	4.57	4.63
LO	*Nitzschia marina*	24X-4, 80-82	24X-5, 80-82	219.02	220.52	4.57	4.63
B	C3n.2n			220.50	220.50	4.62	4.62
LO	*Stichocorys peregrina*	24X-CC	25X-CC	223.20	232.70	5.42	6.22
LO	*Amphymenium challengerae*	24X-CC	25X-CC	223.20	232.70	5.42	6.22
LO	*Thalassiosira praeconvexa*	25X-1, 80-82	25X-2, 80-82	224.02	225.52	5.67	5.92
T	C3An.1n			224.76	224.76	5.89	5.89
FO	*Thalassiosira oestrupii*	25X-2, 80-82	25X-3, 80-82	227.02	228.52	5.99	6.04
B	C3An.1n			231.05	231.05	6.14	6.14
T	C3An.2n			233.85	233.85	6.27	6.27
LO	*Thalassiosira convexa v. aspinosa*	26X-4, 82	27X-2, 110	238.02	244.80	6.43	6.69

(continued)

Table 22. (*continued*)

	Event	Sample (top)	Sample (bottom)	Depth (t)	Depth (b)	Age (t)	Age (b)
B	C3An.2n			241.62	241.62	6.57	6.57
FO	*Amphymenium challengerae*	27X-2, 180	27X-4, 50	244.97	247.20	6.70	6.78
LO	*Lamprocyclas aegles*	27X-2, 180	27X-4, 50	244.97	247.20	6.70	6.78
T	C3Bn			251.25	251.25	6.94	6.94
FO	*Thalassiosira praeconvexa*	28X-1, 81	28X-2, 79	252.51	253.99	6.97	7.01
B	C3Bn			256.75	256.75	7.09	7.09
LO	*Denticulopsis lauta*	28X-4, 79-81	28X-6, 79-81	257.01	258.60	7.12	7.31
T	C4n.1n			259.50	259.50	7.43	7.43
LO	*Nitzschia porteri*	30X-1, 82	30X-2, 80	271.52	273.00	8.04	8.12
FO	*Lamprocyclas aegles*	30X-2, 45	30X-4, 45	272.65	275.65	8.10	8.25
B	C4r.1n			275.85	275.85	8.26	8.26
LO	*Diartus hughesi*	30X-CC	31X-2, 45	280.20	282.15	8.39	8.46
FO	*Stichocorys peregrina*	30X-CC	31X-2, 45	280.20	282.15	8.39	8.46
T	C4An			290.06	290.06	8.70	8.70
FO	*Thalassiosira insigna*	32X-6, 80-82	33X-1, 80-82	298.02	300.02	8.79	8.81
FO	*Nitzschia marina*	32X-6, 82	33X-1, 80	298.02	300.00	8.79	8.81
FO	*Nitzschia reinholdii*	32X-6, 80-82	33X-1, 80-82	298.02	300.02	8.79	8.81
FO	*Cosmiodiscus intersectus*	32X-6, 80-82	33X-1, 80-82	298.02	300.02	8.79	8.81
B	C4An			318.55	318.55	9.03	9.03
LO	*Denticulopsis dimorpha*	35X-1, 80-82	35X-3, 80-82	319.02	322.02	9.05	9.37
FO	*Nitzschia fossilis*	35X-2, 82	35X-5, 80	322.02	325.00	9.16	9.25
T	C4Ar.1n			324.10	324.10	9.23	9.23
B	C4Ar.1n			327.05	327.05	9.31	9.31
T	C4Ar.2n			338.55	338.55	9.58	9.58
FO	*Diartus hughesi*	39X-1, 1	39X-2, 80	356.20	358.10	9.92	9.96
FO	*Hemidiscus cuneiformis*	39X-5, 82	40X-2, 80	363.02	368.00	10.05	10.15
FO	*Thalassiosira miocenica*	40X-2, 82	41X-2, 80	368.02	376.02	10.15	10.30
FO	*Asteromphalus kenettii*	42X-2, 80-82	42X-6, 80-82	387.02	391.52	10.51	10.60
LO	*Cyrtocapsella japonica*	42X-4, 56	42X-6, 13	389.76	392.33	10.57	10.62
B	C5n.2n			409.64	409.64	10.95	10.95
FO	*Simonsenella barboi*	46X-2, 80-82	46X-3, 80-82	425.00	426.50	11.90	11.99
FO	*Cyrtocapsella japonica*	45X-2, 61	45X-CC	415.31	422.70	11.30	11.76
FO	*Cyrtocapsella tetrapera*	45X-2, 59	45X-4, 70	424.79	427.29	11.89	12.04
LO	*Actinocyclus ingens v. nodus*	46X-2, 82	46X-5, 80	425.00	429.50	11.90	12.18
FO	*Denticulopsis dimorpha*	46X-2, 82	46X-5, 80	425.02	429.50	11.90	12.17
LO	*Crucidenticula nicobarica*	46X-5, 82	47X-2, 81	429.52	434.51	12.18	?
LO	*Crucidenticula kanayae*	46X-5, 80	47X-2, 81	429.50	434.51	12.18	?
T	C5An.2n			429.55	429.55	12.18	?
	Unconformity			***	***	***	***
LO	*Denticulopsis maccollumnii*	47X-2, 81	47X-5, 79	434.50	438.99	?	15.05
FO	*Actinocyclus ingens v. nodus*	47X-5, 79	47X-2, 81	434.51	438.99	?	15.05
LO	*Coscinodiscus lewisianus*	47X-2, 81	47X-5, 79	434.51	438.99	?	15.05
LO	*Nitzschia grossepunctata*	47X-2, 81	47X-5, 79	434.51	438.99	?	15.05
B	C5ADn			435.65	435.65	14.61	14.61
T	C5Bn.1n			436.60	436.60	14.80	14.80
B	C5Bn.1n			437.40	437.40	14.89	14.89
FO	*Nitzschia denticuloides*	47X-5, 81	48X-1, 79	439.00	442.49	15.06	15.13
T	C5Bn.2n			439.72	439.72	15.03	15.03
LO	*Nitzschia maleinterpretaria*	48X-1, 79	48X-3, 81	442.59	445.51	15.14	?
B	C5Bn.2n			443.00	443.00	15.16	15.16
FO	*Denticulopsis hustedtii*	48X-3, 81	49X-1, 47	445.51	451.67	?	?
FO	*Cestodiscus peplum*	48X-3, 81	49X-1, 47	445.51	451.67	?	?
LO	*Thalassiosira fraga*	48X-3, 81	49X-1, 47	445.51	451.67	?	?
FO	*Denticulopsis lauta*	48X-3, 79-81	49X-1, 47-49	445.51	451.69	?	?
FO	*Actinocyclus ingens*	48X-3, 79-81	49X-1, 47-49	445.51	451.69	?	?
FO	*Calcidiscus macintyrei*	48X-CC	49X-CC	451.20	460.70	?	16.84
FO	*Simonsensiella barboi*	49X-1, 47	49X-4, 49	451.69	456.19	?	?
	Unconformity			***	***	***	***
B	C5Cn.3n			453.60	453.60	16.73	16.73
LO	*Crucidenticula kanayae*	49X-4, 49	50X-1, 41	456.19	461.15	16.77	16.85
FO	*Denticulopsis maccollumnii*	49X-4, 49	50X-1, 45	456.19	461.15	16.77	16.85
FO	*Nitzschia grossepunctata*	50X-1, 45	51X-1, 45	461.15	463.15	16.85	16.88
FO	*Raphidodiscus marylandicus*	50X-1, 45	51X-1, 45	461.15	463.15	16.85	16.88
FO	*Crucidenticula kanayae*	52X-1, 80-82	52X-4, 42	473.02	477.12	17.03	17.09
LO	*Coscinodiscus rhombicus*	52X-4, 42	52X-CC	477.12	481.70	17.09	17.16
LO	*Thalassiosira aspinosa*	53X-1, 42	53X-3, 42	482.12	485.12	17.17	17.21
FO	*Coscinodiscus lewisianus*	53X-3, 42	54X-2, 40	485.12	493.10	17.21	17.51
T	C5Dn			489.50	489.50	17.28	17.28
FO	*Crucidenticula nicobarica*	54X-2, 42	54X-5, 40	493.10	494.40	17.51	17.60
FO	*Nitzschia maleinterpretaria*	56X-1, 87	57X-2, 82	511.07	522.02	18.68	19.28
LO	*Rocella schraderi*	56X-1, 87	57X-2, 82	511.07	522.02	18.68	19.28

Table 22. (*continued*)

	Event	Sample (top)	Sample (bottom)	Depth (t)	Depth (b)	Age (t)	Age (b)
LO	*Rocella vigilans*	56X-1, 87	57X-2, 82	511.07	522.02	18.68	19.28
LO	*Rocella gelida*	56X-1, 87	57X-2, 82	511.07	522.02	18.68	19.28
B	C5En			512.59	512.59	18.78	18.78
T	C6n			513.94	513.94	19.05	19.05
FO	*Thalassiosira aspinosa*	58X-2, 70-72	58X-5, 70-72	531.42	534.45	19.86	20.05
LO	*Thalassiosira spumellaroides*	58X-5, 70-72	59X-2, 80-82	534.45	541.02	20.05	20.45
B	C6n			535.79	535.79	20.13	20.13
T	C6An.1n			538.29	538.29	20.52	20.52
FO	*Thalassiosira fraga*	59X-2, 82	60X-2, 80	541.02	550.50	20.75	21.54
LO	*Bogorovia veniamini*	59X-2, 82	60X-2, 80	541.02	550.50	20.75	21.54
B	C6AAr.1n			558.99	558.99	22.25	22.25
FO	*Rocella gelida*	65X-CC	66X-CC	605.20	614.70	?	?

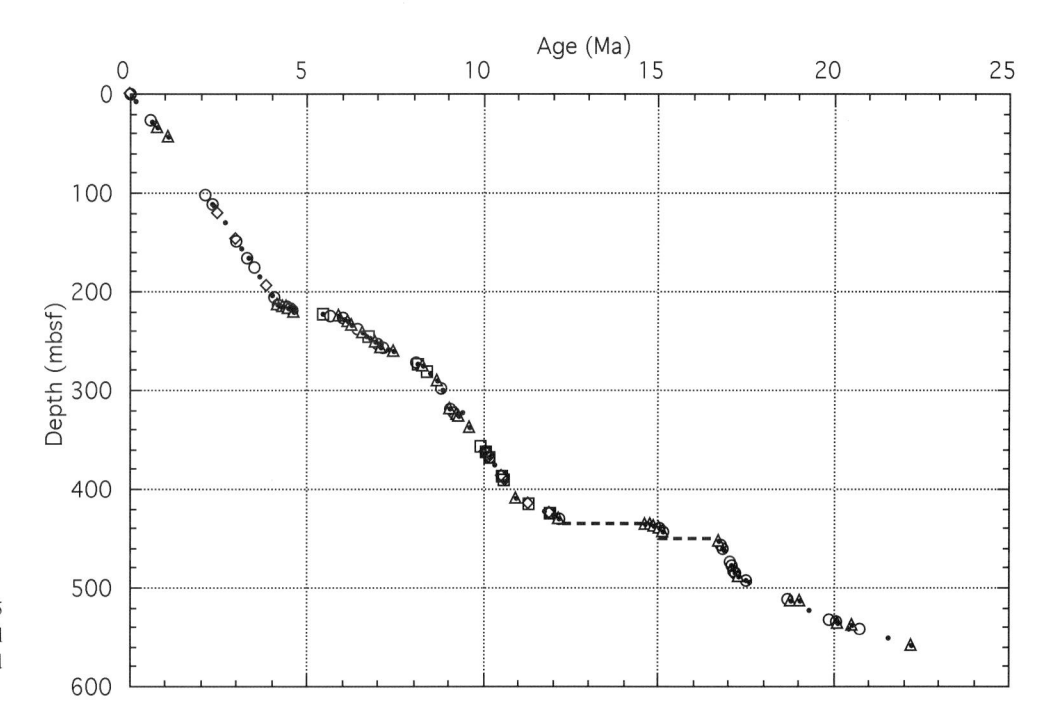

Data sets are integrated from the work of numerous authors (see Table 2).

Fig. 30. An age–depth plot for DSDP Hole 704B utilizing the previously used biostratigraphical datums and polarity reversal interpretation calibrated to the GPTS of Cande & Kent (1995). See Fig. 5 for legend of symbols. One stratigraphical break (dashed line) is based on Hailwood & Clement (1991*b*).

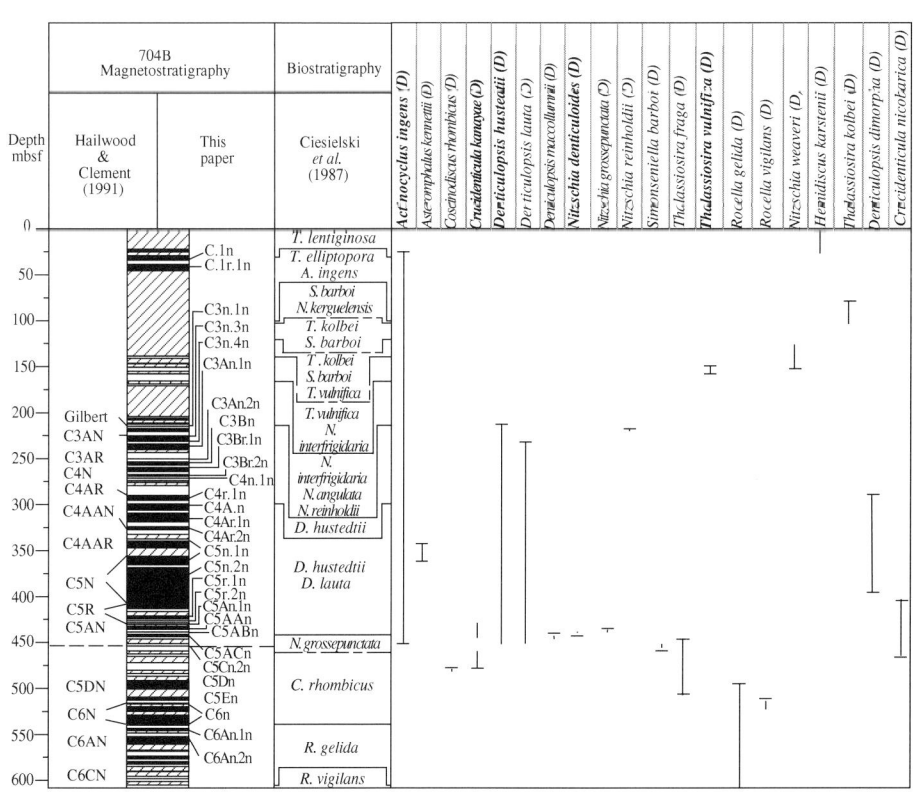

Fig. 31. Comparison of the ODP Hole 704B polarity chronozones and diatom zonal assignments by previous workers with the chronozonal assignments of this study. The stratigraphical ranges of selected stratigraphical markers are based on the original data sets (see text for references). D, diatoms; R, radiolaria; N, calcareous nannofossils. Emboldened events indicate primary biostratigraphical events recognized in this study. The polarity record shown represents the original interpretation calibrated to the GPTS of Cande & Kent (1995) and the interpretation from this study correlated with the GPTS of Cande & Kent (1995). Diagonal lines indicate intervals of uncertainty.

Table 23. *Composite dataset for ODP Hole 704B containing the diatom, radiolarian, calcareous nannofossil and palaeomagnetic data calibrated to the GPTS of Cande & Kent (1995) using the biochronological interpretation based on this study*

	Event	Sample (top)	Sample (bottom)	Depth (t)	Depth (b)	Age (t)	Age (b)
LO	*Hemidiscus karstenii*	1H-1, 0	1H-2, 150	0.00	3.00	0.00	0.07
LO	*Pseudoemiliana lacunosa*	1H-1, 0	1H-CC	0.00	6.70	0.00	0.15
LO	*Actinocyclus ingens*	3H-CC	4H-2, 95	25.70	28.15	0.58	0.64
B	C1n			34.51	34.51	0.78	0.78
B	C1r.1n			44.27	44.27	1.07	1.07
LO	*Simonseniella barboi*	11H-CC	12H-CC	101.70	111.20	2.12	2.30
LO	*Thalassiosira kolbei*	12H-CC	13H-2, 125	111.20	113.50	2.30	2.34
LO	*Helicosphaera sellii*	13H-CC	14H-CC	120.70	130.20	2.47	2.65
LO	*Calcidiscus macintyrei*	16H-CC	17H-CC	147.20	156.70	2.96	3.13
LO	*Discoaster browerii*	16X-CC	17X-CC	147.20	156.70	2.96	3.13
LO	*Thalassiosira vulnifica*	17X-2, 100	17X-CC	149.70	156.70	3.00	3.13
FO	*Thalassiosira vulnifica*	19X-1, 21	19X-1, 75	166.41	166.95	3.31	3.32
FO	*Nitzschia weaveri*	19X-CC	20X-CC	175.70	185.20	3.48	3.66
LO	*Reticulofenestra pseudoumbilica*	21X-CC	22X-CC	194.70	204.20	3.83	4.01
FO	*Nitzschia angulata*	23X-2, 70	23X-CC	206.40	213.70	4.05	4.18
T	C3n.1n			213.70	213.70	4.18	4.18
LO	*Hemidiscus cuneiformis*	23X-CC	24X-1, 80-82	213.70	214.52	4.18	4.25
LO	*Simonseniella barboi*	23X-CC	24X-1, 80-82	213.70	214.52	4.18	4.25
LO	*Thalassiosira miocenica*	23X-CC	24X-1, 80-82	213.70	214.52	4.18	4.25
LO	*Denticulopsis hustedtii*	23X-CC	24X-CC	213.70	223.20	4.18	5.42
B	C3n.1n			215.00	215.00	4.29	4.29
LO	*Thalassiosira insigna*	24X-2, 80-82	24X-3, 80-82	216.02	217.52	4.42	4.57
LO	*Nitzschia reinholdii*	24X-2, 80-82	24X-3, 80-82	216.02	217.52	4.42	4.57
T	C3n.2n			216.50	216.50	4.48	4.48
LO	*Cosmiodiscus intersectus*	24X-3, 80-82	24X-4, 80-82	217.52	219.02	4.52	4.57
LO	*Nitzschia fossilis*	24X-4, 80-82	24X-5, 80-82	219.02	220.52	4.57	4.62
LO	*Nitzschia marina*	24X-4, 80-82	24X-5, 80-82	219.02	220.52	4.57	4.62
B	C3n.2n			220.50	220.50	4.62	4.62
LO	*Stichocorys peregrina*	24X-CC	25X-CC	223.20	232.70	4.73	5.75
LO	*Amphymenium challengerae*	24X-CC	25X-CC	223.20	232.70	4.73	5.75
LO	*Thalassiosira praeconvexa*	25X-1, 80-82	25X-2, 80-82	224.02	225.52	4.77	4.89
T	C3n.3n			224.76	224.76	4.80	4.80
FO	*Thalassiosira oestrupii*	25X-2, 80-82	25X-3, 80-82	227.02	228.52	4.95	5.06
B	C3n.4n			231.05	231.05	5.23	5.23
T	C3An.1n			233.85	233.85	5.89	5.89
LO	*Thalassiosra convexa v. aspinosa*	26X-4, 82	27X-2, 110	238.02	244.80	6.02	6.19
B	C3An.1n			241.62	241.62	6.14	6.14
FO	*Amphymenium challengerae*	27X-2, 180	27X-4, 50	244.97	247.20	6.20	6.24
LO	*Lamprocyclas aegles*	27X-2, 180	27X-4, 50	244.97	247.20	6.20	6.24
T	C3An.2n			248.80	248.80	6.27	6.27
B	C3An.2n			249.45	249.45	6.57	6.57
T	C3Bn			251.25	251.25	6.94	6.94
FO	*Thalassiosira praeconvexa*	28X-1, 81	28X-2, 79	252.51	253.99	6.97	7.01
B	C3Bn			256.75	256.75	7.09	7.09
LO	*Denticulopsis lauta*	28X-4, 79-81	28X-6, 79-81	257.01	258.60	7.09	7.12
T	C3Br.1n			259.50	259.50	7.14	7.14
B	C3Br.1n			264.29	264.29	7.17	7.17
T	C3Br.2n			265.79	265.79	7.34	7.34
B	C3Br.2n			267.79	295.60	7.38	7.38
T	C4n.1n			269.55	269.55	7.43	7.43
B	C4n.1n			270.10	270.10	7.56	7.56
T	C4n.2n			270.82	270.82	7.65	7.65
LO	*Nitzschia porteri*	30X-1, 82	30X-2, 80	271.52	273.00	7.71	7.83
FO	*Lamprocyclas aegles*	30X-2, 45	30X-4, 45	272.65	275.65	7.80	8.05
B	C4n.2n			275.85	275.85	8.07	8.07
LO	*Diartus hughesi*	30X-CC	31X-2, 45	280.20	282.15	8.12	8.14
FO	*Stichocorys peregrina*	30X-CC	31X-2, 45	280.20	282.15	8.12	8.14
T	C4r.1n			290.06	290.06	8.23	8.23
B	C4r.1n			295.60	295.60	8.26	8.26
FO	*Thalassiosira insigna*	32X-6, 80-82	33X-1, 80-82	298.02	300.02	8.54	8.73
FO	*Nitzschia marina*	32X-6, 82	33X-1, 80	298.02	300.02	8.54	8.73
FO	*Nitzschia reinholdii*	32X-6, 80-82	33X-1, 80-82	298.02	300.02	8.54	8.73
FO	*Cosmiodiscus intersectus*	32X-6, 80-82	33X-1, 80-82	298.02	300.02	8.54	8.73
T	C4An			299.37	299.37	8.70	8.70
B	C4An			306.00	306.00	9.03	9.03
T	C4Ar.1n			309.10	309.10	9.23	9.23
B	C4Ar.1n			318.55	318.55	9.31	9.31
LO	*Denticulopsis dimorpha*	35X-1, 80-82	35X-3, 80-82	319.02	322.02	9.33	9.48
FO	*Nitzschia fossilis*	35X-2, 82	35X-5, 80	322.02	325.00	9.48	9.59
T	C4Ar.2n			324.10	324.10	9.58	9.58

Table 23. (*continued*)

	Event	Sample (top)	Sample (bottom)	Depth (t)	Depth (b)	Age (t)	Age (b)
B	C4Ar.2n			331.00	331.00	9.64	9.64
T	C5n.1n			338.55	338.55	9.74	9.74
FO	*Diartus hughesi*	39X-1, 1	39X-2, 80	356.20	358.10	9.83	9.84
FO	*Hemidiscus cuneiformis*	39X-5, 82	40X-2, 80	363.02	368.00	9.87	9.90
B	C5n.1n			365.00	365.00	9.88	9.88
T	C5n.2n			366.00	366.00	9.92	9.92
FO	*Thalassiosira miocenica*	40X-2, 82	41X-2, 80	368.02	376.02	9.97	10.16
LO	*Denticulopsis dimorpha*	40X-5, 82	41X-5, 80	375.52	382.00	10.15	10.30
FO	*Asteromphalus kenettii*	42X-2, 80-82	42X-6, 80-82	387.02	391.52	10.42	10.52
LO	*Cyrtocapsella japonica*	42X-4, 56	42X-6, 13	389.76	392.33	10.48	10.54
B	C5n.2n			409.64	409.64	10.95	10.95
FO	*Cyrtocapsella japonica*	45X-2, 61	45X-CC	415.31	422.70	10.99	11.05
T	C5r.1n			423.10	423.10	11.05	11.05
FO	*Simonsenella barboi*	46X-2, 80-82	46X-3, 80-82	425.00	426.50	11.39	11.66
FO	*Cyrtocapsella tetrapera*	45X-2, 59	45X-4, 70	424.79	427.29	11.35	11.83
LO	*Actinocyclus ingens v. nodus*	46X-2, 82	46X-5, 80	425.00	429.50	11.06	12.95
FO	*Denticulopsis dimorpha*	46X-2, 82	46X-5, 80	425.02	429.50	11.06	12.95
T	C5r.2n			425.50	425.50	11.48	11.48
B	C5r.2n			426.20	426.20	11.53	11.53
T	C5An.1n			427.70	427.70	11.94	11.94
B	C5An.1n			428.50	428.50	12.08	12.08
LO	*Crucidenticula kanayae*	46X-5, 82	47X-2, 81	429.50	434.51	12.95	13.11
LO	*Crucidenticula nicobarica*	46X-5, 82	47X-2, 81	429.50	434.51	12.95	13.11
T	C5AAn			429.55	429.55	12.99	12.99
LO	*Denticulopsis maccollumnii*	47X-2, 81	47X-5, 79	434.50	438.99	13.11	13.22
FO	*Actinocyclus ingens v. nodus*	47X-5, 79	47X-2, 81	434.51	438.99	13.11	13.64
LO	*Coscinodiscus lewisianus*	47X-2, 81	47X-5, 79	434.51	438.99	13.11	13.64
LO	*Nitzschia grossepunctata*	47X-2, 81	47X-5, 79	434.51	438.99	13.11	13.64
B	C5AAn			435.65	435.65	13.14	13.14
T	C5ABn			436.60	436.60	13.30	13.30
B	C5ABn			437.40	437.40	13.51	13.51
FO	*Nitzschia denticuloides*	47X-5, 81	48X-1, 79	439.00	442.49	13.64	14.02
T	C5ACn			439.72	439.72	13.70	13.70
LO	*Nitzschia maleinterpretaria*	48X-1, 79	48X-3, 81	442.59	445.51	14.03	?
B	C5ACn			443.00	443.00	14.08	14.08
FO	*Denticulopsis hustedtii*	48X-3, 81	49X-1, 47	445.51	451.67	?	?
FO	*Cestodiscus peplum*	48X-3, 81	49X-1, 47	445.51	451.67	?	?
LO	*Thalassiosira fraga*	48X-3, 81	49X-1, 47	445.51	451.67	?	?
LO	*Denticulopsis lauta*	48X-3, 79-81	49X-1, 47-49	445.51	451.69	?	?
FO	*Actinocyclus ingens*	48X-3, 79-81	49X-1, 47-49	445.51	451.69	?	?
FO	*Calcidiscus macintyrei*	48X-CC	49X-CC	451.20	460.70	?	16.52
FO	*Simonseniella barboi*	49X-1, 47	49X-4, 49	451.69	456.19	?	16.50
	Unconformity			***	***	***	***
T	C5Cn.2n			452.15	452.15	16.33	16.33
B	C5Cn.2n			453.60	453.60	16.49	16.49
FO	*Denticulopsis maccollumnii*	49X-4, 49	50X-1, 45	456.19	461.15	16.50	16.52
FO	*Nitzschia grossepunctata*	50X-1, 45	51X-1, 45	461.15	463.15	16.52	16.52
FO	*Raphidodiscus marylandicus*	50X-1, 45	51X-1, 45	461.15	463.15	16.64	16.68
FO	*Crucidenticula kanayae*	52X-1, 80-82	52X-4, 42	473.02	477.12	16.56	16.81
T	C5Cn.3n			473.80	473.80	16.56	16.56
B	C5Cn.3n			475.05	475.05	16.73	16.73
LO	*Coscinodiscus rhombicus*	52X-4, 42	52X-CC	477.12	481.70	16.81	16.99
LO	*Thalassiosira aspinosa*	53X-1, 42	53X-3, 42	482.12	485.12	17.00	17.12
FO	*Coscinodiscus lewisianus*	53X-3, 42	54X-2, 40	485.12	493.10	17.12	17.42
FO	*Crucidenticula nicobarica*	54X-2, 42	54X-5, 40	493.10	494.40	17.42	17.47
LO	*Rocella gelida*	54X-5, 40	55X-1, 40	494.40	501.10	17.47	17.85
B	C5Dn			498.20	498.20	17.62	17.62
FO	*Nitzschia maleinterpretaria*	56X-1, 87	57X-2, 82	511.07	522.02	18.66	19.37
LO	*Rocella schraderi*	56X-1, 87	57X-2, 82	511.07	522.02	18.66	19.37
LO	*Rocella vigilans*	56X-1, 87	57X-2, 82	511.07	522.02	18.66	19.37
LO	*Rocella gelida*	56X-1, 87	57X-2, 82	511.07	522.02	18.66	19.37
B	C5En			512.59	512.59	18.78	18.78
T	C6n			513.94	513.94	19.05	19.05
FO	*Thalassiosira aspinosa*	58X-2, 70-72	58X-5, 70-72	531.42	534.45	19.74	19.86
LO	*Thalassiosira spumellaroides*	58X-5, 70-72	59X-2, 80-82	534.45	541.02	19.86	20.12
FO	*Thalassiosira fraga*	59X-2, 82	60X-2, 80	541.02	550.50	20.12	21.07
LO	*Bogorovia veniamini*	59X-2, 82	60X-2, 80	541.02	550.50	20.12	21.07
B	C6n			541.24	541.24	20.13	20.13
T	C6An.2n			548.34	548.34	21.00	21.00
B	C6An.2n			558.99	558.99	21.32	21.32
FO	*Rocella gelida*	65X-CC	66X-CC	605.20	614.70	?	?

Data sets are integrated from the work of numerous authors (see Table 2).

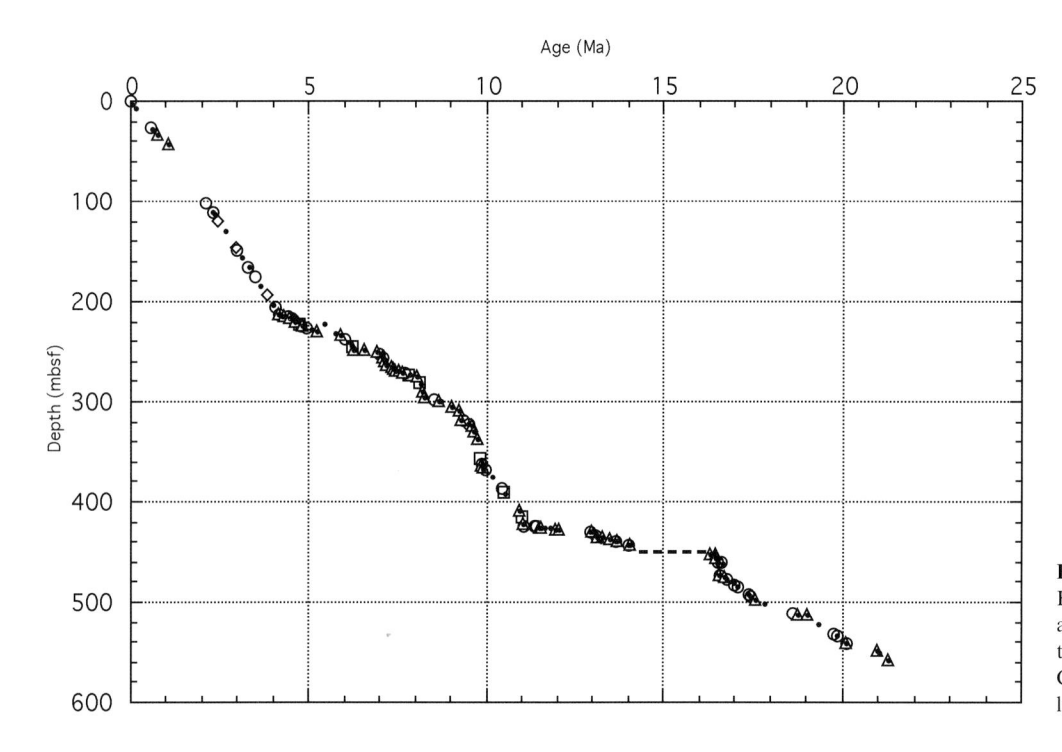

Fig. 32. An age–depth plot for DSDP Hole 704B utilizing the biostratigraphical and polarity reversal events identified in this study and calibrated to the GPTS of Cande & Kent (1995). See Fig. 5 for legend of symbols.

Hole 737A

Twenty-nine cores were recovered from a total penetration of 273.2 mbsf representing a recovery of 66% of the cored interval. Sediments recovered consist of Quaternary to upper Miocene diatom ooze, upper Miocene to middle Miocene calcareous nannofossil ooze with a variable terrigenous and volcanic component, upper Oligocene middle Eocene claystone and limestone (Barron *et al.* 1989).

A partial palaeomagnetic polarity reversal record was interpreted by Barron *et al.* (1991) based on shipboard and shorebase analysis. Diatoms and radiolarians are generally present in the Quaternary to upper Oligocene section and absent from the lower Oligocene to middle Eocene interval. Calcareous nannofossils are generally present throughout the sequence except for a barren Pliocene/Miocene interval. Well-preserved and abundant diatoms are present in the Quaternary to middle Miocene interval. Both abundance and preservation declines with rare, poorly to moderately preserved specimens present in upper to lower upper Oligocene sediments. Diatoms are absent from the lower Oligocene to middle Eocene interval. Calcareous nannofossils are common to abundant with moderate preservation for the Quaternary to lower Pliocene, absent from the lower Pliocene to upper Miocene and abundant with variable preservation in the upper Miocene to middle Eocene. Radiolarians are rare to abundant with poor to good preservation for the Quaternary to upper Oligocene section. Radiolarians are absent from lower upper Oligocene to middle Eocene (Barron *et al.* 1989). The integrated biostratigraphical and magnetostratigraphical data previously used (palaeomagnetics: Barron *et al.* 1991; diatoms: Baldauf & Barron 1991; calcareous nannofossils: none; radiolarians: Caulet (1991) and calibrated to the GPTS of Berggren *et al.* (1985*a,b*) is presented in Table 24. An age–depth graph for Hole 119-737A using this data is shown in Fig. 33.

Figure 34 compares the polarity chronozones and diatom zones assigned by previous workers (Barron *et al.* 1991; Baldauf & Barron 1991) with the chronozonal assignments in this study. The stratigraphical ranges of diatoms and radiolarians reflect the placement of FO and LO of stratigraphically useful species as recognized by Baldauf & Barron (1991), Barron *et al.* (1991) and Caulet (1991), respectively. The revised chronozones and diatom zones illustrated are based on reassessing the data sets of these workers (Table 25). Comparison of the chronozonal assignments of Barron *et al.* (1991; Table 24) with those of this study (Table 25) indicates no changes to the polarity assignments.

Figure 35 shows an age–depth graph for Hole 119-737A using the stratigraphical markers determined in this study. The stratigraphical sequence is continuous for the interval examined (3.0 to about 7.5 Ma). Stratigraphical control is based on both biostratigraphical and palaeomagnetic markers. The stratigraphical markers used include only those markers proven to be useful based on this study. These include the FO of *Nitzschia interfrigidaria* (43.60–46.90 mbsf), FO of *Nitzschia barronii* (62.00–65.6 mbsf), the FO of *Helotholus vema* (62.00–71.50 mbsf)) the FO of *Nitzschia praeinterfrigidaria* (78.10–81.00 mbsf) and the FO of *Denticulopsis hustedtii* (254.04–263.50 mbsf). Barron *et al.* (1991) recognize the occurrence of a hiatus at 1.5 mbsf and of a possible hiatus between 62 and 71 mbsf. The younger hiatus, which coincides with a boundary between glauconitic sand and diatom ooze, is inferred on the basis of diatom biostratigraphy and is adhered to in this study. The possible existence of an older hiatus is based on radiolarian biostratigraphy, but Barron *et al.* (1991) also indicate that the reduced sedimentation rate within the 62–71 mbsf interval may represent a condensed sequence. This study also records the occurrence of reduced sedimentation rates within this interval.

Table 24. *Composite dataset for ODP Hole 737A containing the diatom, radiolarian, calcareous nannofossil and palaeomagnetic data calibrated to the GPTS of Cande & Kent (1995)*

	Event	Sample (top)	Sample (bottom)	Depth (t)	Depth (b)	Age (t)	Age (b)
LO	*Nitzschia interfrigidaria*	1H-1, 57-59	1H-2, 57-59	0.59	2.10	?	?
LO	*Nitzschia praeinterfrigidaria*	1H-1, 57-59	1H-2, 57-59	0.59	2.10	?	?
LO	*Nitzschia weaveri*	1H-1, 57-59	1H-2, 57-59	0.59	2.10	?	?
	Unconformity			***	***	***	***
FO	*Nitzschia kerguelensis*	1H-2, 57-59	1H-CC	2.10	5.00	?	?
T	C2An.2n			11.42	11.42	3.11	3.11
FO	*Thalassiosira kolbei*	2H-5, 57-59	2H-CC	11.60	14.50	3.12	3.23
B	C2An.2n			13.75	13.75	3.22	3.22
LO	*Desmospyris spongiosa*			14.50	15.05	3.24	3.25
LO	*Thalassiosira insigna*	3H-2, 57-59	3H-5, 57-59	16.59	21.10	3.28	3.60
FO	*Nitzschia weaveri*	3H-2, 57-59	3H-5, 57-59	16.59	21.10	3.28	3.60
T	C2An.3n			19.00	19.00	3.33	3.33
B	C2An.3n			19.95	19.95	3.58	3.58
LO	*Simonseniella barboi*	4H-2, 57-59	4H-5, 57-59	26.10	30.60	3.68	3.76
LO	*Hemidiscus karstenii*	4H-CC	5H-2, 57-59	33.50	35.60	3.81	3.84
FO	*Nitzschia interfrigidaria*	6H-1, 57-59	6H-5, 57-59	43.60	46.90	3.98	4.03
FO	*Thalassiosra lentiginosa*	6H-5, 57-59	6H-CC	49.60	52.50	4.08	4.13
T	C3n.1n			55.41	55.41	4.18	4.18
B	C3n.1n			61.32	61.32	4.29	4.29
FO	*Nitzschia barronii*	7H-CC	8H-3, 57-59	62.00	65.60	4.32	4.57
FO	*Helotholus vema*			62.00	71.50	4.32	4.66
T	C3n.2n			65.08	65.08	4.48	4.48
B	C3n.2n			65.88	65.88	4.62	4.62
FO	*Nitzschia praeinterfrigidaria*	9H-5, 57-59	9H-CC	78.10	81.00	4.70	4.72
FO	*Thalassiosira inura*	9H-CC	10H-4, 57-59	81.00	87.60	4.72	4.77
FO	*Thalassiosira torokina*	10H-5, 57-59	10H-6, 57-59	87.60	89.10	4.77	4.77
T	C3n.3n			92.22	92.22	4.80	4.80
B	C3n.3n			97.90	97.90	4.89	4.89
T	C3n.4n			98.90	98.90	4.98	4.98
FO	*Thalassiosira oestrupii*	11H-CC	12H-2, 57-59	100.00	102.00	5.01	5.08

(*continued*)

Table 24. (*continued*)

	Event	Sample (top)	Sample (bottom)	Depth (t)	Depth (b)	Age (t)	Age (b)
LO	*Stichocorys peregrina*			101.90	103.40	5.07	5.12
B	C3n.4n			106.81	106.81	5.23	5.23
LO	*Thalassiosira miocenica*	13H-2, 57-59	13H-CC	111.60	119.00	5.41	5.70
T	C3An.1n			123.91	123.91	5.89	5.89
FO	*Rouxia heteropolara*	17H-5, 57-59	17H-CC	154.10	157.00	6.36	6.41
B	C3An.2n			167.21	167.21	6.57	6.57
FO	*Thalassiosira miocenica*	19-1, 57-59	19X-CC	167.10	176.20	6.57	6.68
LO	*Thalassiosira burckliana*	21X-CC	23X-2, 57-59	195.60	207.40	6.93	7.08
LO	*Actinocyclus ingens*	23X-CC	25X-2, 57-59	215.00	228.90	7.17	7.35
T	C4n.1n			235.28	235.28	7.43	7.43
FO	*Actinocyclus fryxellae*	27X-3, 57-59	27X-CC	247.70	253.80	?	?
FO	*Nitzschia reinholdii*	27X-3, 57-59	27X-CC	247.70	253.80	?	?
FO	*Cosmiodiscus intersectus*			247.70	253.80	?	?
FO	*Denticulopsis hustedtii*	28X-1, 57-59	28X-CC	254.40	263.50	?	?

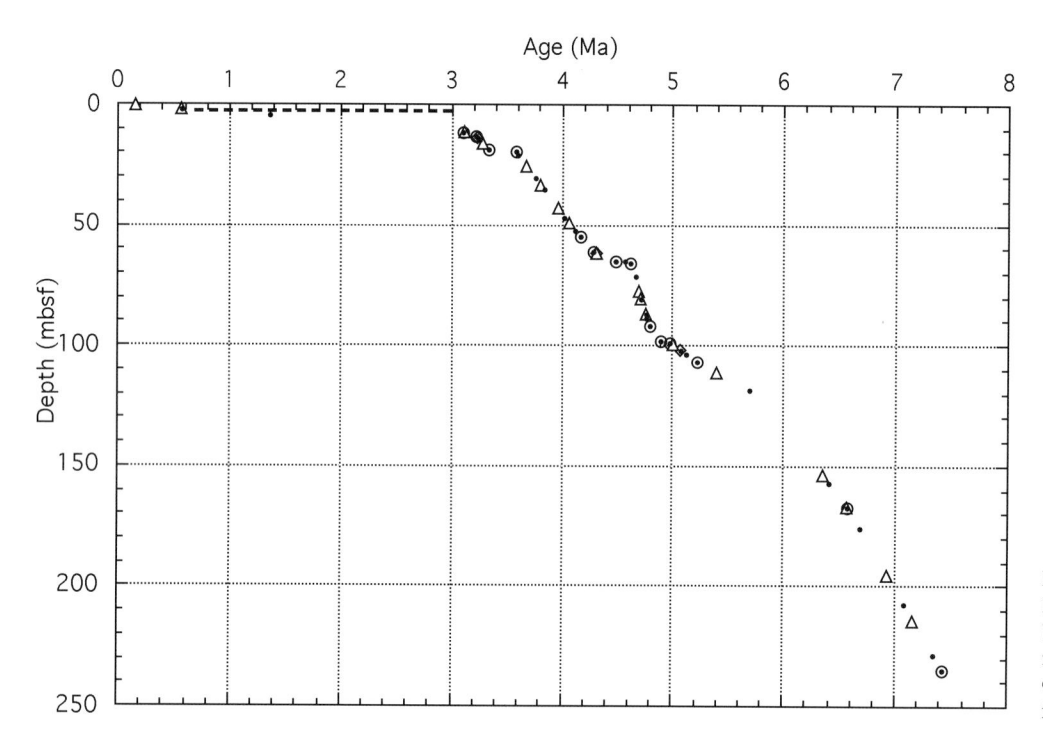

Data sets are integrated from the work of numerous authors (see Table 2).

Fig. 33. An age–depth plot for DSDP Hole 737A utilizing the previously used biostratigraphical datums and polarity reversal interpretation calibrated to the GPTS of Cande & Kent (1995). See Fig. 5 for legend of symbols.

HOLE 737A

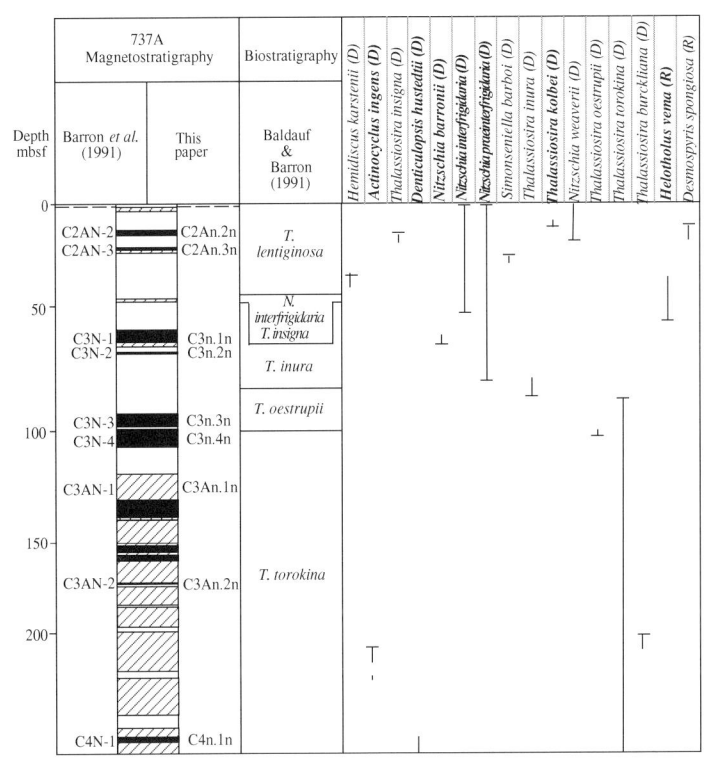

Fig. 34. Comparison of the ODP Hole 737A polarity chronozones and diatom zonal assignments by previous workers with the chronozonal assignments of this study. The stratigraphical ranges of selected stratigraphical markers is based on the original data sets (see text for references). D, diatoms; R, radiolaria; N, calcareous nannofossils. Emboldened events indicate primary biostratigraphical events recognized in this study. The polarity record shown represents the original interpretation calibrated to the GPTS of Cande & Kent (1995) and the interpretation from this study correlated with the GPTS of Cande & Kent (1995). Diagonal lines indicate intervals of no core recovery, dashed lines indicate intervals of uncertainty.

Table 25. *Composite dataset for ODP Hole 737A containing the diatom, radiolarian, calcareous nannofossil and palaeomagnetic data calibrated to the GPTS of Cande & Kent (1995) using the biochronological interpretation based on this study*

	Event	Sample (top)	Sample (bottom)	Depth (t)	Depth (b)	Age (t)	Age (b)
LO	*Nitzschia interfrigidaria*	1H-1, 57-59	1H-2, 57-59	0.59	2.10	?	?
LO	*Nitzschia praeinterfrigidaria*	1H-1, 57-59	1H-2, 57-59	0.59	2.10	?	?
LO	*Nitzschia weaveri*	1H-1, 57-59	1H-2, 57-59	0.59	2.10	?	?
	Unconformity			***	***	***	***
FO	*Nitzschia kerguelensis*	1H-2, 57-59	1H-CC	2.10	5.00	?	?
T	C2An.2n			11.42	11.42	3.11	3.11
FO	*Thalassiosira kolbei*	2H-5, 57-59	2H-CC	11.60	14.50	3.12	3.23
B	C2An.2n			13.91	13.91	3.22	3.22
LO	*Desmospyris spongiosa*			14.50	15.05	3.23	3.22
LO	*Thalassiosira insigna*	3H-2, 57-59	3H-5, 57-59	16.59	21.10	3.28	3.60
FO	*Nitzschia weaveri*	3H-2, 57-59	3H-5, 57-59	16.59	21.10	3.28	3.60
T	C2An.3n			19.00	19.00	3.33	3.33
B	C2An.3n			19.95	19.95	3.58	3.58
LO	*Simonseniella barboi*	4H-2, 57-59	4H-5, 57-59	26.10	30.60	3.68	3.76
LO	*Hemidiscus karstenii*	4H-CC	5H-2, 57-59	33.50	35.60	3.81	3.84
FO	*Nitzschia interfrigidaria*	6H-1, 57-59	6H-5, 57-59	43.60	46.90	3.98	4.03
FO	*Thalassiosira lentiginosa*	6H-5, 57-59	6H-CC	49.60	52.50	4.08	4.13
T	C3n.1n			55.41	55.41	4.18	4.18
B	C3n.1n			61.32	61.32	4.29	4.29
FO	*Nitzschia barronii*	7H-CC	8H-3, 57-59	62.00	65.60	4.32	4.62
FO	*Helotholus vema*			62.00	71.50	4.32	4.66
T	C3n.2n			65.08	65.08	4.48	4.48
B	C3n.2n			65.88	65.88	4.62	4.62
FO	*Nitzschia praeinterfrigidaria*	9H-5, 57-59	9H-CC	78.10	81.00	4.70	4.72
FO	*Thalassiosira inura*	9H-CC	10H-4, 57-59	81.00	87.60	4.72	4.77
LO	*Thalassiosira torokina*	10H-5, 57-59	10H-6, 57-59	87.60	89.10	4.77	4.78
T	C3n.3n			92.22	92.22	4.80	4.80
B	C3n.3n			97.90	97.90	4.89	4.89

(*continued*)

Table 25. (*continued*)

	Event	Sample (top)	Sample (bottom)	Depth (t)	Depth (b)	Age (t)	Age (b)
T	C3n.4n			98.90	98.90	4.98	4.98
FO	*Thalassiosira oestrupii*	11H-CC	12H-2, 57-59	100.00	102.00	5.01	5.08
LO	*Stichocorys peregrina*			101.90	103.40	5.07	5.12
B	C3n.4n			106.81	106.81	5.23	5.23
LO	*Thalassiosira miocenica*	13H-2, 57-59	13H-CC	111.60	119.00	5.41	5.70
T	C3An.1n			123.91	123.91	5.89	5.89
FO	*Rouxia heteropolara*	17H-5, 57-59	17H-CC	154.10	157.00	6.36	6.40
B	C3An.2n			167.21	167.21	6.56	6.56
FO	*Thalassiosira miocenica*	19-1, 57-59	19X-CC	167.10	176.20	6.56	6.67
LO	*Thalassiosira burckliana*	21X-CC	23X-2, 57-59	195.60	207.40	6.92	7.07
LO	*Actinocyclus ingens*	23X-CC	25X-2, 57-59	215.00	228.90	7.17	7.35
T	C4n.1n			235.28	235.28	7.43	7.43
FO	*Actinocyclus fryxellae*	27X-3, 57-59	27X-CC	247.70	253.80	?	?
FO	*Nitzschia reinholdii*	27X-3, 57-59	27X-CC	247.70	253.80	?	?
FO	*Cosmiodiscus intersectus*			247.70	253.80	?	?
FO	*Denticulopsis hustedtii*	28X-1, 57-59	28X-CC	254.40	263.50	?	?

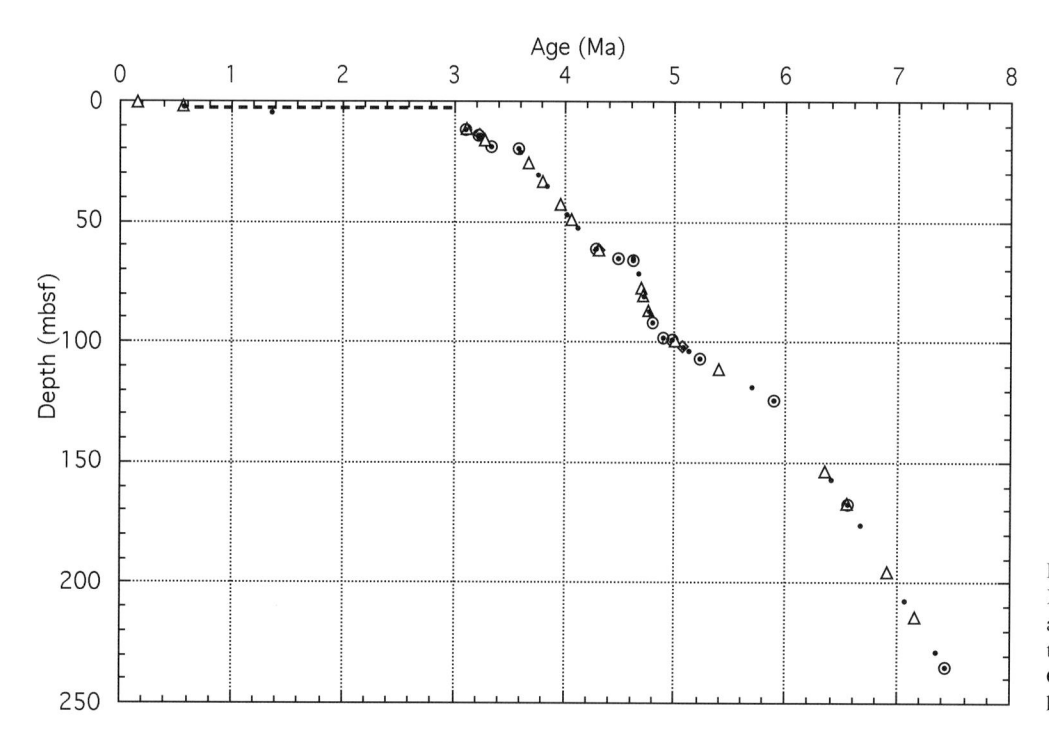

Data sets are integrated from the work of numerous authors (see Table 2).

Fig. 35. An age–depth plot for DSDP Hole 737A utilizing the biostratigraphical and polarity reversal events identified in this study and calibrated to the GPTS of Cande & Kent (1995). See Fig. 5 for legend of symbols.

Hole 744A

Twenty cores were recovered from a total penetration of 176.1 mbsf representing a recovery of 82% of the cored interval. Sediments recovered consist of Quaternary to lower Pliocene diatom ooze and upper Miocene to upper Eocene calcareous nannofossil ooze (Barron et al. 1989). A good palaeomagnetic polarity reversal record was interpreted by Barron et al. (1991) for the lowermost portion of the section. All three microfossil groups are recorded from the sediments examined (Barron et al. 1989). Diatoms are well preserved and abundant in the Quaternary to lower Oligocene interval. Abundance and preservation deteriorates in the lowermost Oligocene and diatoms are absent in upper Eocene sediments. Calcareous nannofossils are sporadic in the Quaternary to Pliocene and abundant in the upper Miocene to upper Eocene with moderate to good preservation. Radiolarians are common to abundant with moderate to good preservation for the Quaternary to lower Oligocene interval and are absent from upper Eocene sediments. The integrated biostratigraphical and magnetostratigraphical data previously used (palaeomagnetics: Barron et al. 1991; diatoms: Baldauf & Barron 1991; calcareous nannofossils: Wei & Thierstein 1991; radiolarians: none) and calibrated to the GPTS of Berggren et al. (1985a, b) is shown in Table 26. An age–depth plot for Hole 119-744A using this data is shown in Fig. 36.

Figure 37 compares the polarity chronozones and diatom zones assigned by previous workers. Barron et al. (1991) and Baldauf & Barron (1991) with the chronozonal assignments in this study for the interval containing polarity reversals from 80 to 170 mbsf. In addition, this figure illustrates the stratigraphical ranges of diatoms, radiolarians and calcareous nannofossils. The stratigraphical ranges reflect the placement of FO and LO of stratigraphically useful species as recognized by Baldauf & Barron (1991) and Wei & Thierstein (1991) respectively. The revised chronozones and diatom zones illustrated are based on reassessing the data sets of these workers (Table 27). Comparison of the chronozonal assignments of Barron et al. (1991; Table 26) with those of this study (Table 27) indicates minimal reinterpretation which affects the placement of one hiatus. Nomenclature changes also reflect in part differences in the GPTS used in these studies.

Figure 38 shows an age–depth graph for Hole 119-744A using the stratigraphical markers determined in this study. Two hiatuses are recognized, the younger between 98.9 and 99.20 mbsf and the older between 115.29 and 118.37 mbsf. The upper boundary of the younger hiatus is constrained by the top of C6AAn at 98.9 mbsf. The lower boundary of this break is constrained by a sequence of polarity events from the base of C7An at 100.24 mbsf to the top of C9n at 115.29 mbsf. This interval lacks primary biostratigraphical events. The LO of *Chiasmolithus altus* (106.92–107.62 mbsf) provides a potentially useful marker within this sequence. The upper boundary of the older hiatus is constrained by the base of C9n mbsf. The lower boundary is constrained by the base of C11n.2n at 118.37 mbsf. Placement of the boundaries for the older hiatus is tenuous and is based on the polarity reversal succession directly above and below the break. This break is constrained by the FO of *Isthmolithus recurvus* (168.85–170.35 mbsf) and the LO of *Reticulofenestra reticulata* (168.86–170.36 mbsf) .and the polarity reversals identified.

Barron et al. (1991) identified three hiatuses in this sequence. The younger two breaks occur at 97 and 99 mbsf. The placement of these unconformities is based on the basis of strontium isotope studies, diatom biostratigraphy and/or the interpretation of magnetostratigraphy. We combine these two unconformities into a single unconformity at 99 mbsf based on our interpretation of the magnetostratigraphy. The older hiatus identified by Barron et al. (1991) approximates a depth of 117 mbsf and corresponds in depth and age to the placement of the older hiatus in this study.

Table 26. *Composite dataset for ODP Hole 744A containing the diatom, radiolarian, calcareous nannofossil and palaeomagnetic data calibrated to the GPTS of Cande & Kent (1995)*

	Event	Sample (top)	Sample (bottom)	Depth (t)	Depth (b)	Age (t)	Age (b)
B	C5En			81.70	81.70	18.78	18.78
T	C6n			83.18	83.18	19.05	19.05
LO	*Thalassiosira spumellaroides*	10HCC	11H-1, 63-65	88.16	89.70	19.61	19.78
FO	*Cyrtocapsella longithorax*	10HCC	11H-1, 63-65	88.25	90.30	19.61	19.85
FO	*Thalassiosira fraga*	11H-1, 63-65	11H-2, 63-65	90.30	91.80	19.85	20.02
LO	*Rocella schraderi*	11H-2, 63-65	11H-3, 63-65	91.80	93.30	20.01	20.24
B	C6n			92.80	92.80	20.13	20.13
T	C6An.1n			94.63	94.63	20.52	20.52
LO	*Rocella gelida*	11H-4, 63-65	11H-5, 63-65	94.80	96.33	20.55	20.70
FO	*Thalassiosira spumellaroides*	11H-4, 63-65	11H-5, 63-65	94.90	96.33	20.55	20.70
LO	*Bogorovia veniamini*	11H-5, 63-65	11H-6, 63-65	96.33	97.83	20.69	?
FO	*Raphidodiscus marylandicus*	11H-5, 63-65	11H-6, 63-65	96.40	97.83	20.69	?
B	C6An.1n			96.68	96.68	20.73	20.73
	Unconformity			***	***	***	***
B	C6Cn.3n			97.77	97.77	24.12	24.12
	Unconformity			***	***	***	***
T	C7An			98.90	98.90	25.50	25.50
LO	*Thalassiosira primalabiata*	11H-CC	12H-1, 60-62	99.20	99.80	25.53	25.60
LO	*Reticulofenestra bisecta*	11H-CC	12H-1, 92.93	99.20	101.22	25.53	25.76
B	C7An			100.24	100.24	25.65	25.65
LO	*Lisitzina ornata*	12H-4, 60-62	12H-5, 60-62	104.30	106.00	25.80	25.86
T	C8n.1n			104.79	104.79	25.82	25.82
LO	*Chiasmolithus altus*	13H-1, 60-62	13H-2, 60-62	106.92	107.62	26.04	26.11
FO	*Nitzschia maleinterpretaria*	12H-6, 60-62	12H-7, 61-63	106.30	108.70	25.98	26.22
FO	*Rocella gelida*	13H-1, 60-62	13H-2, 60-62	108.70	110.20	26.23	26.38
FO	*Lisitzina ornata*	13H-1, 60-62	13H-2, 60-62	109.32	110.80	26.29	26.44
B	C8n.2n			111.81	111.81	26.55	26.55
T	C9n			115.29	115.29	27.03	27.03
	Unconformity			***	***	***	***
B	C11n.2n			118.37	118.37	30.10	30.10

(continued)

Table 26. (*continued*)

	Event	Sample (top)	Sample (bottom)	Depth (t)	Depth (b)	Age (t)	Age (b)
FO	*Synedra jouseana*	14H-2, 60-62	14H-4, 60-62	118.82	120.32	30.27	30.56
T	C12n			119.36	119.36	30.48	30.48
FO	*Rocella vigilans*	14H-2, 60-62	14H-4, 60-62	121.80	123.30	30.69	30.83
B	C12n			124.59	124.59	30.94	30.94
LO	*Reticulofenestra umbilica*	14H-6, 75-76	14H-CC	126.45	127.70	31.21	31.38
FO	*Cyclicargolithus abisectus*			127.70	128.46	31.39	31.50
LO	*Isthmolithus recurvus*	14H-2, 75-76	14H-3, 75-76	129.95	131.45	31.72	31.93
T	C13n			139.25	139.25	33.06	33.06
B	C13n			146.64	146.64	33.55	33.55
FO	*Rhizosolenia oligocenica*	17H-1, 0	17H-1, 68	146.70	147.53	33.56	33.66
LO	*Reticulofenestra oamaruensis*	17H-1, 75-76	17H-2, 75-76	147.60	148.30	33.67	33.75
T	C15n			155.65	155.65	34.66	34.66
B	C15n			158.10	158.10	34.94	34.94
T	C16n.1n			161.30	161.30	35.34	35.34
B	C16n.1n			162.90	162.90	35.53	35.53
T	C16n.2n			164.60	164.60	35.69	35.69
FO	*Reticulofenestra oamaruensis*	19H-CC	20H-1, 75-76	166.60	167.35	35.86	35.92
FO	*Isthmolithus recurvus*	20H-2, 75-76	20H-3, 75-76	168.85	170.35	36.03	36.15
LO	*Reticulofenestra reticulata*	20H-2, 75-76	20H-3, 75-76	168.86	170.36	36.03	36.15
B	C16n.2n			172.85	172.85	36.34	36.34

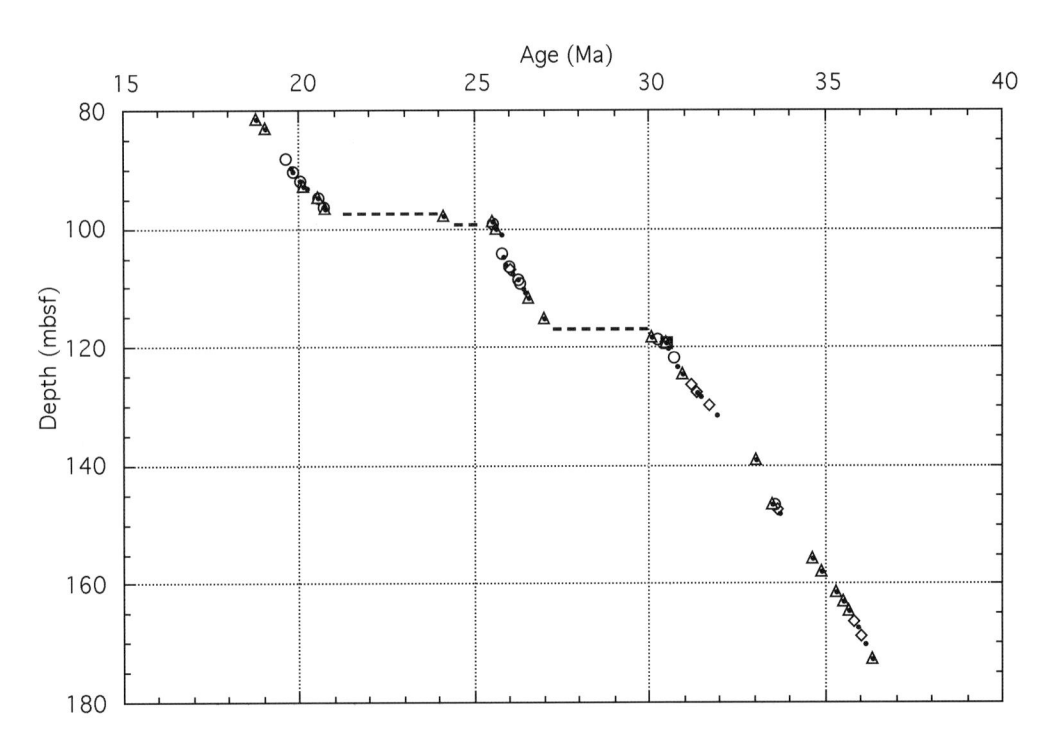

Data sets are integrated from the work of numerous authors (see Table 2).

Fig. 36. An age–depth plot for DSDP Hole 744A utilizing the previously used biostratigraphical datums and polarity reversal interpretation calibrated to the GPTS of Cande & Kent (1995). See Fig. 5 for legend of symbols.

HOLE 744A

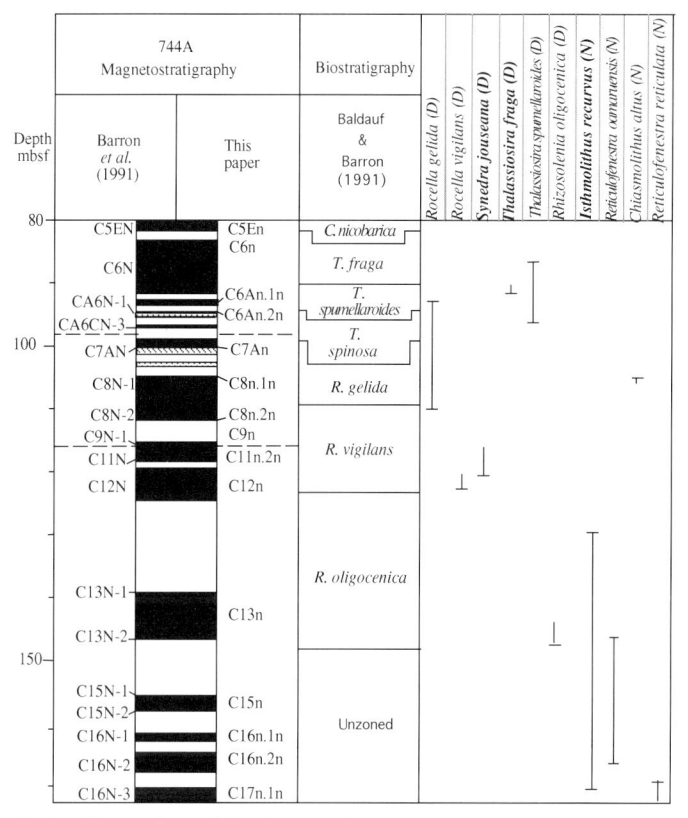

Fig. 37. Comparison of the ODP Hole 744A polarity chronozones and diatom zonal assignments by previous workers with the chronozonal assignments of this study. The stratigraphical ranges of selected species are based on the original data sets (see text for references). D, diatoms; R, radiolaria; N, calcareous nannofossils. Emboldened events indicate primary biostratigraphical events recognized in this study. The polarity record shown represents the original interpretation calibrated to the GPTS of Cande & Kent (1995) and the interpretation from this study correlated with the GPTS of Cande & Kent (1995). Diagonal lines indicate intervals of uncertainty.

Table 27. *Composite dataset for ODP Hole 744A containing the diatom, radiolarian, calcareous nannofossil and palaeomagnetic data calibrated to the GPTS of Cande & Kent (1995) using the biochronological interpretation based on this study*

	Event	Sample (top)	Sample (bottom)	Depth (t)	Depth (b)	Age (t)	Age (b)
B	C5En			81.70	81.70	18.78	18.78
T	C6n			83.18	83.18	19.05	19.05
LO	*Thalassiosira spumellaroides*	10HCC	11H-1, 63-65	88.16	89.70	19.61	19.85
FO	*Cyrtocapsella longithorax*	10HCC	11H-1, 63-65	88.25	89.40	19.61	19.85
FO	*Thalassiosira fraga*	11H-1, 63-65	11H-2, 63-65	90.30	91.80	19.85	20.02
LO	*Rocella schraderi*	11H-2, 63-65	11H-3, 63-65	91.80	93.30	20.01	20.24
B	C6n			92.80	92.80	20.13	20.13
T	C6An.1n			94.63	94.63	20.52	20.52
LO	*Rocella gelida*	11H-4, 63-65	11H-5, 63-65	94.80	96.40	20.54	20.70
FO	*Thalassiosira spumellaroides*	11H-4, 63-65	11H-5, 63-65	94.90	96.33	20.55	20.70
LO	*Bogorovia veniamini*	11H-5, 63-65	11H-6, 63-65	96.33	97.83	20.69	21.34
FO	*Raphidodiscus marylandicus*	11H-5, 63-65	11H-6, 63-65	96.33	97.83	20.69	21.34
B	C6An.1n			96.68	96.68	20.73	20.73
B	C6An.2n			97.77	97.77	21.32	21.32
B	C6 AAn			98.90	98.90	21.86	21.86
	Unconformity			***	***	***	***
LO	*Thalassiosira primalabiata*	11H-CC	12H-1, 60-62	99.20	99.80	?	?
LO	*Reticulofenestra bisecta*	11H-CC	12H-1, 92.93	99.20	99.80	?	?
B	C7An			100.24	100.24	25.65	25.65
LO	*Lisitzina ornata*	12H-4, 60-62	12H-5, 60-62	104.30	106.00	24.94	25.80
T	C8n.1n			104.79	104.79	25.82	25.82
LO	*Chiasmolithus altus*	13H-1, 60-62	13H-2, 60-62	106.92	107.62	26.04	26.11
FO	*Nitzschia maleinterpretaria*	12H-6, 60-62	12H-7, 61-63	106.30	108.70	25.98	26.23
FO	*Rocella gelida*	13H-1, 60-62	13H-2, 60-62	108.70	110.20	26.23	26.38
FO	*Lisitzina ornata*	13H-1, 60-62	13H-2, 60-62	109.32	110.80	26.29	26.45
B	C8n.2n			111.81	111.81	26.55	26.55
T	C9n			115.29	115.29	27.03	27.03

(*continued*)

Table 26. (*continued*)

	Event	Sample (top)	Sample (bottom)	Depth (t)	Depth (b)	Age (t)	Age (b)
	Unconformity			***	***	***	***
B	C11n.2n			118.37	118.37	30.10	30.10
FO	*Synedra jouseana*	14H-2, 60-62	14H-4, 60-62	118.82	120.32	30.27	30.85
T	C12n			119.36	119.36	30.48	30.48
FO	*Rocella vigilans*	14H-2, 60-62	14H-4, 60-62	121.80	123.30	30.69	30.83
B	C12n			124.59	124.59	30.94	30.94
LO	*Reticulofenestra umbilica*			126.45	127.70	31.21	31.39
FO	*Cyclicargolithus abisectus*			127.70	128.46	31.39	31.50
LO	*Isthmolithus recurvus*			129.95	131.45	31.72	31.93
T	C13n			139.25	139.25	33.06	33.06
B	C13n			146.64	146.64	33.55	33.55
FO	*Rhizosolenia oligocenica*	17H-1, 0	17H-1, 68	146.70	147.53	33.56	33.66
LO	*Reticulofenestra oamaruensis*	17H-1, 75-76	17H-2, 75-76	147.60	148.30	33.67	33.75
T	C15n			155.65	155.65	34.66	34.66
B	C15n			158.10	158.10	34.94	34.94
T	C16n.1n			161.30	161.30	35.34	35.34
B	C16n.1n			162.90	162.90	35.53	35.53
T	C16n.2n			164.60	164.60	35.69	35.69
FO	*Reticulofenestra oamaruensis*	19H-CC	20H-1, 75-76	166.60	167.35	35.85	35.91
FO	*Isthmolithus recurvus*	20H-2, 75-76	20H-3, 75-76	168.85	170.35	36.02	36.14
LO	*Reticulofenestra reticulata*	20H-2, 75-76	20H-3, 75-76	168.86	170.36	36.03	36.14
B	C16n.2n			172.85	172.85	36.34	36.34

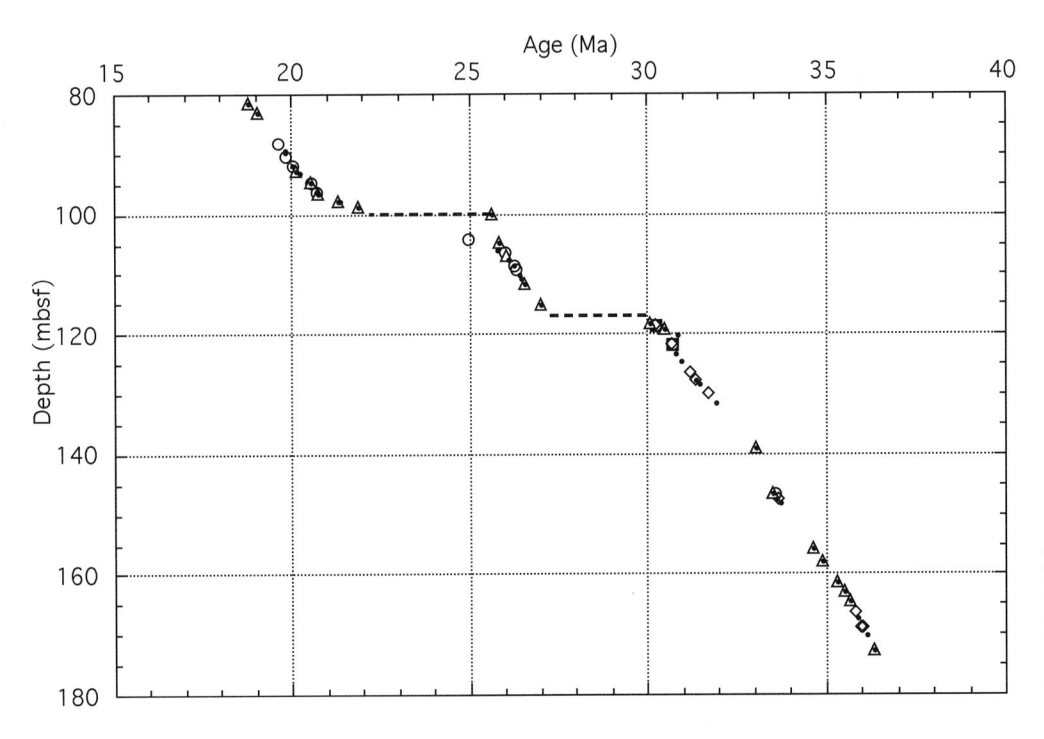

Data sets are integrated from the work of numerous authors (see Table 2).

Fig. 38. An age–depth plot for DSDP Hole 744A utilizing the biostratigraphical and polarity reversal events identified in this study and calibrated to the GPTS of Cande & Kent (1995). See Fig. 5 for legend of symbols.

Hole 744B

Nine cores were recovered from a total penetration of 78.5 mbsf representing a recovery of 101% of the cored interval. Sediments recovered consist of Quaternary to lower Pliocene diatom ooze and upper Miocene to lower Miocene calcareous nannofossil ooze (Barron *et al.* 1989). A good palaeomagnetic polarity reversal record was interpreted by Barron *et al.* (1991) for the entire section. Both diatoms and radiolarians are recorded from the sediments examined (Barron *et al.* 1989). Diatoms are well preserved and abundant in the Quaternary to lower Miocene interval. Calcareous nannofossils are sporadic in the Quaternary to Pliocene and abundant in the Miocene with moderate to good preservation. Radiolarians are common to abundant with moderate to good preservation for the Quaternary to lower Miocene section (Barron *et al.* 1989). The integrated biostratigraphical and magnetostratigraphical data previously used (palaeomagnetics: Barron *et al.* 1991; diatoms: Baldauf & Barron 1991; calcareous nannofossils: none; radiolarians: Caulet 1991; Lazarus 1992) is presented in Table 28. An age–depth graph for Hole 119-744B using this data is shown in Fig. 39.

Figure 40 compares the polarity chronozones and diatom zones assigned by previous workers (Barron *et al.* 1991; Baldauf & Barron 1991) with the chronozonal assignments in this study. In addition, this figure illustrates the stratigraphical ranges of diatoms and radiolarians. The stratigraphical ranges reflect the placement of FO and LO of stratigraphically useful species as recognized by Baldauf & Barron (1991) and Caulet (1991), respectively. The revised chronozones and diatom zones illustrated are based on reassessing the data sets of these workers (Table 29). Comparison of the chronozonal assignments of Barron *et al.* (1991; Table 28) with those of this study (Table 29) indicates that nomenclature changes reflect in part different GPTS used in these studies.

An age–depth graph for Hole 119-744B using the stratigraphical markers determined in this study is shown in Fig. 41. Three hiatuses are recognized in the section. The younger hiatus is placed at 21.5 mbsf, the intermediate break occurs between 23.90 and 24.55 mbsf, and the older hiatus is recognized between 52.20 and 53.70 mbsf. The upper boundary of the younger hiatus is constrained by the base of C3n.1n at 20.6 mbsf and the co-occurrence of the FO of *Nitzschia interfrigidaria*, *Nitzschia barronii* and *Nitzschia praeinterfrigidaria* at 21.5 mbsf. The lower boundary is constrained by a truncated C3An.1n at 23.55 mbsf. The occurrence of upper Miocene diatoms in the interval 21.5–23.9 mbsf indicates the occurrence of an upper Miocene section between these two younger hiatuses. The upper boundary of the intermediate hiatus is constrained by the base of C3An.1n at 23.55 mbsf. The lower boundary is constrained the top of C5n.2n at 24.55 mbsf. The upper boundary of the older hiatus is constrained by the base of C5r.2n at 51.72 mbsf and the lower boundary is constrained by the top C5ADn at 54.14 mbsf and the FO of *Denticulopsis praedimorpha* (53.7–55.2 mbsf), LO of *Synedra jouseana* (53.7–55.2 mbsf) Three additional primary events occur within the lower sequence, the FO of *Denticulopsis hustedtii* (56.70–58.20 mbsf), the FO of *Nitzshia denticuloides* (61.60–63.10 mbsf) and the FO of *Crucidenticula kanayae* (74.10–75.60 mbsf).

Barron *et al.* (1991) identify three hiatuses and discuss the probability of a fourth. The presence of the youngest hiatus, at approximately 22 mbsf, is based on the biostratigraphy of diatoms, radiolarians and benthic foraminifera and the interpretation of the magnetostratigraphy. This hiatus corresponds to the break recorded at 21.5 mbsf in this study. The recognition of the hiatus at approximately 24 mbsf, which corresponds to the intermediate hiatus of this study, is based on biostratigraphy (diatoms, radiolarians and benthic foraminifera) and the occurrence of an erosional contact between diatom ooze and calcareous ooze. The hiatus at 53 mbsf is inferred from the interpretation of the magnetostratigraphy and the presence of a colour change within the sequence. The occurrence of the earliest hiatus between 54.5 and 55.2 mbsf is based on integrating biostratigraphy with carbon isotope and magnetostratigraphy. The older hiatus of this study incorporates the lower two hiatuses of Barron *et al.* (1991).

Table 28. *Composite dataset for ODP Hole 744B containing the diatom, radiolarian, calcareous nannofossil and palaeomagnetic data calibrated to the GPTS of Cande & Kent (1995)*

	Event	Sample (top)	Sample (bottom)	Depth (t)	Depth (b)	Age (t)	Age (b)
FO	*Actinocyclus actinochilus*	1H-1, 8	1H-CC	0.80	9.50	0.11	1.76
LO	*Actinocyclus ingens*	1H-1, 8	1H-CC	0.80	9.50	0.11	1.76
LO	*Thalassiosira elliptopora*	1H-1, 8	1H-CC	0.80	9.50	0.11	1.76
B	C1n			5.50	5.50	0.78	0.78
LO	*Cycladophora pliocenica*	1H-5, 123	1H-7, 23	7.73	9.23	1.33	1.69
LO	*Coscinodiscus lewisianus*	1H-CC	2H-3, 50	9.50	13.00	1.76	2.32
LO	*Thalassiosira vulnifica*	1H-CC	2H-3, 50	9.50	13.00	1.76	2.32
LO	*Thalassiosira insigna*	1H-CC	2H-3, 50	9.50	13.00	1.76	2.32
LO	*Denticulopsis dimorpha*	1H-CC	2H-3, 50	9.50	13.00	1.76	2.32
LO	*Hemidiscus karstenii*	1H-CC	2H-3, 50	9.50	13.00	1.76	2.32
LO	*Nitzschia interfrigidaria*	1H-CC	2H-3, 50	9.50	13.00	1.76	2.32
FO	*Nitzschia kerguelensis*	1H-CC	2H-3, 50	9.50	13.00	1.76	2.32
LO	*Thalassiosira kolbei*	1H-CC	2H-3, 50	9.50	13.00	1.76	2.32
FO	*Thalassiosira lentiginosa*	1H-CC	2H-3, 50	9.50	13.00	1.75	2.32
LO	*Prunopyle titan*	1H-CC	2H-CC	9.50	13.00	1.76	3.46
LO	*Eucyrtidium calvertense*	1H-CC	2H-CC	9.50	13.00	1.76	3.46
LO	*Desmospyris spongiosa*	1H-CC	2H-CC	9.50	13.00	1.76	3.46
LO	*Helotholus vema*	1H-CC	2H-CC	9.50	13.00	1.76	3.46
T	C2n			9.55	9.55	1.77	1.77
B	C2n			10.05	10.05	1.95	1.95
LO	*Denticulopsis dimorpha*	2H-3, 50	2H-CC	13.00	19.00	2.32	3.46
FO	*Thalassiosira elliptopora*	2H-3, 50	2H-CC	13.00	19.00	2.32	3.46
FO	*Thalassiosira vulnifica*	2H-3, 50	2H-CC	13.00	19.00	2.32	3.46
LO	*Nitzschia praeinterfrigidaria*	2H-3, 50	2H-CC	13.00	19.00	2.32	3.46
LO	*Simonseniella barboi*	2H-3, 50	2H-CC	13.00	19.00	2.32	3.46
FO	*Thalassiosira inura*	2H-3, 50	2H-CC	13.00	19.00	2.32	3.46

(*continued*)

Table 28. (continued)

	Event	Sample (top)	Sample (bottom)	Depth (t)	Depth (b)	Age (t)	Age (b)
FO	Thalassiosira kolbei	2H-3, 50	2H-CC	13.00	19.00	2.32	3.46
T	C2An.1n			15.05	15.05	2.58	2.58
LO	Nitzschia weaveri	2H-CC	3H-2, 33	19.00	20.80	3.46	?
FO	Thalassiosira oestrupii	2H-CC	3H-2, 33	19.00	20.80	3.46	?
B	C2An.3n			19.55	19.55	3.58	3.58
T	C3n.1n			20.40	20.40	4.18	4.18
B	C3n.1n			20.60	20.60	4.29	4.29
FO	Nitzschia interfrigidaria			20.80	21.50	?	?
	Unconformity			***	***	***	***
FO	Nitzschia weaveri	3H-2, 38	3H-CC	20.80	21.50	?	?
FO	Nitzschia praeinterfrigidaria	3H-CC	4H-1, 70	21.50	22.20	?	?
FO	Nitzschia barronii	3H-CC	4H-1, 70	21.50	22.20	?	?
LO	Thalassiosira miocenica	3H-CC	4H-1, 70	21.50	22.20	?	?
FO	Nitzschia reinholdii	4H-1, 70	4H-2, 24	22.20	23.24	?	?
LO	Thalassiosira praeconvexa	4H-1, 70	4H-2, 24	22.20	23.24	?	?
FO	Thalassiosira miocenica	4H-2, 37	4H-2, 90	23.37	23.90	?	?
FO	Thalassiosira torokina	4H-2, 37	4H-2, 90	23.37	23.90	?	?
	Unconformity			***	***	***	***
B	C3An.1n			23.55	23.55	6.14	6.14
	Unconformity			***	***	***	***
LO	Asteromphalus kennettii	4H-2, 90	4H-3, 27	23.90	24.77	?	?
T	C5n.1n			24.55	24.55	9.74	9.74
FO	Asteromphalus kennettii	4H-6, 60	4H-CC	29.60	31.00	10.13	10.24
LO	Reticulofenestra pseudoumbilica	5H-1, 110-111	5H-2, 110-111	32.10	33.60	10.33	10.45
LO	Acrosphaera australis	5H-3, 23	5H-5, 23	33.23	36.23	10.41	10.65
LO	Calcidiscus macintyrei	5H-2, 110-111	5H-3, 110-111	33.60	35.10	10.45	10.56
FO	Hemidiscus cuneiformis			34.50	36.10	10.52	10.64
LO	Denticulopsis lauta	5H-5, 60	5H-6, 60	37.60	39.10	10.76	10.88
LO	Actinomma golownini	5H-6, 10	6H-2, 123	37.60	42.73	10.76	11.12
B	C5n.2n			40.05	40.05	10.95	10.95
FO	Actinocyclus fryxellae	5H-CC	6H-1, 60	40.50	41.10	10.99	11.05
T	C5r.1n			41.12	41.12	11.05	11.05
B	C5r.1n			42.25	42.25	11.10	11.10
FO	Cycladophora spongothorax	6H-5, 123	7H-2, 125	47.23	52.65	11.32	?
LO	Denticulopsis praedimorpha	6H-6, 60	6H-CC	48.60	50.00	11.38	11.44
T	C5r.2n			50.90	50.90	11.48	11.48
B	C5r.2n			51.72	51.72	11.53	11.53
	Unconformity			***	***	***	***
FO	Denticulopsis dimorpha	7H-2, 60	7H-3, 70	52.20	53.70	?	?
LO	Denticulopsis maccollumnii	7H-2, 60	7H-3, 70	52.20	53.70	?	?
LO	Nitzschia denticuloides	7H-2, 60	7H-3, 70	52.20	53.70	?	?
LO	Nitzschia grossepunctata	7H-2, 60	7H-3, 70	52.20	53.70	?	?
FO	Actinomma golownini	7H-2, 125	7H-4, 123	52.65	55.23	?	?
FO	Dendrospyris megalocephalis	7H-2, 125	7H-4, 123	52.65	55.23	?	?
FO	Denticulopsis praedimorpha	7H-3, 70	7H-4, 70	53.70	55.20	?	?
LO	Synedra jouseana			53.70	55.20	?	?
T	C5ACn			54.14	54.14	13.70	13.70
LO	Actinocyclus ingens v. nodus	5H-3, 60	5H-4, 60	55.20	58.20	13.91	14.51
FO	Cycladophora humerus	7H-4, 123	7H-5, 73	55.23	56.73	13.92	14.22
FO	Denticulopsis hustedtii	7H-5, 70	7H-6, 50	56.70	58.20	14.21	14.51
B	C5ADn			58.68	58.68	14.61	14.61
T	C5Bn.2n			60.07	60.07	15.03	15.03
B	C5Bn.2n			61.36	61.36	15.16	15.16
FO	Nitzschia denticuloides	8H-1, 60	8H-2, 60	61.60	63.10	15.21	15.55
FO	Actinocyclus ingens v. nodus	8H-3, 60	8H-4, 60	63.10	64.60	15.55	15.89
FO	Actinocyclus ingens	8H-4, 60	8H-5, 60	64.60	66.10	15.89	16.40
T	C5Cn.1n			65.11	65.11	16.01	16.01
LO	Crucidenticula kanayae	8H-5, 60	8H-6, 60	66.10	67.60	16.40	16.80
LO	Nitzschia maleinterpretaria	8H-5, 60	8H-6, 60	66.10	67.60	16.40	16.80
FO	Eucyrtidium pseudoinflatum	8H-5, 73	9H-2, 53	66.23	71.30	16.45	17.20
B	C5Cn.3n			66.95	66.95	16.73	16.73
LO	Thalassiosira fraga	8H-6, 60	8H-CC	67.60	69.00	16.80	16.95
FO	Denticulopsis lauta	8H-CC	9H-1, 58	69.00	69.60	16.95	17.02
FO	Denticulopsis maccollumnii	8H-CC	9H-1, 58	69.00	69.60	16.95	17.02
FO	Nitzschia grossepunctata	8H-CC	9H-1, 58	69.00	69.60	16.95	17.02
FO	Raphidodiscus marylandicus	9H-1, 58	9H-2, 70	69.60	71.10	17.01	17.18
FO	Cycladophora golli regipileus	9H-2, 90	9H-6, CC	71.30	80.73	17.20	?
T	C5Dn			72.00	72.00	17.28	17.28
FO	Crucidenticula kanayae	9H-4, 58	9H-5, 58	74.10	75.60	17.49	17.69

Table 28. (*continued*)

Event		Sample (top)	Sample (bottom)	Depth (t)	Depth (b)	Age (t)	Age (b)
FO	*Crucidenticula nicobarica*	9H-4, 58	9H-5, 58	74.10	75.60	17.49	17.69
LO	*Reticulofenestra hesslandii*	9H-4, 118-119	9H-5, 118-119	74.70	76.20	17.55	17.92
B	C5Dn			75.43	75.43	17.62	17.62
T	C5En			77.10	77.10	18.28	18.28
B	C5En			77.50	77.50	18.78	18.78
FO	*Calcidiscus leptoporus*	9H-6, 118-119	9H-CC	77.70	78.50	?	?
FO	*Calcidiscus macintyrei*	9H-6, 118-119	9H-CC	77.70	78.50	?	?

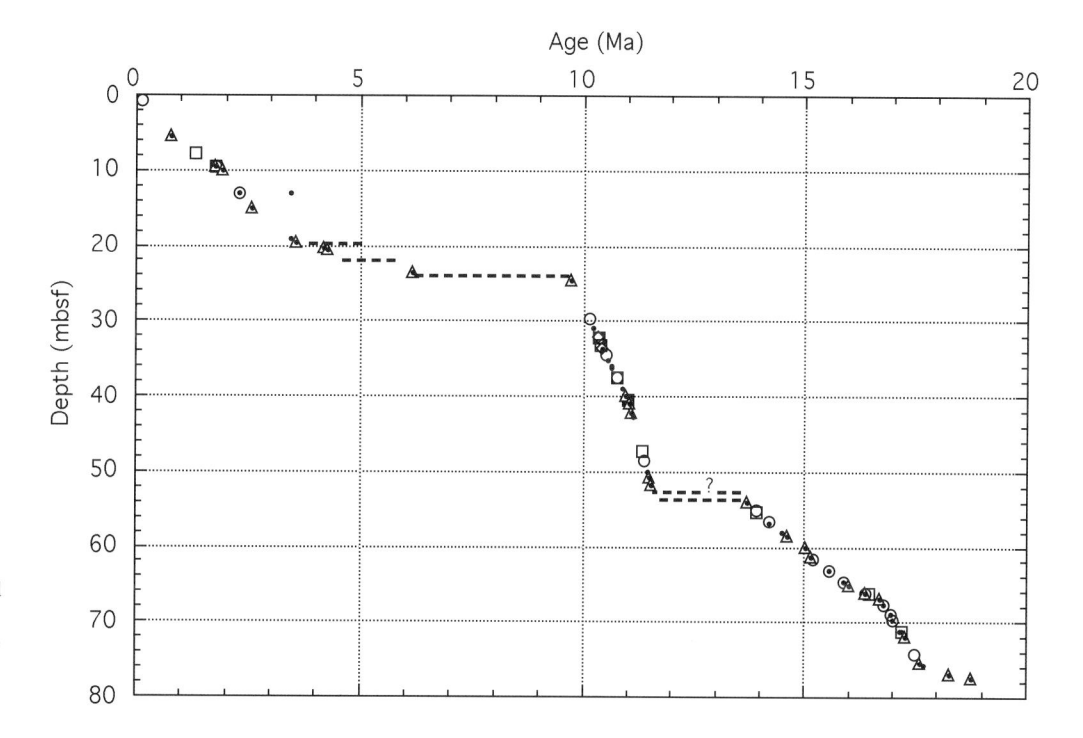

Data sets are integrated from the work of numerous authors (see Table 2).

Fig. 39. An age–depth plot for DSDP Hole 744B utilizing the previously used biostratigraphical datums and polarity reversal interpretation calibrated to the GPTS of Cande & Kent (1995). See Fig. 5 for legend of symbols.

Fig. 40. Comparison of the ODP Hole 744B polarity chronozones and diatom zonal assignments by previous workers with the chronozonal assignments of this study. The stratigraphical ranges of selected species are based on the original data sets (see text for references). D, diatoms; R, radiolaria; N, calcareous nannofossils. Emboldened events indicate primary biostratigraphical events recognized in this study. The polarity record shown represents the original interpretation calibrated to the GPTS of Cande & Kent (1995) and the interpretation from this study correlated with the GPTS of Cande & Kent (1995). Diagonal lines indicate intervals of uncertainty.

Table 29. *Composite dataset for ODP Hole 744B containing the diatom, radiolarian, calcareous nannofossil and palaeomagnetic data calibrated to the GPTS of Cande & Kent (1995) using the biochronological interpretation based on this study*

	Event	Sample (top)	Sample (bottom)	Depth (t)	Depth (b)	Age (t)	Age (b)
FO	Actinocyclus actinochilus	1H-1, 8	1H-CC	0.80	9.50	0.89	1.76
LO	Actinocyclus ingens	1H-1, 8	1H-CC	0.80	9.50	0.89	1.76
LO	Thalassiosira elliptopora	1H-1, 8	1H-CC	0.80	9.50	0.89	1.76
B	C1n			5.50	5.50	0.78	0.78
LO	Cycladophora pliocenica	1H-5, 123	1H-7, 23	7.73	9.23	1.32	1.69
LO	Coscinodiscus lewisianus	1H-CC	2H-3, 50	9.50	13.00	1.76	2.32
LO	Thalassiosira vulnifica	1H-CC	2H-3, 50	9.50	13.00	1.76	2.32
LO	Thalassiosira insigna	1H-CC	2H-3, 50	9.50	13.00	1.76	2.32
LO	Denticulopsis dimorpha	1H-CC	2H-3, 50	9.50	13.00	1.76	2.32
LO	Hemidiscus karstenii	1H-CC	2H-3, 50	9.50	13.00	1.76	2.32
LO	Nitzschia interfrigidaria	1H-CC	2H-3, 50	9.50	13.00	1.76	2.32
FO	Nitzschia kerguelensis	1H-CC	2H-3, 50	9.50	13.00	1.76	2.32
LO	Thalassiosira kolbei	1H-CC	2H-3, 50	9.50	13.00	1.76	2.32
FO	Thalassiosira lentiginosa	1H-CC	2H-3, 50	9.50	13.00	1.76	2.32
LO	Prunopyle titan	1H-CC	2H-CC	9.50	13.00	1.76	3.46
LO	Eucyrtidium calvertense	1H-CC	2H-CC	9.50	13.00	1.76	3.46
LO	Desmospyris spongiosa	1H-CC	2H-CC	9.50	13.00	1.76	3.46
LO	Helotholus vema	1H-CC	2H-CC	9.50	13.00	1.76	3.46
T	C2n			9.55	9.55	1.77	1.77
B	C2n			10.05	10.05	1.95	1.95
LO	Denticulopsis dimorpha	2H-3, 50	2H-CC	13.00	19.00	2.32	3.46
FO	Thalassiosira elliptopora	2H-3, 50	2H-CC	13.00	19.00	2.32	3.46
FO	Thalassiosira vulnifica	2H-3, 50	2H-CC	13.00	19.00	2.32	3.46
LO	Nitzschia praeinterfrigidaria	2H-3, 50	2H-CC	13.00	19.00	2.32	3.46
LO	Simonseniella barboi	2H-3, 50	2H-CC	13.00	19.00	2.32	3.46
FO	Thalassiosira inura	2H-3, 50	2H-CC	13.00	19.00	2.32	3.46
FO	Thalassiosira kolbei	2H-3, 50	2H-CC	13.00	19.00	2.32	3.46
T	C2An.1n			15.05	15.05	2.58	2.58
LO	Nitzschia weaveri	2H-CC	3H-2, 33	19.00	20.80	3.46	?
FO	Thalassiosira oestrupii	2H-CC	3H-2, 33	19.00	20.80	3.46	?
B	C2An.3n			19.55	19.55	3.58	3.58
T	C3n.1n			20.40	20.40	4.18	4.18
B	C3n.1n			20.60	20.60	4.29	4.29
FO	Nitzschia interfrigidaria	3H-2, 38	3H-CC	20.80	21.50	?	?
FO	Nitzschia weaveri	3H-2, 38	3H-CC	20.80	21.50	?	?
FO	Nitzschia praeinterfrigidaria	3H-CC	4H-1, 70	21.50	22.20	?	?
FO	Nitzschia barronii	3H-CC	4H-1, 70	21.50	22.20	?	?
	Unconformity			***	***	***	***
LO	Thalassiosira miocenica	3H-CC	4H-1, 70	21.50	22.20	?	?
FO	Nitzschia reinholdii	4H-1, 70	4H-2, 24	22.20	23.24	?	?
LO	Thalassiosira praeconvexa	4H-1, 70	4H-2, 24	22.20	23.24	?	?
FO	Thalassiosira miocenica	4H-2, 37	4H-2, 90	23.37	23.90	?	?
FO	Thalassiosira torokina	4H-2, 37	4H-2, 90	23.37	23.90	?	?
B	C3An.1n			23.55	23.55	6.14	6.14
LO	Asteromphalus kennettii	4H-2, 90	4H-3, 27	23.90	24.77	?	?
	Unconformity			***	***	***	***
T	C5n.2n			24.55	24.55	9.92	9.92
FO	Asteromphalus kennettii	4H-6, 60	4H-CC	29.60	31.00	10.25	10.35
LO	Reticulofenestra pseudoumbilica	5H-1, 110-111	5H-2, 110-111	32.10	33.60	10.42	10.52
LO	Acrosphaera australis	5H-3, 23	5H-5, 23	33.23	36.23	10.50	10.70
LO	Calcidiscus macintyrei	5H-2, 110-111	5H-3, 110-111	33.60	35.10	10.52	10.62
FO	Hemidiscus cuneiformis			34.50	36.10	10.58	10.69
LO	Denticulopsis lauta	5H-5, 60	5H-6, 60	37.60	39.10	10.79	10.89
LO	Actinomma golownini	5H-6, 10	6H-2, 123	37.60	42.73	10.79	11.12
B	C5n.2n			40.05	40.05	10.95	10.95
FO	Actinocyclus fryxellae	5H-CC	6H-1, 60	40.50	41.10	10.99	11.05
T	C5r.1n			41.12	41.12	11.05	11.05
B	C5r.1n			42.25	42.25	11.10	11.10
FO	Cycladophora spongothorax	6H-5, 123	7H-2, 125	47.23	52.65	11.32	?
LO	Denticulopsis praedimorpha	6H-6, 60	6H-CC	48.60	50.00	11.38	11.44
T	C5r.2n			50.90	50.90	11.48	11.48
B	C5r.2n			51.72	51.72	11.53	11.53
FO	Denticulopsis dimorpha	7H-2, 60	7H-3, 70	52.20	53.70	?	?
	Unconformity			***	***	***	***
LO	Denticulopsis maccollumnii	7H-2, 60	7H-3, 70	52.20	53.70	?	?
LO	Nitzschia denticuloides	7H-2, 60	7H-3, 70	52.20	53.70	?	?
LO	Nitzschia grossepunctata	7H-2, 60	7H-3, 70	52.20	53.70	?	?
FO	Actinomma golownini	7H-2, 125	7H-4, 123	52.65	55.23	?	?

Table 29. (*continued*)

	Event	Sample (top)	Sample (bottom)	Depth (t)	Depth (b)	Age (t)	Age (b)
FO	*Dendrospyris megalocephalis*	7H-2, 125	7H-4, 123	52.65	55.23	?	?
FO	*Denticulopsis praedimorpha*	7H-3, 70	7H-4, 70	53.70	55.20	?	?
LO	*Synedra jouseana*			53.70	55.20	?	?
T	C5ADn			54.14	54.14	14.18	14.18
LO	*Actinocyclus ingens v. nodus*	7H-3, 60	5H-4, 60	55.20	58.20	14.28	14.56
FO	*Cycladophora humerus*	7H-4, 123	7H-5, 73	55.23	56.73	14.28	14.42
FO	*Denticulopsis hustedtii*	7H-5, 70	7H-6, 50	56.70	58.20	14.42	14.56
B	C5ADn			58.68	58.68	14.61	14.61
T	C5Bn.2n			60.07	60.07	15.03	15.03
B	C5Bn.2n			61.36	61.36	15.16	15.16
FO	*Nitzschia denticuloides*	8H-1, 60	8H-2, 60	61.60	63.10	15.21	15.55
FO	*Actinocyclus ingens v. nodus*	8H-3, 60	8H-4, 60	63.10	64.60	15.55	15.89
FO	*Actinocyclus ingens*	8H-4, 60	8H-5, 60	64.60	66.10	15.89	16.40
T	C5Cn.1n			65.11	65.11	16.01	16.01
LO	*Crucidenticula kanayae*	8H-5, 60	8H-6, 60	66.10	67.60	16.40	16.80
LO	*Nitzschia maleinterpretaria*	8H-5, 60	8H-6, 60	66.10	67.60	16.40	16.80
FO	*Eucyrtidium pseudoinflatum*	8H-5, 73	9H-2, 53	66.23	71.30	16.45	17.20
B	C5Cn.3n			66.95	66.95	16.73	16.73
LO	*Thalassiosira fraga*	8H-6, 60	8H-CC	67.60	69.00	16.80	16.95
FO	*Denticulopsis lauta*	8H-CC	9H-1, 58	69.00	69.60	16.95	17.02
FO	*Denticulopsis maccollumii*	8H-CC	9H-1, 58	69.00	69.60	16.95	17.02
FO	*Nitzschia grossepunctata*	8H-CC	9H-1, 58	69.00	69.60	16.95	17.02
FO	*Raphidodiscus marylandicus*	9H-1, 58	9H-2, 70	69.60	71.10	17.01	17.18
FO	*Cycladophora golli regipileus*	9H-2, 90	9H-6, CC	71.30	80.73	17.20	?
T	C5Dn			72.00	72.00	17.28	17.28
FO	*Crucidenticula kanayae*	9H-4, 58	9H-5, 58	74.10	75.60	17.49	17.69
FO	*Crucidenticula nicobarica*	9H-4, 58	9H-5, 58	74.10	75.60	17.49	17.69
LO	*Reticulofenestra hesslandii*	9H-4, 118-119	9H-5, 118-119	74.70	76.20	17.55	17.92
B	C5Dn			75.43	75.43	17.62	17.62
T	C5En			77.10	77.10	18.28	18.28
B	C5En			77.50	77.50	18.78	18.78
FO	*Calcidiscus leptoporus*	9H-6, 118-119	9H-CC	77.70	78.50	?	?
FO	*Calcidiscus macintyrei*	9H-6, 118-119	9H-CC	77.70	78.50	?	?

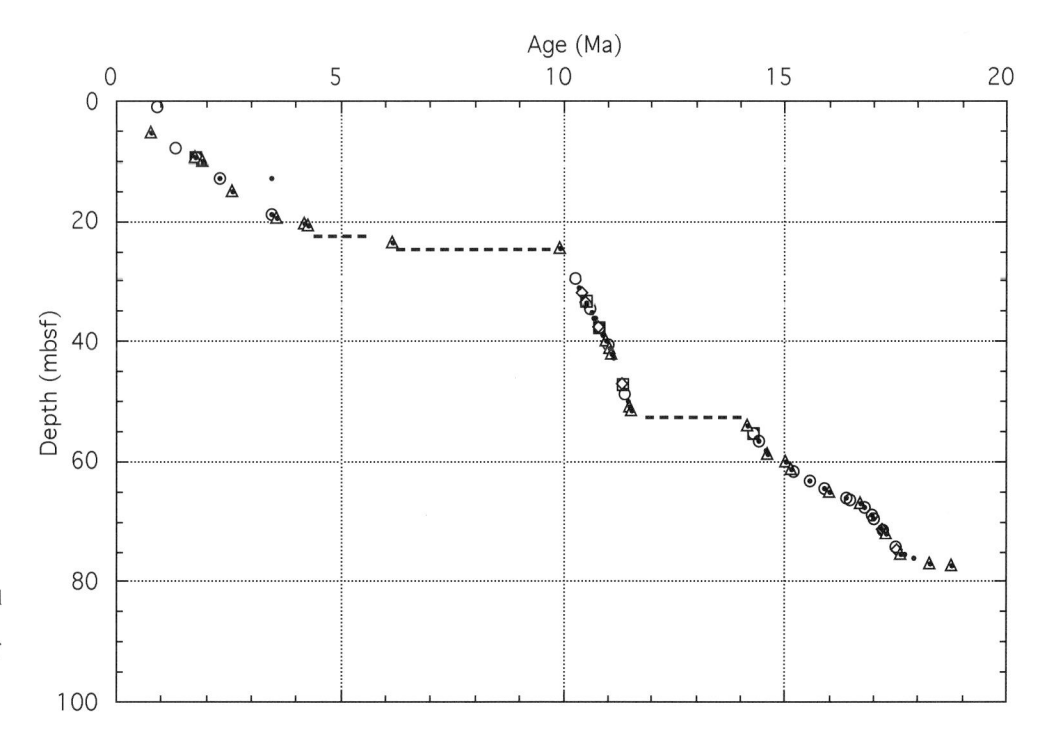

Data sets are integrated from the work of numerous authors (see Table 2).

Fig. 41. An age–depth plot for DSDP Hole 744B utilizing the biostratigraphical and polarity reversal events identified in this study and calibrated to the GPTS of Cande & Kent (1995). See Fig. 5 for legend of symbols.

Hole 745B

Twenty-four cores were recovered from a total penetration of 215 mbsf representing a recovery of 102% of the cored interval. Sediments recovered consist of Quaternary to upper Miocene diatom ooze with various amounts of clay (Barron et al. 1989). A palaeomagnetic polarity reversal record was interpreted by Barron et al. (1991) for the entire section. Diatoms and radiolarians are generally abundant and well preserved throughout the entire stratigraphical section. Calcareous nannofossils are generally absent, but have a sporadic occurrence in the Quaternary and upper Pliocene (Barron et al. 1989). The integrated biostratigraphical and magnetostratigraphical data previously used (palaeomagnetics: Barron et al. 1991; diatoms: Baldauf & Barron 1991; calcareous nannofossils: none; radiolarians: Caulet 1991; Lazarus 1992) and calibrated to the GPTS of Berggren et al. (1985a,b) is shown in Table 30. An age–depth graph for Hole 119-745B using this data is shown in Fig. 42.

Figure 43 compares the polarity chronozones and diatom zones assigned by previous workers (Barron et al. 1991; Baldauf & Barron 1991) with the chronozonal assignments in this study. In addition, this figure illustrates the stratigraphical ranges of diatoms and radiolarians. The stratigraphical ranges reflect the placement of FO and LO of stratigraphically useful species as recognized by Baldauf & Barron (1991); Caulet (1991), and Lazarus (1992), respectively. The revised chronozones and diatom zones illustrated are based on reassessing the data sets of these workers (Table 31). Comparison of the chronozonal assignments of Barron et al. (1991; Table 30) with those of this study (Table 31) indicates that only minor changes, associated with differences in the GPTS were required.

Figure 44 shows an age–depth graph for Hole 119-745B using the stratigraphical markers determined in this study. The hole represents a continuous stratigraphical sequence to 205.66 mbsf as indicated by the continuous nature of the polarity record from the base of C1n.1n at 42.80 mbsf to the top of C3An.2n at 205.66 mbsf. Six primary biostratigraphical datums are recognized in this site, the LO of *Thallassiosira kolbei* (90.5–94.10 mbsf), the LO of *Thallassiosira vulnifica* (102.10–105.10 mbsf), the FO of *Nitzschia interfrigidaria* (135.10–136.60 mbsf), the FO of *Nitzschia barronii* (144.10–145.60 mbsf), the FO of *Helotholus vema* (153.50–155.00 mbsf) and the FO of *Nitzschia praeinterfrigidaria* (158.60–166.10 mbsf). These events are used to calibrate the polarity reversals and the results correspond with those of Barron et al. (1991).

Table 30. *Composite dataset for ODP Hole 745B containing the diatom, radiolarian, calcareous nannofossil and palaeomagnetic data calibrated to the GPTS of Cande & Kent (1995)*

	Event	Sample (top)	Sample (bottom)	Depth (t)	Depth (b)	Age (t)	Age (b)
LO	*Triceraspyris antarctica*	1H-1, 0	1H-CC	0.00	5.00	0.00	0.09
LO	*Hemidiscus karstenii*	2H-CC	3H-CC	14.50	24.00	0.26	0.44
LO	*Actinocyclus ingens*	3H-CC	4H-CC	24.00	33.50	0.44	0.61
LO	*Denticulopsis dimorpha*	4H-CC	5H-CC	33.50	43.00	0.61	0.79
LO	*Eucyrtidium calvertense*	4H-CC	5H-1, 53-55	33.50	34.00	0.61	0.62
LO	*Antarctissa cylindrica*			34.00	35.50	0.61	0.65
B	C1n			42.80	42.80	0.78	0.78
LO	*Thalassiosira elliptopora*	5H-CC	6H-CC	43.00	52.30	0.79	1.27
FO	*Denticulopsis dimorpha*	5H-CC	6H-CC	43.00	52.30	0.79	1.27
T	C1r.1n			50.30	50.30	0.99	0.99
B	C1r.1n			54.64	54.64	1.07	1.07
FO	*Thalassiosira elliptopora*	9H-CC	10H-CC	81.00	90.50	1.57	1.75
LO	*Hemidiscus cuneiformis*	9H-CC	10H-CC	81.00	90.50	1.57	1.75
LO	*Simonseniella barboi*	9H-CC	10H-CC	81.00	90.50	1.57	1.75
LO	*Cycladophora pliocenica*	9H-CC	10H-1, 53-55	81.00	81.50	1.57	1.58
LO	*Nitzschia weaveri*	10H-CC	11H-3, 60-62	90.50	94.10	1.75	1.97
LO	*Thalassiosira kolbei*	10H-CC	11H-3, 60-62	90.50	94.10	1.75	1.97
T	C2n			91.55	91.55	1.77	1.77
B	C2n			93.55	93.55	1.95	1.95
LO	*Nitzschia interfrigidaria*	11H-3, 60-62	11H-6, 60-62	94.10	98.60	1.97	2.12
LO	*Thalassiosira torokina*	11H-6, 60-62	12H-1, 60-62	98.60	100.60	2.12	2.18
FO	*Nitzschia kerguelensis*	11H-6, 60-62	12H-1, 60-62	98.60	100.60	2.12	2.18
LO	*Thalassiosira insigna*	12H-1, 60-62	12H-2, 60-62	100.60	102.10	2.18	2.23
LO	*Rouxia heteropolara*	12H-1, 60-62	12H-2, 60-62	100.60	102.10	2.18	2.23
FO	*Actinocyclus ingens*	12H-2, 60-62	12H-4, 60-62	102.10	105.10	2.23	2.33
LO	*Thalassiosira vulnifica*	12H-2, 60-62	12H-4, 60-62	102.10	105.10	2.23	2.33
FO	*Nitzschia cylindrica*	12H-2, 60-62	12H-4, 60-62	102.10	105.10	2.23	2.33
LO	*Rouxia californica*	12H-4, 60-62	12H-6, 60-62	105.10	108.10	2.33	2.43
LO	*Helotholus vema*	12H-CC	13H-1, 53-55	109.50	110.00	2.48	2.50
LO	*Desmospyris spongiosa*	13H-1, 53-55	13H-2, 53-55	110.00	111.00	2.50	2.53
T	C2An.1n			112.55	112.55	2.58	2.58
FO	*Cycladophora davisiana*	13H-4, 53-55	13H-5, 53-55	114.50	116.00	2.66	2.72
FO	*Thalassiosira vulnifica*	13H-4, 60-62	13H-7, 60-62	115.60	118.60	2.71	2.83
LO	*Nitzschia praeinterfrigidaria*	13H-4, 60-62	13H-7, 60-62	115.60	118.60	2.71	2.83
LO	*Nitzschia reinholdii*	13H-4, 60-62	13H-7, 60-62	115.60	118.60	2.71	2.83
LO	*Stephanopyxis grunowii*	13H-4, 60-62	13H-7, 60-62	115.60	118.60	2.71	2.83
FO	*Thalassiosira insigna*	14H-1, 64-66	14H-2, 60-62	120.20	121.70	2.90	2.96
FO	*Thalassiosira kolbei*	14H-1, 64-66	14H-2, 60-62	120.20	121.70	2.90	2.96
B	C2An.1n			123.55	123.55	3.04	3.04
FO	*Nitzschia weaveri*	14H-4, 64-66	14H-5, 64-66	124.20	125.70	3.06	3.13
T	C2An.2n			125.55	125.55	3.11	3.11
B	C2An.2n			126.55	126.55	3.22	3.22

Table 30. (*continued*)

	Event	Sample (top)	Sample (bottom)	Depth (t)	Depth (b)	Age (t)	Age (b)
T	C2An.3n			128.55	128.55	3.33	3.33
LO	*Prunopyle titan*	15H-1, 53-55	15H-2, 53-55	130.50	132.00	3.44	3.52
B	C2An.3n			133.05	133.05	3.58	3.58
FO	*Nitzschia interfrigidaria*	15H-5, 60-62	15H-6, 60-62	135.10	136.60	3.71	3.80
LO	*Nitzschia fossilis*	16H-CC	17H-2, 60-62	139.00	141.10	3.95	4.08
FO	*Rouxia heteropolara*	17H-2, 60-62	17H-3, 60-62	139.00	141.10	3.95	4.08
LO	*Rhizosolenia miocenica*	17H-2, 60-62	17H-3, 60-62	139.00	141.10	3.95	4.08
T	C3n.1n			142.73	142.73	4.18	4.18
FO	*Nitzschia barronii*	17H-4, 60-62	17H-5, 60-62	144.10	145.60	4.22	4.27
B	C3n.1n			146.28	146.28	4.29	4.29
T	C3n.2n			149.30	149.30	4.48	4.48
FO	*Helotholus vema*	18H-4, 53-55	18H-5, 53-55	153.50	155.00	4.57	4.60
B	C3n.2n			155.93	155.93	4.62	4.62
FO	*Nitzschia praeinterfrigidaria*	19H-1, 60-62	19H-6, 60-62	158.60	166.10	4.72	4.87
T	C3n.3n			160.82	160.82	4.80	4.80
B	C3n.3n			167.90	167.90	4.89	4.89
FO	*Thalassiosira inura*	20H-1, 60-62	20H-4, 60-62	168.10	172.60	4.80	4.89
T	C3n.4n			177.95	177.95	4.98	4.98
B	C3n.4n			180.85	180.85	5.23	5.23
LO	*Rocella vigilans*	21H-2, 60-62	21H-5, 60-62	182.10	183.60	5.38	5.56
FO	*Thalassiosira oestrupii*	21H-2, 60-62	21H-5, 60-62	182.10	183.60	5.38	5.56
T	C3An.1n			186.37	186.37	5.89	5.89
LO	*Thalassiosira convexa*	22H-3, 60-62	22H-6, 60-62	190.10	194.60	5.98	6.09
FO	*Thalassiosira convexa v. aspinosa*	22H-6, 60-62	23H-3, 60-62	194.60	199.60	6.09	6.18
LO	*Trinacria excavata*	21H-5, 60-62	22H-3, 60-62	194.60	199.60	6.09	6.18
B	C3An.1n			196.63	196.63	6.14	6.14
T	C3An.2n			205.66	205.66	6.27	6.27
FO	*Rocella vigilans*	24H-2, 60-62	24H-6, 60-62	207.60	13.60	?	?
LO	*Thalassiosira miocenica*	24H-3, 60-62	24H-6, 60	209.10	213.60	?	?
FO	*Thalassiosira inura*	24H-1, 60-62	24H-2, 60-62	210.60	212.10	?	?

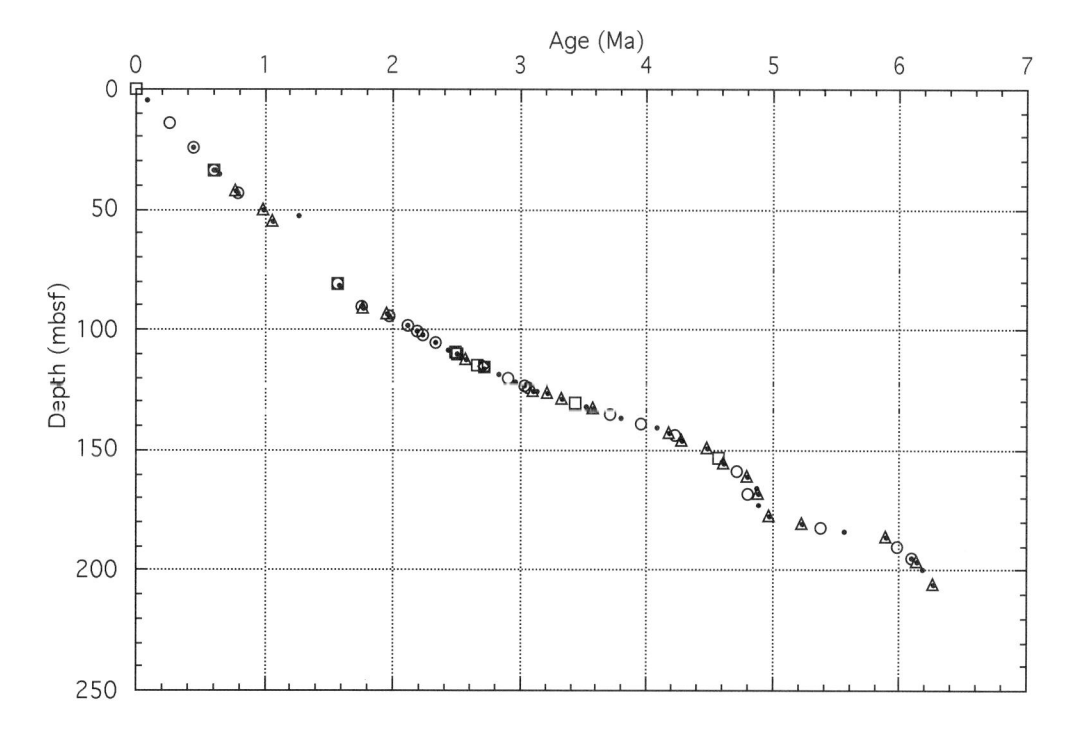

Data sets are integrated from the work of numerous authors (see Table 2).

Fig. 42. An age–depth plot for DSDP Hole 745B utilizing the previously used biostratigraphical datums and polarity reversal interpretation calibrated to the GPTS of Cande & Kent (1995). See Fig. 5 for legend of symbols.

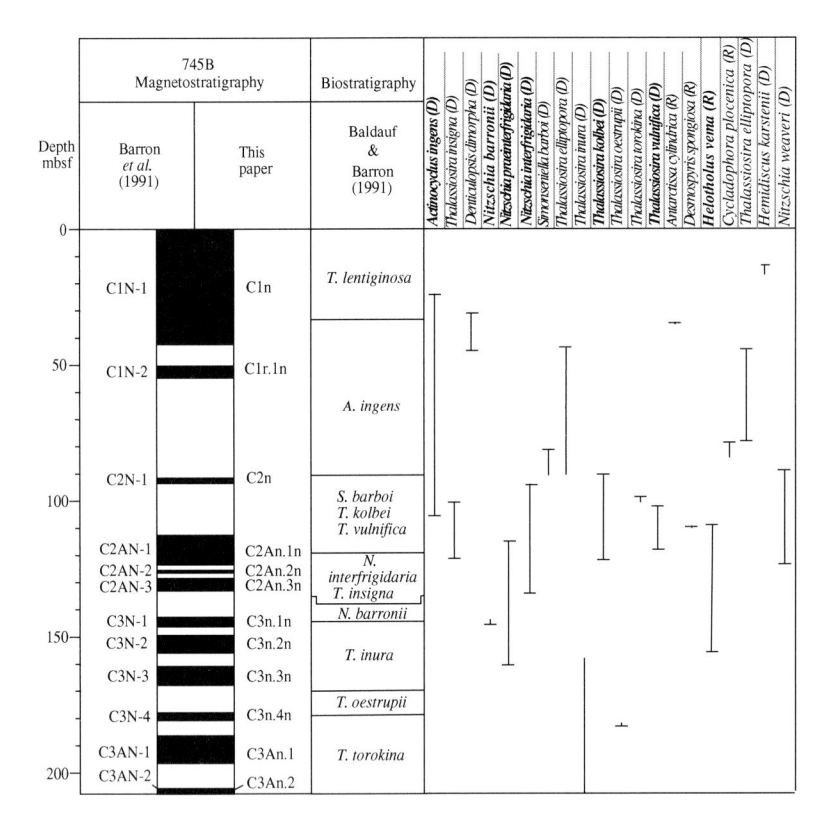

Fig. 43. Comparison of the ODP Hole 745B polarity chronozones and diatom zonal assignments by previous workers with the chronozonal assignments of this study. The stratigraphical ranges of selected stratigraphical markers are based on the original data sets (see text for references). D, diatoms; R, radiolaria; N, calcareous nannofossils. Emboldened events indicate primary biostratigraphical events recognized in this study. The polarity record shown represents the original interpretation calibrated to the GPTS of Cande & Kent (1995) and the interpretation from this study correlated with the GPTS of Cande & Kent (1995).

Table 31. *Composite dataset for ODP Hole 745B containing the diatom, radiolarian, calcareous nannofossil and palaeomagnetic data calibrated to the GPTS of Cande & Kent (1995) using the biochronological interpretation based on this study*

	Event	Sample (top)	Sample (bottom)	Depth (t)	Depth (b)	Age (t)	Age (b)
LO	Triceraspyris antarctica	1H-1, 0	1H-CC	0.00	5.00	0.00	0.09
LO	Hemidiscus karstenii	2H-CC	3H-CC	14.50	24.00	0.26	0.44
LO	Actinocyclus ingens	3H-CC	4H-CC	24.00	33.50	0.44	0.61
LO	Denticulopsis dimorpha	4H-CC	5H-CC	33.50	43.00	0.61	0.79
LO	Eucyrtidium calvertense	4H-CC	5H-CC	33.50	34.00	0.61	0.62
LO	Antarctissa cylindrica	4H-CC	5H-1, 53-55	34.00	35.50	0.61	0.65
B	C1n			42.80	42.80	0.78	0.78
LO	Thalassiosira elliptopora	5H-CC	6H-CC	43.00	52.30	0.79	1.27
FO	Denticulopsis dimorpha	5H-CC	6H-CC	43.00	52.30	0.79	1.27
T	C1r.1n			50.30	50.30	0.99	0.99
B	C1r.1n			54.64	54.64	1.07	1.07
FO	Thalassiosira elliptopora	9H-CC	10H-CC	81.00	90.50	1.57	1.75
LO	Hemidiscus cuneiformis	9H-CC	10H-CC	81.00	90.50	1.57	1.75
LO	Simonseniella barboi	9H-CC	10H-CC	81.00	90.50	1.57	1.75
LO	Cycladophora pliocenica	9H-CC	10H-1, 53-55	81.00	81.50	1.57	1.58
LO	Nitzschia weaveri	10H-CC	11H-3, 60-62	90.50	94.10	1.75	1.97
LO	Thalassiosira kolbei	10H-CC	11H-3, 60-62	90.50	94.10	1.75	1.97
T	C2n			91.55	91.55	1.77	1.77
B	C2n			93.55	93.55	1.95	1.95
LO	Nitzschia interfrigidaria	11H-3, 60-62	11H-6, 60-62	94.10	98.60	1.97	2.12
LO	Thalassiosira torokina	11H-6, 60-62	12H-1, 60-62	98.60	100.60	2.12	2.18
FO	Nitzschia kerguelensis	11H-6, 60-62	12H-1, 60-62	98.60	100.60	2.12	2.18
LO	Thalassiosira insigna	12H-1, 60-62	12H-2, 60-62	100.60	102.10	2.18	2.23
LO	Rouxia heteropolara	12H-1, 60-62	12H-2, 60-62	100.60	102.10	2.18	2.23
FO	Actinocyclus ingens	12H-2, 60-62	12H-4, 60-62	102.10	105.10	2.23	2.33
LO	Thalassiosira vulnifica	12H-2, 60-62	12H-4, 60-62	102.10	105.10	2.23	2.33
FO	Nitzschia cylindrica	12H-2, 60-62	12H-4, 60-62	102.10	105.10	2.23	2.33
LO	Rouxia californica	12H-4, 60-62	12H-6, 60-62	105.10	108.10	2.33	2.43
LO	Helotholus vema	12H-CC	13H-1, 53-55	109.50	110.00	2.48	2.50
LO	Desmospyris spongiosa	13H-1, 53-55	13H-2, 53-55	110.00	111.00	2.50	2.53
T	C2An.1n			112.55	112.55	2.58	2.58
FO	Cycladophora davisiana	13H-4, 53-55	13H-5, 53-55	114.50	116.00	2.66	2.72
FO	Thalassiosira vulnifica	13H-4, 60-62	13H-7, 60-62	115.60	118.60	2.71	2.83
LO	Nitzschia praeinterfrigidaria	13H-4, 60-62	13H-7, 60-62	115.60	118.60	2.71	2.83
LO	Nitzschia reinholdii	13H-4, 60-62	13H-7, 60-62	115.60	118.60	2.71	2.83
LO	Stephanopyxis grunowii	13H-4, 60-62	13H-7, 60-62	115.60	118.60	2.71	2.83
FO	Thalassiosira insigna	14H-1, 64-66	14H-2, 60-62	120.20	121.70	2.90	2.96

Table 31. (*continued*)

	Event	Sample (top)	Sample (bottom)	Depth (t)	Depth (b)	Age (t)	Age (b)
FO	*Thalassiosira kolbei*	14H-1, 64-66	14H-2, 60-62	120.20	121.70	2.90	2.96
B	C2An.1n			123.55	123.55	3.04	3.04
FO	*Nitzschia weaveri*	14H-4, 64-66	14H-5, 64-66	124.20	125.70	3.06	3.13
T	C2An.2n			125.55	125.55	2.99	3.11
B	C2An.2n			126.55	126.55	3.22	3.22
T	C2An.3n			128.55	128.55	3.33	3.33
LO	*Prunopyle titan*	15H-1, 53-55	15H-2, 53-55	130.50	132.00	3.44	3.52
B	C2An.3n			133.05	133.05	3.58	3.58
FO	*Nitzschia interfrigidaria*	15H-5, 60-62	15H-6, 60-62	135.10	136.60	3.71	3.80
LO	*Nitzschia fossilis*	16H-CC	17H-2, 60-62	139.00	141.10	3.95	4.08
FO	*Rouxia heteropolara*	17H-2, 60-62	17H-3, 60-62	139.00	141.10	3.95	4.08
LO	*Rhizosolenia miocenica*	17H-2, 60-62	17H-3, 60-62	139.00	141.10	3.95	4.08
T	C3n.1n			142.73	142.73	4.18	4.18
FO	*Nitzschia barronii*	17H-4, 60-62	17H-5, 60-62	144.10	145.60	4.22	4.27
B	C3n.1n			146.28	146.28	4.29	4.29
T	C3n.2n			149.30	149.30	4.48	4.48
FO	*Helotholus vema*	18H-4, 53-55	18H-5, 53-55	153.50	155.00	4.57	4.60
B	C3n.2n			155.93	155.93	4.62	4.62
FO	*Nitzschia praeinterfrigidaria*	19H-1, 60-62	19H-6, 60-62	158.60	166.10	4.72	4.87
T	C3n.3n			160.82	160.82	4.80	4.80
B	C3n.3n			167.90	167.90	4.89	4.89
FO	*Thalassiosira inura*	20H-1, 60-62	20H-4, 60-62	168.10	172.60	4.80	4.89
T	C3n.4n			177.95	177.95	4.98	4.98
B	C3n.4n			180.85	180.85	5.23	5.23
LO	*Rocella vigilans*	21H-2, 60-62	21H-5, 60-62	182.10	183.60	5.38	5.56
FO	*Thalassiosira oestrupii*	21H-2, 60-62	21H-5, 60-62	182.10	183.60	5.38	5.56
T	C3An.1n			186.37	186.37	5.89	5.89
LO	*Thalassiosira convexa*	22H-3, 60-62	22H-6, 60-62	190.10	194.60	5.98	6.09
FO	*Thalassiosira convexa v. aspinosa*	22H-6, 60-62	23H-3, 60-62	194.60	199.60	6.09	6.18
LO	*Trinacria excavata*	21H-5, 60-62	22H-3, 60-62	194.60	199.60	6.09	6.18
B	C3An.1n			196.63	196.63	6.14	6.14
T	C3An.2n			205.66	205.66	6.27	6.27
FO	*Rocella vigilans*	24H-2, 60-62	24H-6, 60-62	207.60	13.60	?	?
LO	*Thalassiosira miocenica*	24H-3, 60-62	24H-6, 60	209.10	213.60	?	?
FO	*Thalassiosira inura*	24H-1, 60-62	24H-2, 60-62	210.60	212.10	?	?

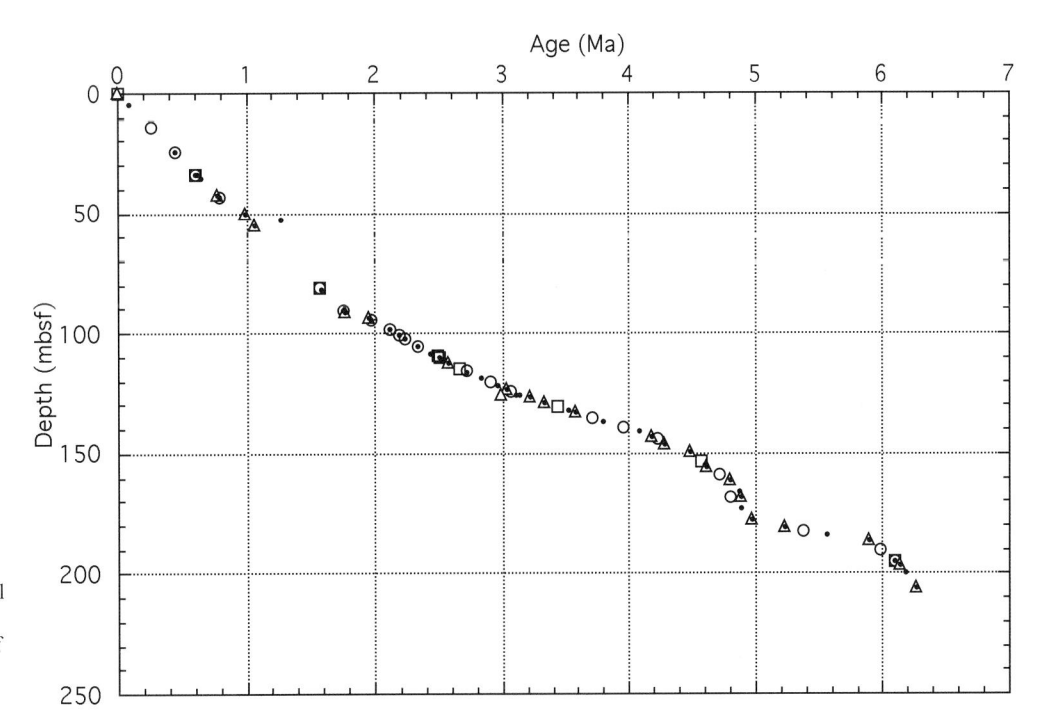

Data sets are integrated from the work of numerous authors (see Table 2).

Fig. 44. An age–depth plot for DSDP Hole 745B utilizing the biostratigraphical and polarity reversal events identified in this study and calibrated to the GPTS of Cande & Kent (1995). See Fig. 5 for legend of symbols.

Hole 746A

Sixteen cores were recovered from a total penetration of 280.8 mbsf representing a recovery of 33% of the cored interval. Sediments recovered consist of lower Pliocene and upper Miocene diatom ooze (Barron *et al.* 1989). A palaeomagnetic polarity reversal record was interpreted by Barron *et al.* (1991) for the upper Miocene section. Diatoms and radiolarians are generally abundant and well preserved throughout the entire stratigraphical section. Calcareous nannofossils are recorded in a single sample from the upper Miocene. (Barron *et al.* 1989). The integrated biostratigraphical and magnetostratigraphical data previously used (palaeomagnetics: Barron *et al.* 1991; diatoms: Baldauf & Barron 1991; calcareous nannofossils: none; radiolarians: Caulet 1991; Lazarus 1992) and calibrated to the GPTS of Berggren *et al.* (1985*a,b*) is shown in Table 32. An age–depth graph for Hole 119-46A using this data is shown in Fig. 45.

Figure 46 compares the polarity chronozones and diatom zones assigned by previous workers (Barron *et al.* 1991) with the chronozonal assignments in this study. In addition, this figure illustrates the stratigraphical ranges of diatoms and radiolarians. The stratigraphical ranges reflect the placement of FO and LO of stratigraphically useful species as recognized by Baldauf & Barron (1991), Caulet (1991) and Lazarus (1992). The revised chronozones and diatom zones illustrated are based on reassessing the data sets of these workers (Table 33). Comparison of the chronozonal assignments of Barron *et al.* (1991) with those of this study indicate only nomenclatural changes.

Figure 47 shows an age–depth graph for Hole 119-746A for the interval 180–250 mbsf, which contains a record of polarity reversals. The interval from 168 to 236 mbsf contains a continuous sequence of polarity reversals from the base of C3An.2n to the top of C4Ar.2n. These findings compare with those of Barron *et al.* (1991). There are no primary biostratigraphical events within this sequence.

Table 32. *Composite dataset for ODP Hole 746A containing the diatom, radiolarian, calcareous nannofossil and palaeomagnetic data calibrated to the GPTS of Cande & Kent (1995)*

	Event	Sample (top)	Sample (bottom)	Depth (t)	Depth (b)	Age (t)	Age (b)
FO	*Thalassiosira miocenica*	4H-1, 60	4H-2, 60	165.40	166.90	?	?
B	C3An.2n			168.86	168.86	6.57	6.57
FO	*Thalassiosira praeconvexa*	4H-4, 60	4H-5, 60	169.90	171.40	6.60	6.65
LO	*Actinocyclus ingens* v. *nodus*	4H-4, 60	4H-5, 60	169.90	171.40	6.60	6.65
LO	*Denticulopsis dimorpha*	4H-4, 60	4H-5, 60	169.90	171.40	6.60	6.65
FO	*Nitzschia miocenica*	4H-6, 60	4H-CC	172.90	174.30	6.70	6.74
LO	*Thalassiosira burckliana*	5H-2, 60	5H-3, 60	176.40	177.90	6.81	6.86
T	C3Bn			180.37	180.37	6.94	6.94
T	C4n.1n			189.26	189.26	7.43	7.43
T	C4n.2n			193.47	193.47	7.65	7.65
LO	*Hemidiscus karstenii*	7H-CC	8H-1, 60	202.80	203.40	7.99	8.01
FO	*Actinocyclus ingens* v. *nodus*	8H-2, 60	?8H-3, 60	204.90	206.40	8.07	8.12
B	C4n.2n			204.97	204.97	8.07	8.07
FO	*Nitzschia marina*	8H-1, 60	?8H-2, 60	206.40	208.00	8.12	8.18
LO	*Cycladophora humerus*	8H-CC	9H-4, 48-50	209.30	214.30	8.23	8.27
T	C4r.1n			209.30	209.30	8.23	8.23
B	C4r.1n			213.80	213.80	8.26	8.26
LO	*Actinocyclus ingens*	9H-4, 45	?9H-5, 45	214.30	215.80	8.27	8.32
FO	*Thalassiosira torokina*	9H-CC	10H-1, 60	217.80	218.40	8.38	8.39
FO	*Cosmiodiscus intersectus*	10H-2, 60	?10H-3, 60	219.90	221.40	8.44	8.48
T	C4An			228.85	228.85	8.70	8.70
B	C4An			231.35	231.35	9.03	9.03
LO	*Cycladophora spongothorax*	11X-4, 53-55	?11X-5, 53-55	233.30	233.80	9.11	9.13
LO	*Asteromphalus kennettii*	11X-CC	13X-1, 60	234.80	242.80	9.17	?
FO	*Thalassiosira burckliana*	13X-6, 60	13X-CC	234.80	242.80	9.17	?
T	C4Ar.1n			236.30	236.30	9.23	9.23
FO	*Hemidiscus karstenii*	13H-5, 60	?13H-6, 60	249.00	250.30	?	?
FO	*Cycladophora humerus*	13X-6, 53-55	13X-CC	250.20	251.80	?	?
FO	*Asteromphalus kennettii*	13H-CC	14H-1, 60	251.80	252.40	?	?
FO	*Eucyrtidium pseudoinflatum*	14X-CC	15X-CC	261.50	271.00	?	?

Data sets are integrated from the work of numerous authors (see Table 2).

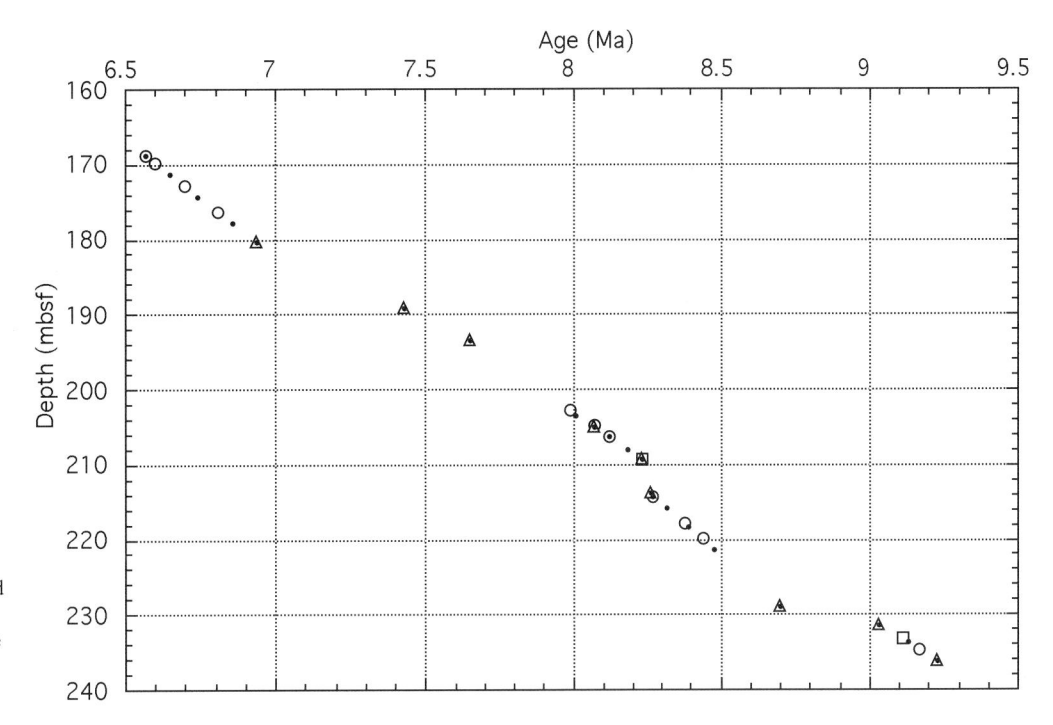

Fig. 45. An age–depth plot for DSDP Hole 746A utilizing the previously used biostratigraphical datums and polarity reversal interpretation calibrated to the GPTS of Cande & Kent (1995). See Fig. 5 for legend of symbols.

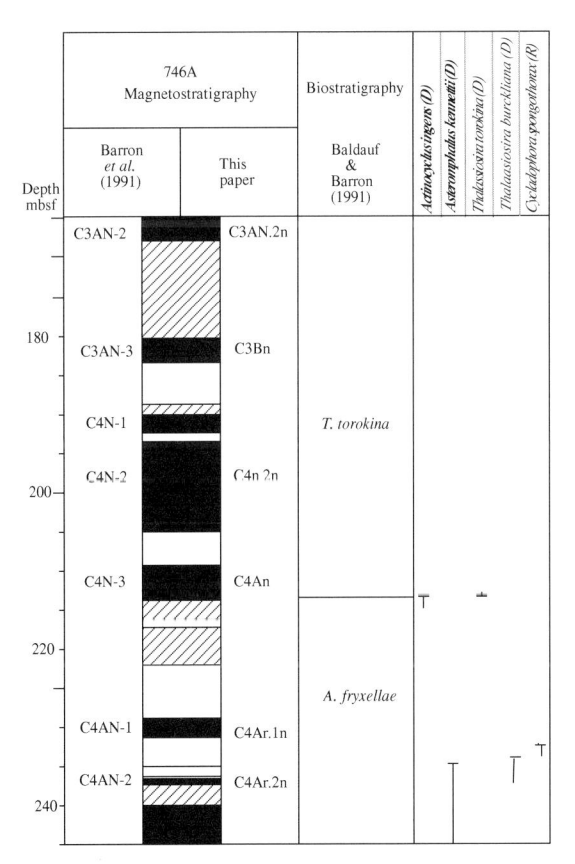

Fig. 46. Comparison of the ODP Hole 746A polarity chronozones and diatom zonal assignments by previous workers with the chronozonal assignments of this study. The stratigraphical ranges of selected stratigraphical markers are based on the original data sets (see text for references). D, diatoms; R, radiolaria; N, calcareous nannofossils. Emboldened events indicate primary biostratigraphical events recognized in this study. The polarity record shown represents the original interpretation calibrated to the GPTS of Cande & Kent (1995) and the interpretation from this study correlated with the GPTS of Cande & Kent (1995). Diagonal lines indicate intervals of uncertainty.

Table 33. *Composite dataset for ODP Hole 746A containing the diatom, radiolarian, calcareous nannofossil and palaeomagnetic data calibrated to the GPTS of Cande & Kent (1995) using the biochronological interpretation based on this study*

	Event	Sample (top)	Sample (bottom)	Depth (t)	Depth (b)	Age (t)	Age (b)
FO	*Thalassiosira miocenica*	4H-1, 60	?4H-2, 60	165.40	166.90	?	?
B	C3An.2n			168.86	168.86	6.57	6.57
FO	*Thalassiosira praeconvexa*	4H-4, 60	?4H-5, 60	169.90	171.40	6.60	6.65
LO	*Actinocyclus ingens v. nodus*	4H-4, 60	?4H-5, 60	169.90	171.40	6.60	6.65
LO	*Denticulopsis dimorpha*	4H-4, 60	?4H-5, 60	169.90	171.40	6.60	6.65
FO	*Nitzschia miocenica*	4H-6, 60	4H-CC	172.90	174.30	6.70	6.74
LO	*Thalassiosira burckliana*	5H-2, 60	?5H-3, 60	176.40	177.90	6.81	6.86
T	C3Bn			180.37	180.37	6.94	6.94
B	C3Bn			183.31	183.31	7.09	7.09
T	C4n.1n			189.26	189.26	7.43	7.43
B	C4n.1n			192.29	192.29	7.56	7.56
T	C4n.2n			193.47	193.47	7.65	7.65
LO	*Hemidiscus karstenii*	7H-CC	8H-1, 60	202.80	203.40	7.99	8.01
FO	*Actinocyclus ingens v. nodus*	8H-2, 60	?8H-3, 60	204.90	206.40	8.07	8.12
B	C4n.2n			204.97	204.97	8.07	8.07
FO	*Nitzschia marina*	8H-1, 60	?8H-2, 60	206.40	208.00	8.12	8.18
LO	*Cycladophora humerus*	8H-CC	9H-4, 48-50	209.30	214.30	8.23	8.29
T	C4r.1n			209.30	209.30	8.23	8.23
B	C4r.1n			213.80	213.80	8.26	8.26
LO	*Actinocyclus ingens*	9H-4, 45	?9H-5, 45	214.30	215.80	8.29	8.39
FO	*Thalassiosira torokina*	9H-CC	10H-1, 60	217.80	218.40	8.52	8.56
FO	*Cosmiodiscus intersectus*	10H-2, 60	?10H-3, 60	219.90	221.40	8.65	8.75
T	C4Ar.1n			228.85	228.85	9.23	9.23
B	C4Ar.1n			231.35	231.35	9.31	9.31
LO	*Cycladophora spongothorax*	11X-4, 53-55	?11X-5, 53-55	233.30	233.80	9.41	9.44
LO	*Asteromphalus kennettii*	11X-CC	13X-1, 60	234.80	242.80	9.50	?
FO	*Thalassiosira burckliana*	13X-6, 60	13X-CC	234.80	242.80	9.50	?
T	C4Ar.2n			236.30	236.30	9.58	9.58
FO	*Hemidiscus karstenii*	13H-5, 60	?13H-6, 60	249.00	250.30	?	?
FO	*Cycladophora humerus*	13X-6, 53-55	13X-CC	250.20	251.80	?	?
FO	*Asteromphalus kennettii*	13H-CC	14H-1, 60	251.80	252.40	?	?
FO	*Eucyrtidium pseudoinflatum*	14X-CC	15X-CC	261.50	271.00	?	?

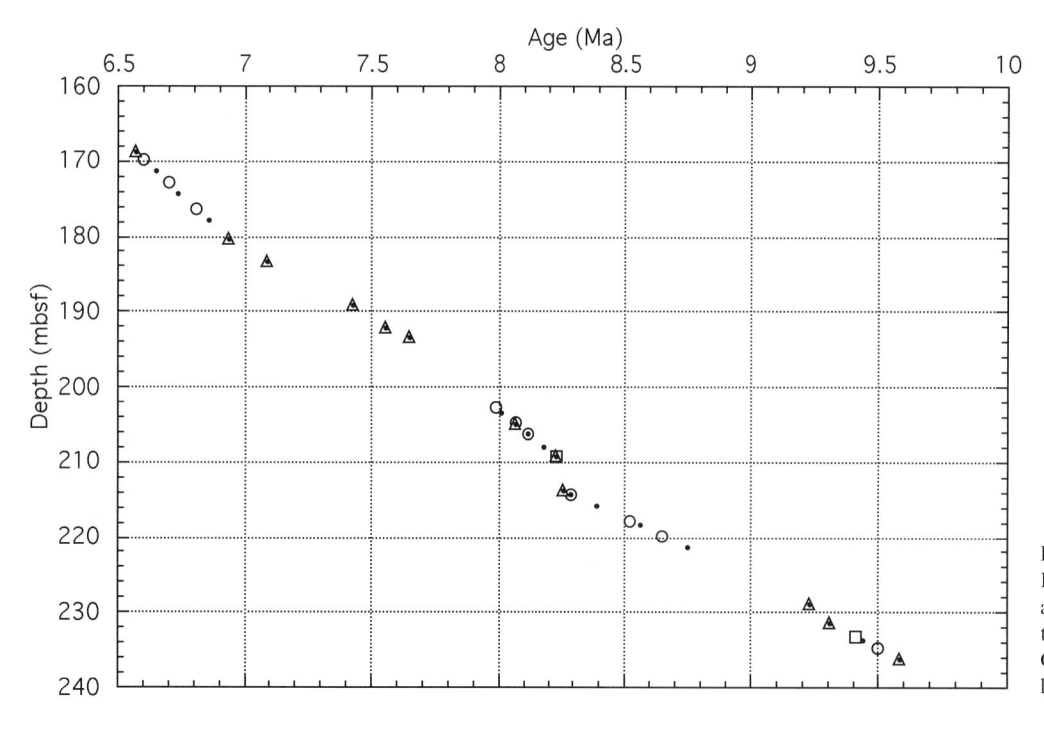

Data sets are integrated from the work of numerous authors (see Table 2).

Fig. 47. An age–depth plot for DSDP Hole 746A utilizing the biostratigraphical and polarity reversal events identified in this study and calibrated to the GPTS of Cande & Kent (1995). See Fig. 5 for legend of symbols.

Hole 747A

Twenty-seven cores were recovered from a total penetration of 256 mbsf representing a recovery of 88% of the cored interval. Sediments recovered consist of Quaternary to lower Pliocene diatom ooze with calcareous nannofossils and planktonic foraminifers, upper Pliocene to upper Oligocene calcareous nannofossil ooze, upper Oligocene to upper Cretaceous chalk with calcareous nannofossils. Basaltic pebbles and chert occur in the Cretaceous portion of the section (Wise et al. 1989).

A palaeomagnetic polarity record was obtained for the Quaternary to Cretaceous sequence. Shipboard workers resolved their inability to recognize short polarity reversals, by shorebase measurements of discrete samples (Wise et al. 1989; Heider et al. 1992a). Diatoms calcareous nannofossils and radiolarians are observed in the sequence recovered. Diatoms are abundant and well preserved in the Quaternary and Pliocene with few moderately well-preserved valves recorded in the Miocene and upper Oligocene. Calcareous nannofossils are few to abundant and well preserved throughout the sedimentary sequence (Aubry 1992; Wei & Wise 1992a). Radiolarians are common to abundant and generally well preserved in the Quaternary to middle Miocene sediments. Abundance and preservation declines in the lower Miocene and Oligocene. Radiolarians were not observed in sediments older than Oligocene (Wise et al. 1989). The integrated biostratigraphical and magnetostratigraphical data previously used (palaeomagnetics: Heider et al. 1992; Harwood et al. 1992; diatoms: Harwood & Maruyama 1992; calcareous nannofossils: Wei & Wise 1992; radiolarians: Abelmann 1992; Lazarus 1992) and calibrated to the GPTS of Berggren et al. (1985a, b) are presented in Table 34. An age–depth graph for Hole 120-747A using these data is presented in Fig. 48.

Figure 49 compares the polarity chronozones and diatom zones assigned by previous workers (Heider et al. 1992; Harwood et al. 1992; Harwood & Maruyama 1992) with the chronozonal assignments in this study. In addition, this figure illustrates the stratigraphical ranges of diatoms, radiolarians and calcareous nannofossils. The stratigraphical ranges reflect the placement of FO and LO of stratigraphically useful species as recognized by Harwood & Maruyama (1992), Abelmann (1992), Lazarus (1992), and Wei & Wise (1992). The revised chronozones and diatom zones illustrated are based on reassessing the data sets of these workers (Table 34). Comparison of the chronozonal assignments of Harwood et al. (1992; Table 34) with those of this study (Table 35) indicates that whereas the chronozonal placements are somewhat similar, the interpretation of hiatuses and the continuity of the section differs significantly.

An age–depth graph for Hole 120-747A using the stratigraphical markers determined in this study is presented in Fig. 50. Two hiatuses are placed in the section. The younger break is placed at approximately 24 mbsf and the older unconformity occurs between 30.26 and 33.0 mbsf. The upper boundary of the younger break is constrained by the continuous sequence of polarity reversals from the top of C1r.1n (14 mbsf) to the top of C2An.3n at 22.30 mbsf. The last occurrence of *Thallassiosira kolbei* (18.50–19.00 mbsf) and the LO of *Thallassiosira vulnifica* (21.50–22.00 mbsf). The lower boundary is constrained by the base of C3n.1n (25.90 mbsf), the FO of *Nitzschia barronii* (28.00–28.50 mbsf), the FO of *Helotholus vema* (28.00–29.50 mbsf) and the FO of *Nitzschia praeinterfrigidaria* (28.50–30.00 mbsf). The upper boundary of the older hiatus is constrained by the top of C3n.3n (30.26 mbsf) and the biostratigraphical events identified above. The lower boundary is constrained by the top of C4An (33.0 mbsf) and the top C4Ar.1n (34.41 mbsf). This study suggests that the stratigraphical section is significantly more continuous than previously recognized. Harwood & Maruyama (1992) and Harwood et al. (1992) identified ten hiatuses for the interval representing the last 25 million years based on both biostratigraphical and palaeomagnetic data. The two unconformities identified in this study approximate the placement of two of the unconformities previously recognized.

Harwood & Maruyama (1992) and Harwood et al. (1992) identified the youngest hiatus between 18 and 20 mbsf. The placement of this break is based on the co-occurrence of the LO of *Thalassiosira insigna*, LO of *Nitzschia weaveri* and the occurrence of *Thalassiosira vulnifica*. In addition, the previously used markers, such as the LO of *Thallassiosira inura*, FO of *Nitzschia kerguelensis* and the LO of *Nitzschia interfrigidaria* have shortened ranges. The stratigraphical relationship between the LO of *Thalassiosira insigna*, LO of *Nitzschia weaveri* and the FO of *Thalassiosira vulnifica* differs in holes where they have been identified for Legs 113, 119 and 120. The stratigraphical occurrence of *Nitzschia weaveri* is sporadic and therefore is not stratigraphically reliable. *Thalassiosira vulnifica* and *Thalassiosira insigna* co-occur in several holes with the FO of *Thalassiosira insigna* generally placed beneath the FO of *Thalassiosira vulnifica*. The LO of *Thalassiosira vulnifica* generally supersedes the LO of *Thalassiosira insigna*. These datums have a sporadic record in the Southern Ocean and we question their reliability as stratigraphical markers based on the currently available data discussed above.

Harwood & Maruyama (1992) and Harwood et al. (1992) place a second hiatus between 28.0 and 28.47 mbsf based on the LO of *Rhizosolenia costata* at 28 mbsf, below the FO of *Nitzschia barronii* (28.00–28.50 mbsf). The LO of *Nitzschia praeinterfrigidaria* is restricted based on previous stratigraphical interpretations. The recognition of a hiatus suggests a premature FO of *Nitzschia barronii* which is based on non overlapping ranges of *Nitzschia barronii* and *Rhizololenia costata*. We, however, incorporate the FO of *Nitzschia barronii* into our biostratigraphical model based on the consistent chronstratigraphical occurrence of this datum at sites throughout the Southern Ocean, including Hole 747A. The FO of *Helotholus vema* in Hole 747A at 28.0 mbsf approximates the FO of *Nitzschia barronii* in this hole. A similar relationship is observed throughout the Southern Ocean and supports our interpretation. This interpretation suggests that the LO of *Rhizoselenia costata* is an unreliable stratigraphical marker.

A third hiatus was identified by Harwood & Maruyama (1992) and Harwood et al. (1992) between 28.47 and 29.98 mbsf based on the co-occurrence of the LO of *Lychocanium grande*, the FO of *Thallassiosira inura*, the FO of *Nitzschia praeintefrigidaria*, the LO of *Thalassiosira compacta* and the absence of *Thalassiosira oestrupii* in this interval. Based on this study, the most reliable stratigraphical marker from this interval is the FO of *Nitzschia praeinterfrigidaria* at 28.0–30.00 mbsf. This event has an estimated age range of 4.46–4.76 Ma at Hole 747A which is within the ranges estimated for this event at other Southern Ocean sites.

Harwood & Maruyama (1992) and Harwood et al. (1992) identify an additional hiatus between 34.5 and c. 36 mbsf based on the premature FO of *Acrosphaera labrata* at 33.6 mbsf and the LO of *Nitzschia reinholdii* without the occurrence of *Thalassiosira miocenica* from this interval. The FO of *Acrosphaera labrata* at 33.6 mbsf and the LO of *Cycladophora spongothorax* at 35.95 mbsf are potential stratigraphical indicators based on then results of this study. The presence of these events within this interval suggests that the sequence is continuous.

Harwood & Maruyama (1992) and Harwood et al. (1992) place another hiatus between 47.47 and 48.97 mbsf based on an interpreted premature FO of *Asteromphalus kennetii* at 49.0 mbsf. This interpretation was based on the absence of a gap between the placement of the FO of *Asteromphalus kennettii* and the last common occurrence (LCO) of *Denticulopsis dimorpha* in this hole. This relationship was previously recognized by Baldauf & Barron (1991) at Site 744. In addition, Baldauf & Barron (1991) questioned the stratigraphical integrity of *Asteromphalus kennettii* because its finely silicified skeleton may have been prone to dissolution and relied on *Denticulopsis dimorpha* as a stratigraphical marker. The usefulness of *Asteromphalus kennettii* as a stratigraphical indicator requires additional analysis.

Harwood & Maruyama (1992) and Harwood et al. (1992) interpret the occurrence of an additional hiatus at 59.97 and

Table 34. Composite dataset for ODP Hole 747A containing the diatom, radiolarian, calcareous nannofossil and palaeomagnetic data calibrated to the GPTS of Cande & Kent (1995)

	Event	Sample (top)	Sample (bottom)	Depth (t)	Depth (b)	Age (t)	Age (b)
FO	*Emiliana huxleyii*	1H-2, 10-11	1H-3, 58-62	1.60	3.58	1.70	0.38
LO	*Actinocyclus ingens*	1H-3, 47-48	1H-4, 47-48	3.50	5.00	0.37	0.53
LO	*Thalassiosira inura*	1H-5, 47-48	1H-6, 47-48	6.50	8.00	0.69	0.85
LO	*Antarctissa cylindrica*	1H-6, 45	1H-CC	7.95	9.00	0.84	0.95
LO	*Pterocanium trilobium*	1H-6, 45	1H-CC	7.95	9.00	0.84	0.95
LO	*Thalassiosira elliptopora*	1H-6, 47-48	1H-CC	8.00	9.00	0.85	0.95
LO	*Nitzschia barronii*	2H-2, 47-50	2H-3, 47-50	11.00	12.50	1.17	1.32
FO	*Thalassiosira elliptopora*	2H-3, 47-48	2H-4, 47-50	12.50	14.00	1.32	1.48
LO	*Cycladophora pliocenica*	2H-5, 45	2H-6, 45	15.45	16.95	1.64	1.80
LO	*Simonseniella barboi*	2H-6, 47-48	2H-CC	17.00	18.50	1.80	?
B	2n			18.40	18.40	1.95	1.95
LO	*Triceraspyris antarctica*	2H-7, 45	2H-CC	18.45	18.50	?	?
LO	*Eucyrtidium calvertense*	2H-7, 45	2H-CC	18.45	18.50	?	?
LO	*Thalassiosira kolbei*	2H-CC	3H-1, 47-48	18.50	19.00	?	?
	Unconformity			***	***	***	***
LO	*Desmospyris spongiosa*	3H-1, 45	3H-2, 45	18.95	20.45	?	?
LO	*Helotholus vema*	3H-1, 45	3H-2, 45	18.95	20.45	?	?
FO	*Actinocyclus actinochilus*	3H-1, 47-48	3H-2, 47-48	19.00	20.50	?	?
LO	*Nitzschia interfrigidaria*	3H-1, 47-48	3H-2, 47-48	19.00	20.50	?	?
FO	*Nitzschia kerguelensis*	3H-1, 47-48	3H-2, 47-48	19.00	20.50	?	?
LO	*Nitzschia weaveri*	3H-1, 47-48	3H-2, 47-48	19.00	20.50	?	?
LO	*Thalassiosira insigna*	3H-1, 47-48	3H-2, 47-48	19.00	21.50	?	?
LO	*Thalassiosira vulnifica*	3H-1, 47-48	3H-2, 47-48	19.00	21.50	?	?
LO	*Reticulofenestra hesslandii*	3H-2, 58-62	3H-3, 34-36	20.58	22.58	?	?
T	C2An.1n			21.30	21.30	2.58	2.58
FO	*Thalassiosira vulnifica*	3H-2, 47-48	3H-3, 47-48	21.50	22.00	?	?
FO	*Cycladophora davisiana*	3H-3, 45	3H-4, 45	21.95	23.45	?	?
LO	*Prunopyle titan*	3H-3, 45	3H-4, 45	21.95	23.45	?	?
FO	*Nitzschia weaveri*	3H-4, 47-48	3H-5, 47-48	23.50	25.00	?	?
LO	*Thalassiosira complicata*	3H-4, 47-48	3H-5, 47-48	23.50	25.00	?	?
LO	*Lampromitra coronata*	3H-5, 45	3H-6, 45	24.95	26.45	?	?
FO	*Nitzschia interfrigidaria*	3H-5, 47-48	3H-6, 47-48	25.00	26.50	?	?
LO	*Nitzschia praeinterfrigidaria*	3H-5, 47-48	3H-6, 47-48	25.00	26.50	?	?
LO	*Rouxia diploneides*	3H-5, 47-48	3H-6, 47-48	25.00	26.50	?	?
FO	*Thalassiosira kolbei*	3H-6, 47-48	3H-CC	26.50	28.00	?	?
FO	*Thalassiosira oestrupii*	3H-6, 47-48	3H-CC	26.50	28.00	?	?
	Unconformity			***	***	***	***
FO	*Nitzschia barronii*	3H-CC	4H-1, 47-50	28.00	28.50	?	?
FO	*Thalassiosra lentiginosa*	3H-CC	4H-1, 47-50	28.00	28.50	?	?
FO	*Helotholus vema*	3H-CC	4H-2, 45	28.00	29.50	?	?
LO	*Denticulopsis dimorpha*	4H-1, 47-50	4H-2, 47-50	28.50	30.00	?	?
FO	*Nitzschia praeinterfrigidaria*	4H-1, 47-50	4H-2, 47-50	28.50	30.00	?	?
FO	*Rouxia heteropolara*	4H-1, 47-50	4H-2 47-50	28.50	30.00	?	?
FO	*Thalassiosira complicata*	4H-1, 47-50	4H-2 47-50	28.50	30.00	?	?
B	C3n.2n			29.00	29.00	4.62	4.62
	Unconformity			***	***	***	***
LO	*Asteromphalus kennettii*	4H-2, 47-50	4H-3, 47-50	30.00	31.50	?	?
LO	*Hemidiscus triangularis*	4H-2 47-50	4H-3, 47-50	30.00	31.50	?	?
LO	*Neobrunia miraibilis*	4H-2 47-50	4H-3, 47-50	30.00	31.50	?	?
FO	*Nitzschia praecurta*	4H-2 47-50	4H-3, 47-50	30.00	31.50	?	?
T	C3n.3n			31.00	31.00	4.80	4.80
T	C3An.1n			31.50	31.50	5.89	5.89
FO	*Thalassiosira inura*	4H-3, 47-48	4H-4, 47-50	31.50	33.00	5.89	6.14
B	C3An.1n			33.00	33.00	6.14	6.14
FO	*Thalassiosira torokina*	4H-4, 47-50	4H-5, 47-50	33.00	34.50	6.14	?
FO	*Acrosphaera labrata*	4H-4, 110	4H-5, 45	33.60	34.45	6.20	?
T	C3An.2n			34.41	34.41	6.27	6.27
LO	*Cycladophora spongothorax*	4H-5, 45	4H-6, 45	34.45	35.95	?	?
	Unconformity			***	***	***	***
FO	*Hemidiscus ovalis*	4H-6, 47-50	4H-7, 48-50	36.00	37.50	?	?
B	C4An			40.97	40.97	9.00	9.00
LO	*Calcidiscus macintyrei*	5H-6, 10-11	5H-7, 58-62	45.10	47.08	9.68	9.85
T	C5n.1n			45.47	45.47	9.74	9.74
B	C5n.1n			47.56	47.56	9.88	9.88
	Unconformity			***	***	***	***
FO	*Eucyrtidium pseudoinflatum*	6H-3, 45	6H-4, 45	50.45	51.95	10.20	10.37
LO	*Actinomma golownini*	6H-4, 45	6H-6, 45	51.95	54.95	10.37	10.71
FO	*Hemidiscus cuneiformis*	6H-4, 47-48	6H-5, 47-48	52.00	53.50	10.38	10.55

Table 34. (*continued*)

	Event	Sample (top)	Sample (bottom)	Depth (t)	Depth (b)	Age (t)	Age (b)
B	C5n.2n			57.10	57.10	10.95	10.95
FO	*Cycladophora spongothorax*	7H-2, 45	7H-3, 45	58.45	59.45	?	?
LO	*Denticulopsis praedimorpha*	7H-2, 47-49	7H-3, 47-49	58.50	60.00	?	?
	Unconformity			***	***	***	***
LO	*Nitzschia denticuloides*	7H-3, 47-49	7H-4, 47-49	60.00	61.50	?	?
FO	*Reticulofenestra hesslandii*	7H-4, 10-11	7H-5, 62-63	61.15	63.08	?	16.73
T	C5An.1n			61.30	61.30	11.94	11.94
B	C5An.1n			62.10	62.10	12.08	12.08
B	C5An.2n			65.10	65.10	12.40	12.40
FO	*Denticulopsis dimorpha*	7H-CC	8H-1, 47-49	66.00	66.50	?	?
LO	*Actinocyclus ingens* v. *nodus*	7H-CC	8H-1, 47-49	66.00	66.50	?	?
FO	*Denticulopsis praedimorpha*	7H-CC	8H-1, 47-49	66.00	66.50	?	?
	Unconformity			***	***	***	***
FO	*Actinomma golownini*	7H-CC	8H-1, 45	66.00	66.50	?	?
LO	*Nitzschia grossepunctata*	8H-1, 47-49	8H-2, 47-49	66.50	68.00	?	?
LO	*Coscinodiscus lewisianus*	8H-1, 47-49	8H-2, 47-49	66.50	68.00	?	?
T	C5ABn			66.80	66.80	13.30	13.30
FO	*Nitzschia denticuloides*	9H-6, 47-49	9H-7, 47-49	67.97	69.47	13.41	13.60
B	C5ABn			68.90	68.90	13.51	13.51
T	C5ACn			70.10	70.10	13.70	13.70
FO	*Denticulopsis hustedtii*	8H-5, 47-49	8H-6, 47-49	72.50	74.00	14.13	14.31
T	C5ADn			72.80	72.80	14.18	14.18
LO	*Denticulopsis maccolumnii*	8H-6, 47-49	8H-CC	74.00	75.50	14.31	14.49
FO	*Actinocyclus ingens*	v. nodus 8H-CC	9H-2, 47-49	75.50	77.49	14.49	14.68
LO	*Synedra jouseana*	8H-CC	9H-2, 47-49	75.50	79.10	14.49	14.80
LO	*Crucidenticula kanayae*	9H-2, 47-49	9H-4, 47-49	76.06	79.06	14.56	14.80
B	C5ADn			76.50	76.50	14.61	14.61
T	C5Bn.1n			79.10	79.10	14.80	14.80
B	C5Bn.1n			80.00	80.00	14.89	14.89
B	C5Bn.2n			81.30	81.30	15.16	15.16
FO	*Actinocyclus ingens*	9H-CC	10H-1, 47-48	85.00	85.50	?	?
FO	*Nitzschia grossepunctata*	9H-CC	10H-1, 47-49	85.00	85.50	?	?
FO	*Denticulopsis lauta*	9H-CC	10H-1, 47-49	85.00	85.60	?	?
	Unconformity			***	***	***	***
T	C5Cn.1n			85.47	85.47	16.01	16.01
LO	*Reticulofenestra pseudoumbilica*	10H-4, 10-11	10H5-58-62	89.60	91.58	16.51	16.75
B	C5Cn.3n			91.47	91.47	16.73	16.75
FO	*Denticulopsis maccolumnii*	10H-5, 47-49	10H-6, 47-49	91.50	93.00	16.74	17.14
FO	*Calcidiscus macintyrei*	10H-6, 58-62	10H-7, 20-24	93.08	94.20	17.17	17.38
T	C5Dn			93.50	93.50	17.28	17.28
FO	*Crucidenticula kanayae*	10H-7, 47-49	11H-1, 47-49	94.50	95.00	17.43	17.51
FO	*Calcidiscus leptoporus*	11H-1, 58-62	11-H2, 10-11	95.08	96.10	17.52	17.68
LO	*Thalassiosira fraga*	11H-3, 47-49	11H-5, 47-49	98.00	101.00	17.97	18.38
T	C5En			100.00	100.00	18.28	18.28
T	C6n			107.50	107.50	19.05	19.05
B	C6n			112.90	112.90	20.13	20.13
FO	*Thalassiosira fraga*	13H-1, 47-49	13H-2, 47-49	114.00	115.50	20.33	20.57
T	C6An.1n			115.07	115.07	20.52	20.52
B	C6An.1n			116.80	116.80	20.73	20.73
LO	*Bogorovia veniamini*	13H-3, 47-49	13H-4, 47-49	116.99	118.49	20.79	21.26
B	C6An.2n			118.68	118.68	21.32	21.32
FO	*Raphidodiscus marylandicus*	13H-5, 47-49	13H-6, 47-49	119.97	121.47	21.68	22.19
T	C6AAn			120.30	120.30	21.77	21.77
B	C6AAr.1n			121.65	121.65	22.25	22.25
T	C6Bn.1n			123.10	123.10	22.59	22.59
LO	*Rocella gelida*	13H-CC	14H-1, 47-49	123.47	126.47	22.68	23.50
B	C6Bn.2n			125.05	125.05	23.07	23.07
T	C6Cn.1n			125.70	125.70	23.35	23.35
LO	*Lisitzina ornata*	14H-3, 47-49	14H-4, 47-49	126.00	128.00	23.41	23.78
B	C6Cn.1n			126.70	126.70	23.54	23.54
T	C6n.2n			127.60	127.60	23.68	23.68
LO	*Thalassiosira primalabiata*	14H-4, 47-49	14H-5, 47-49	128.00	129.50	23.78	24.14
LO	*Reticulofenestra bisecta*	14H-4, 58-62	14H-4, 10-12	128.08	129.10	23.80	?
B	C6n.2n			128.10	128.10	23.80	23.80
T	C6n.3n			128.90	128.90	24.00	24.00
	Unconformity			***	***	***	***
B	C7n.2n			133.60	133.60	25.20	25.20
LO	*Rocella vigilans*	15H-3, 47-49	15H-4, 47-49	135.97	137.47	?	?

(*continued*)

Table 34. (*continued*)

	Event	Sample (top)	Sample (bottom)	Depth (t)	Depth (b)	Age (t)	Age (b)
	Unconformity			***	***	***	***
T	C7An			137.50	137.50	25.50	25.50
B	C7An			138.30	138.30	25.65	25.65
T	C8n.1n			138.70	138.70	25.82	25.82
FO	*Rocella gelida*	15H-5, 47-49	15H-6, 47-49	140.00	140.50	25.89	25.92
B	C8n.1n			141.05	141.05	25.95	25.95
LO	*Chiasmolithus altus*	15H-7, 10-12	16H-1, 10-12	141.60	142.10	25.96	25.97
T	C8n.2n			143.17	143.17	25.99	25.99
FO	*Lisitzina ornata*	16H-CC	17H-1, 47-49	151.50	151.97	26.53	?
FO	*Rocella vigilans*	16H-CC	17H-1, 47-49	151.50	151.97	26.53	?
B	C8n.2n			151.84	151.84	26.55	26.55

Data sets are integrated from the work of numerous authors (see Table 2).

61.47 mbsf based on the assumption that the LO of *Nitzschia denticuloides* is premature relatively to the LO of *Denticulopsis praedimorpha*. The LO of *Nitzschia denticuloides* is assigned to sample 7H-4, 47–48 cm and the LO of *Denticulopsis praedimorpha* is assigned either to sample 7H-5, 47–48 cm or to sample 7H-3, 47–48 cm. Barron & Baldauf (1991) have shown that these events are approximately coeval. Harwood & Maruyama (1992) suggest that the LO of *Denticulopsis praedimorpha* is approximately 0.1 million years younger than the LO of *Nitzschia denticuloides*; however we question the stratigraphical usefulness of the LO of *Nitzschia denticuloides* (Table 3).

Harwood & Maruyama (1992) and Harwood *et al.* (1992) use the concurrent FO's of *Denticulopsis praedimorpha* and *Denticulopsis dimorpha* at 66 mbsf and the FO of *Nitzschia denticuloides* at 67.97 mbsf to interpret the occurrence of a hiatus between 66.0 and 66.47 mbsf. This interpretation was based on results from Baldauf & Barron (1991), Hole 744B, which placed the FO of *D. praedimorpha* between the FO of *Nitzschia denticuloides* and the FO of *Denticulopsis dimorpha*. The FO of *Denticulopsis praedimorpha* is identified in this study as a primary marker. The potential usefulness of *Denticulopsis dimorpha* as a stratigraphical indicator needs to be re-examined.

Harwood & Maruyama (1992) and Harwood *et al.* (1992) identify the occurrence of a hiatus between 85 and 85.47 mbsf based on the FO of *Nitzschia grossepunctata*, the FO of *Denticulopsis lauta* and the FO of *Actinocyclas ingens* at 85 mbsf. The succession of these datums is inconsistent within the Southern Ocean (see Gersonde & Burckle 1990, Hole 689B; Baldauf & Barron 1991, Hole 744B), and they are therefore unreliable as stratigraphical markers. The FO of *Nizschia denticuloides* (67.97–69.47 mbsf), the FO of *Denticulopsis hustedtii* (72.50–74.00 mbsf) and the LO of *Synedra jouseana* (76.10–79.10) are primary markers as they have consistent ranges throughout the Southern Ocean. The occurrence of these events within this interval suggests that the sequence is complete.

Harwood *et al.* (1992) record a hiatus between 129.47 and 130.97 and a possible second break from 135.97 to 137.47 mbsf but do not indicate the rational for the placement of these breaks. In this study we infer a continuous polarity sequence throughout this interval which is devoid of primary markers.

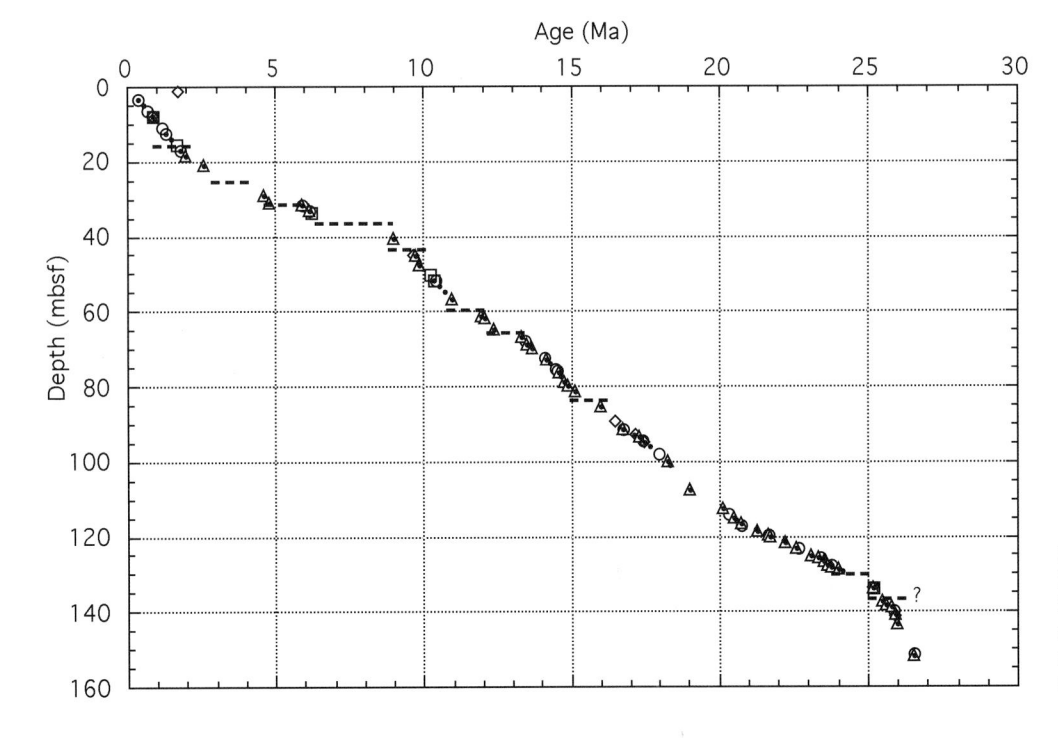

Fig. 48. An age–depth plot for DSDP Hole 747A utilizing the previously used biostratigraphical datums and polarity reversal interpretation calibrated to the GPTS of Cande & Kent (1995). See Fig. 5 for legend of symbols.

HOLE 747A

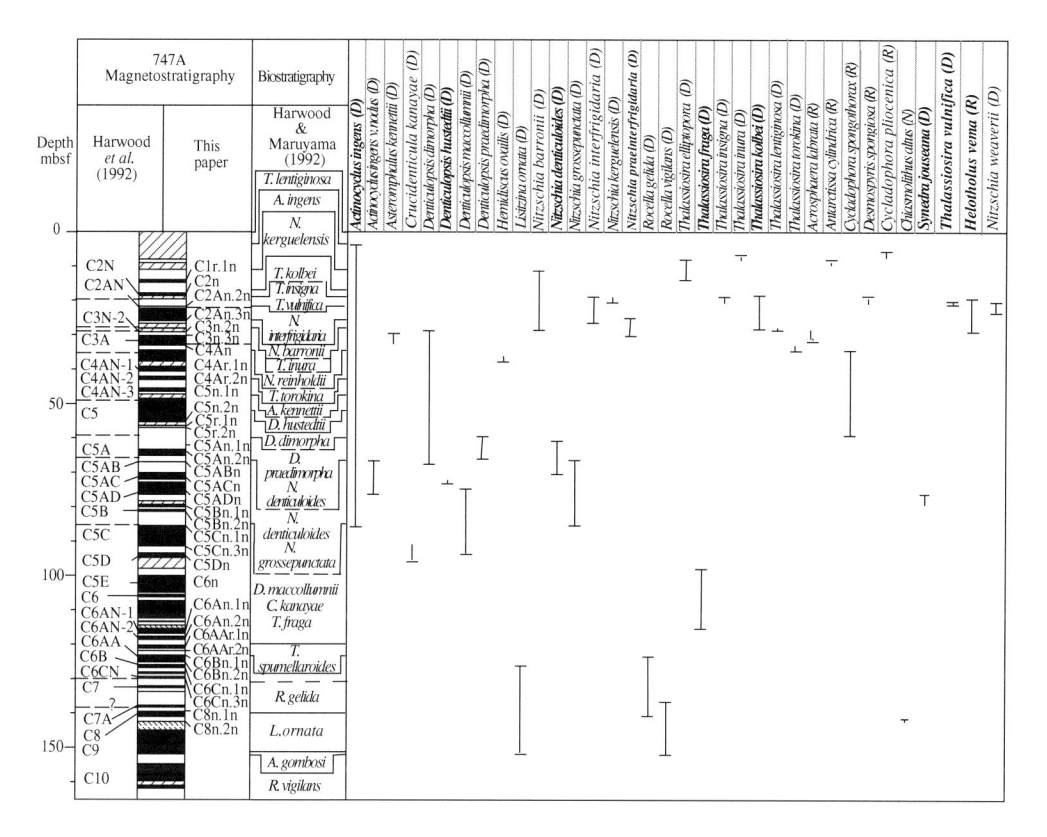

Fig. 49. Comparison of the ODP Hole 747A polarity chronozones and diatom zonal assignments by previous workers with the chronozonal assignments of this study. The stratigraphical ranges of selected stratigraphical markers are based on the original data sets (see text for references). D, diatoms; R, radiolaria; N, calcareous nannofossils. Emboldened events indicate primary biostratigraphical events recognized in this study. The polarity record shown represents the original interpretation calibrated to the GPTS of Cande & Kent (1995) and the interpretation from this study correlated with the GPTS of Cande & Kent (1995). Diagonal lines indicate intervals of uncertainty.

Table 35. *Composite dataset for ODP Hole 747A containing the diatom, radiolarian, calcareous nannofossil and palaeomagnetic data calibrated to the GPTS of Cande & Kent (1995) using the biochronological interpretation based on this study*

	Event	Sample (top)	Sample (bottom)	Depth (t)	Depth (b)	Age (t)	Age (b)
FO	*Emiliania huxleyii*	1H-2, 10-11	1H-3, 58-62	1.60	3.58	0.11	0.25
LO	*Actinocyclus ingens*	1H-3, 47-48	1H-4, 47-48	3.50	5.00	0.25	0.35
LO	*Thalassiosira inura*	1H-5, 47-48	1H-6, 47-48	6.50	8.00	0.46	0.57
LO	*Antarctissa cylindrica*	1H-6, 45	1H-CC	7.95	9.00	0.56	0.64
LO	*Pterocanium trilobium*	1H-6, 45	1H-CC	7.95	9.00	0.56	0.64
LO	*Thalassiosira elliptopora*	1H-6, 47-48	1H-CC	8.00	9.00	0.57	0.64
LO	*Nitzschia barronii*	2H-2, 47-50	2H-3, 47-50	11.00	12.50	0.78	0.88
FO	*Thalassiosira elliptopora*	2H-3, 47-48	2H-4, 47-50	12.50	14.00	0.88	0.99
T	C1r.1n			14.00	14.00	0.99	0.99
B	C1r.1n			14.65	14.65	1.07	1.07
LO	*Cycladophora pliocenica*	2H-5, 45	2H-6, 45	15.45	16.95	1.27	1.63
LO	*Simonseniella barboi*	2H-6, 47-48	2H-CC	17.00	18.50	1.65	1.99
T	C2n			17.50	17.50	1.77	1.77
B	C2n			18.40	18.40	1.95	1.95
LO	*Triceraspyris antarctica*	2H-7, 45	2H-CC	18.45	18.50	1.97	1.99
LO	*Eucyrtidium calvertense*	2H-7, 45	2H-CC	18.45	18.50	1.97	1.99
LO	*Thalassiosira kolbei*	2H-CC	3H-1, 47-48	18.50	19.00	1.99	2.19
LO	*Desmospyris spongiosa*	3H-1, 45	3H-2, 45	18.95	20.45	2.17	2.77
LO	*Helotholus vema*	3H-1, 45	3H-2, 45	18.95	20.45	2.17	2.77
FO	*Actinocyclus actinochilus*	3H-1, 47-48	3H-2, 47-48	19.00	20.50	2.19	2.79
LO	*Nitzschia interfrigidaria*	3H-1, 47-48	3H-2, 47-48	19.00	20.50	2.19	2.79
FO	*Nitzschia kerguelensis*	3H-1, 47-48	3H-2, 47-48	19.00	20.50	2.19	2.79
LO	*Nitzschia weaveri*	3H-1, 47-48	3H-2, 47-48	19.00	20.50	2.19	2.79
LO	*Thalassiosira insigna*	3H-1, 47-48	3H-2, 47-48	19.00	21.50	2.19	3.22
LO	*Thalassiosira vulnifica*	3H-1, 47-48	3H-2, 47-48	19.00	21.50	2.19	3.22
LO	*Reticulofenestra hesslandii*	3H-2, 58-62	3H-3, 34-36	20.58	22.58	2.82	3.56
T	C2An.2n			21.30	21.30	3.11	3.11
B	C2An.2n			21.50	21.50	3.22	3.22
FO	*Thalassiosira vulnifica*	3H-2, 47-48	3H-3, 47-48	21.50	22.00	3.22	3.29
FO	*Cycladophora davisiana*	3H-3, 45	3H-4, 45	21.95	23.45	3.28	?
LO	*Prunopyle titan*	3H-3, 45	3H-4, 45	21.95	23.45	3.28	?
T	C2An.3n			22.30	22.30	3.33	3.33
FO	*Nitzschia weaveri*	3H-4, 47-48	3H-5, 47-48	23.50	25.00	?	?
LO	*Thalassiosira complicata*	3H-4, 47-48	3H-5, 47-48	23.50	25.00	?	?
	Unconformity			***	***	***	***
LO	*Lampromitra coronata*	3H-5, 45	3H-6, 45	24.95	26.45	?	?

(*continued*)

Table 35. (continued)

	Event	Sample (top)	Sample (bottom)	Depth (t)	Depth (b)	Age (t)	Age (b)
FO	*Nitzschia interfrigidaria*	3H-5, 47-48	3H-6, 47-48	25.00	26.50	?	?
LO	*Nitzschia praeinterfrigidaria*	3H-5, 47-48	3H-6, 47-48	25.00	26.50	?	?
LO	*Rouxia diploneides*	3H-5, 47-48	3H-6, 47-48	25.00	26.50	?	?
B	C3n.1n			25.90	25.90	4.29	4.29
FO	*Thalassiosira kolbei*	3H-6, 47-48	3H-CC	26.50	28.00	4.33	4.43
FO	*Thalassiosira oestrupii*	3H-6, 47-48	3H-CC	26.50	28.00	4.33	4.43
FO	*Nitzschia barronii*	3H-CC	4H-1, 47-50	28.00	28.50	4.43	4.43
FO	*Thalassiosira lentiginosa*	3H-CC	4H-1, 47-50	28.00	28.50	4.43	4.43
FO	*Helotholus vema*	3H-CC	4H-2, 45	28.00	29.50	4.43	4.69
LO	*Denticulopsis dimorpha*	4H-1, 47-50	4H-2, 47-50	28.50	30.00	4.46	4.76
FO	*Nitzschia praeinterfrigidaria*	4H-1, 47-50	4H-2, 47-50	28.50	30.00	4.46	4.76
FO	*Rouxia heteropolara*	4H-1, 47-50	4H-2, 47-50	28.50	30.00	4.46	4.76
FO	*Thalassiosira complicata*	4H-1, 47-50	4H-2, 47-50	28.50	30.00	4.46	4.76
T	C3n.2n			28.73	28.73	4.48	4.48
B	C3n.2n			29.00	29.00	4.62	4.62
LO	*Asteromphalus kennettii*	4H-2, 47-50	4H-3, 47-50	30.00	31.50	4.76	?
LO	*Hemidiscus triangularis*	4H-2, 47-50	4H-3, 47-50	30.00	31.50	4.76	?
LO	*Neobrunia miraibilis*	4H-2, 47-50	4H-3, 47-50	30.00	31.50	4.76	?
FO	*Nitzschia praecurta*	4H-2, 47-50	4H-3, 47-50	30.00	31.50	4.76	?
T	C3n.3n			30.26	30.26	4.80	4.80
	Unconformity			***	***	***	***
T	C4An			30.33	30.33	8.70	8.70
FO	*Acrosphaera labrata*	4H-4, 110	4H-5, 45	31.30	31.62	8.82	8.86
FO	*Thalassiosira inura*	4H-3, 47-48	4H-4, 47-50	31.50	33.00	8.84	9.03
B	C4An			33.00	33.00	9.03	9.03
FO	*Thalassiosira torokina*	4H-4, 47-50	4H-5, 47-50	33.00	34.50	9.03	9.24
FO	*Acrosphaera labrata*	4H-4, 110	4H-5, 45	33.60	34.45	9.12	9.24
T	C4Ar.1n			34.41	34.41	9.23	9.23
LO	*Cycladophora spongothorax*	4H-5, 45	4H-6, 45	34.45	35.95	9.23	9.25
FO	*Hemidiscus ovalis*	4H-6, 47-50	4H-7, 48-50	36.00	37.50	9.25	9.27
B	C4Ar.1n			41.20	41.20	9.31	9.31
T	C4Ar.2n			42.10	42.10	9.58	9.58
B	C4Ar.2n			43.02	43.02	9.64	9.64
LO	*Calcidiscus macintyrei*	5H-6, 10-11	5H-7, 58-62	45.10	47.08	9.72	9.80
T	C5n.1n			45.47	45.47	9.74	9.74
B	C5n.1n			47.56	47.56	9.88	9.88
T	C5n.2n			49.61	49.61	9.92	9.92
FO	*Eucyrtidium pseudoinflatum*	6H-3, 45	6H-4, 45	50.45	51.95	10.08	10.37
LO	*Actinomma golownini*	6H-4, 45	6H-6, 45	51.95	54.95	10.37	10.94
FO	*Hemidiscus cuneiformis*	6H-4, 47-48	6H-5, 47-48	52.00	53.50	10.38	10.66
B	C5n.2n			55.00	55.00	10.95	10.95
B	C5r.1n			56.45	56.45	?	?
T	C5r.2n			56.80	56.80	?	?
B	C5r.2n			57.10	57.10	11.53	11.53
FO	*Cycladophora spongothorax*	7H-2, 45	7H-3, 45	58.45	59.45	11.66	11.76
LO	*Denticulopsis praedimorpha*	7H-2, 47-49	7H-3, 47-49	58.50	60.00	11.67	11.81
LO	*Nitzschia denticuloides*	7H-3, 47-49	7H-4, 47-49	60.00	61.50	11.81	11.98
FO	*Reticulofenestra hesslandii*	7H-4, 10-11	7H-5, 62-63	61.15	63.08	11.93	12.18
T	C5An.1n			61.30	61.30	11.94	11.94
B	C5An.1n			62.10	62.10	12.08	12.08
B	C5An.2n			65.10	65.10	12.40	12.40
FO	*Actinomma golownini*	7H-CC	8H-1, 45	66.00	66.50	12.88	13.11
FO	*Denticulopsis dimorpha*	7H-CC	8H-1, 47-49	66.00	66.50	12.88	13.14
LO	*Actinocyclus ingens v. nodus*	7H-CC	8H-1, 47-49	66.00	66.50	12.88	13.14
FO	*Denticulopsis praedimorpha*	7H-CC	8H-1, 47-49	66.00	66.50	12.88	13.14
LO	*Nitzschia grossepunctata*	8H-1, 47-49	8H-2, 47-49	66.50	68.00	13.14	13.57
LO	*Coscinodiscus lewisianus*	8H-1, 47-49	8H-2, 47-49	66.50	68.00	13.14	13.57
T	C5ABn			66.80	66.80	13.30	13.30
B	C5ABn			66.95	66.95	13.51	13.51
FO	*Nitzschia denticuloides*	9H-6, 47-49	9H-7, 47-49	67.97	69.47	13.57	13.66
T	C5ACn			70.10	70.10	13.70	13.70
B	C5ACn			72.00	72.00	14.08	14.08
FO	*Denticulopsis hustedtii*	8H-5, 47-49	8H-6, 47-49	72.50	74.00	14.14	14.32
T	C5ADn			72.80	72.80	14.18	14.18
LO	*Denticulopsis maccolumnii*	8H-6, 47-49	8H-CC	74.00	75.50	14.32	14.49
FO	*Actinocyclus ingens v. nodus*	8H-CC	9H-2, 47-49	75.50	77.49	14.49	14.68
LO	*Synedra jouseana*	8H-CC	9H-2, 47-49	75.50	79.10	14.49	14.80
LO	*Crucidenticula kanayae*	9H-2, 47-49	9H-4, 47-49	76.06	79.06	14.56	14.80
B	C5ADn			76.50	76.50	14.61	14.61

Table 35. (*continued*)

	Event	Sample (top)	Sample (bottom)	Depth (t)	Depth (b)	Age (t)	Age (b)
T	C5Bn.1n			79.10	79.10	14.80	14.80
B	C5Bn.1n			80.00	80.00	14.89	14.89
B	C5Bn.2n			81.30	81.30	15.16	15.16
FO	*Actinocyclus ingens*	9H-CC	10H-1, 47-48	85.00	85.50	15.91	16.01
FO	*Nitzschia grossepunctata*	9H-CC	10H-1, 47-49	85.00	85.50	15.91	16.01
FO	*Denticulopsis lauta*	9H-CC	10H-1, 47-49	85.00	85.60	15.91	16.02
T	C5Cn.1n			85.47	85.47	16.01	16.01
LO	*Reticulofenestra pseudoumbilica*	10H-4, 10-11	10H-5, 58-62	89.60	91.58	16.51	16.75
B	C5Cn.3n			91.47	91.47	16.73	16.73
FO	*Denticulopsis maccollumnii*	10H-5, 47-49	10H-6, 47-49	91.50	93.00	16.74	17.28
FO	*Calcidiscus macintyrei*	10H-6, 58-62	10H-7, 20-24	93.08	94.20	17.17	17.45
T	C5Dn			93.50	93.50	17.28	17.28
FO	*Crucidenticula kanayae*	10H-7, 47-49	11H-1, 47-49	94.50	95.00	17.52	17.65
B	C5Dn			94.90	94.90	17.62	17.62
FO	*Calcidiscus leptoporus*	11H-1, 58-62	11H-2, 10-11	95.08	96.10	17.67	17.96
LO	*Thalassiosira fraga*	11H-3, 47-49	11H-5, 47-49	98.00	101.00	18.49	19.13
T	C6n			100.00	100.00	19.05	19.05
B	C6n			113.83	113.83	20.13	20.13
FO	*Thalassiosira fraga*	13H-1, 47-49	13H-2, 47-49	114.00	115.50	20.18	20.57
T	C6An.1n			115.07	115.07	20.52	20.52
B	C6An.1n			116.80	116.80	20.73	20.73
B	C6An.2n			118.68	118.68	21.32	21.32
LO	*Bogorovia veniamini*	13H-3, 47-49	13H-4, 47-49	119.00	120.00	21.48	22.00
FO	*Raphidodiscus marylandicus*	13H-5, 47-49	13H-6, 47-49	119.97	121.47	21.98	22.42
T	C6AAr.1n			120.30	120.30	22.15	22.15
T	C6AAr.2n			121.65	121.65	22.46	22.46
T	C6Bn.1n			123.10	123.10	22.59	22.59
LO	*Rocella gelida*	13H-CC	14H-1, 47-49	123.47	126.47	22.68	23.50
B	C6Bn.2n			125.05	125.05	23.07	23.07
T	C6Cn.1n			125.70	125.70	23.35	23.35
LO	*Lisitzina ornata*	14H-3, 47-49	14H-4, 47-49	126.00	128.00	23.34	23.80
B	C6Cn.1n			126.70	126.70	23.54	23.54
T	C6Cn.2n			127.60	127.60	23.68	23.68
LO	*Thalassiosira primalabiata*	14H-4, 47-49	14H-5, 47-49	128.00	129.50	23.78	24.11
LO	*Reticulofenestra bisecta*	14H-4, 58-62	14H-4, 10-12	128.08	129.10	23.80	24.03
B	C6Cn.2n			128.10	128.10	23.80	23.80
T	C6Cn.3n			128.90	128.90	24.00	24.00
B	C6Cn.3n			129.70	129.70	24.12	24.12
T	C7n.1n			132.10	132.10	24.73	24.73
B	C7n.1n			132.60	132.60	24.78	24.78
T	C7n.2n			133.10	133.10	24.84	24.84
B	C7n.2n			133.60	133.60	25.18	25.18
LO	*Rocella vigilans*	15H-3, 47-49	15H-4, 47-49	135.97	137.47	25.37	25.50
T	C7An			137.50	137.50	25.50	25.50
B	C7An			138.30	138.30	25.65	25.65
T	C8n.1n			138.70	138.70	25.82	25.82
FO	*Rocella gelida*	15H-5, 47-49	15H-6, 47-49	140.00	140.50	25.89	25.92
B	C8n.1n			141.05	141.05	25.95	25.95
LO	*Chiasmolithus altus*	15H-7, 10-12	16H-1, 10-12	141.60	142.10	25.96	25.97
T	C8n.2n			143.17	143.17	25.99	25.99
FO	*Lisitzina ornata*	16H-CC	17H-1, 47-49	151.50	151.50	26.53	?
FO	*Rocella vigilans*	16H-CC	17H-1, 47-49	151.50	151.97	26.53	?
B	C8n.2n			151.84	151.84	26.55	26.55

Data sets are integrated from the work of numerous authors (see Table 2).

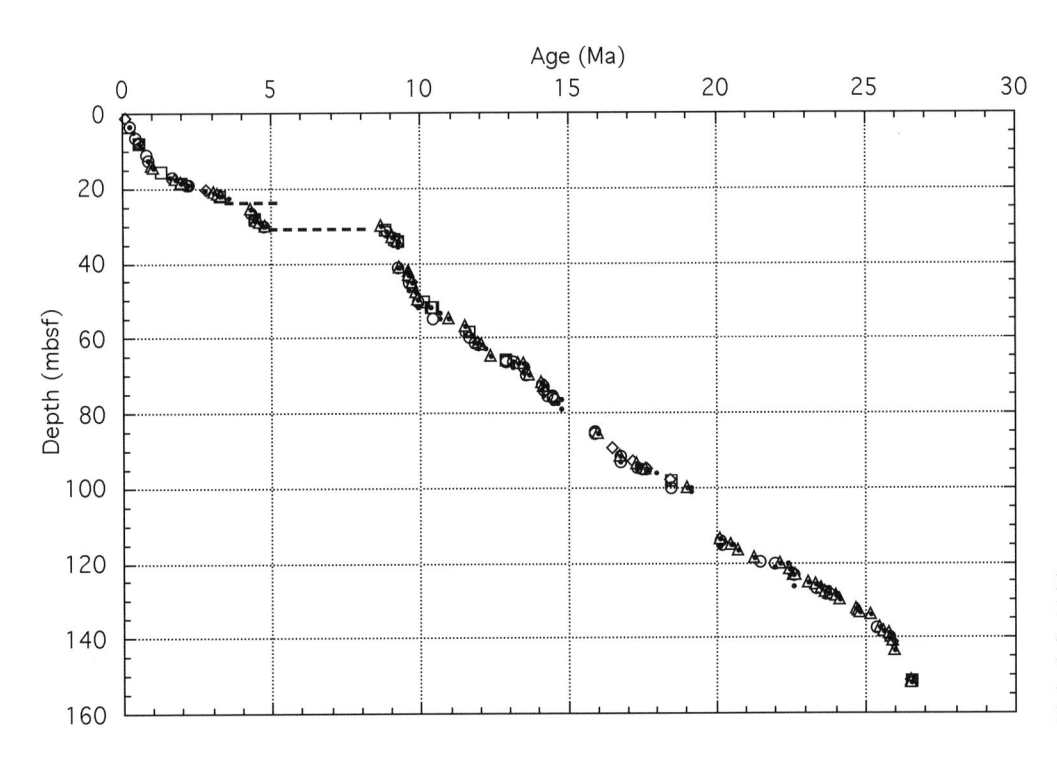

Fig. 50. An age–depth plot for DSDP Hole 747A utilizing the biostratigraphical and polarity reversal events identified in this study and calibrated to the GPTS of Cande & Kent (1995). See Fig. 5 for legend of symbols.

Hole 747B

Six cores were recovered from a total penetration of 53.3 mbsf representing a recovery of 97% of the cored interval. Sediments recovered consist of Quaternary to upper Miocene diatom and nannofossil ooze (Wise *et al.* 1989). A palaeomagnetic polarity record was obtained for the Quaternary to upper Miocene sequence. Shipboard workers resolved their inability to recognize short polarity reversal by shorebase measurements of discrete samples (Wise *et al.* 1989; Heider *et al.* 1992). Diatoms, calcareous nannofossils and radiolarians are observed in the sequence recovered. Diatoms are abundant and well preserved in the Quaternary and Pliocene and are few and moderately preserved in the Miocene. Calcareous nannofossils are recorded from the sedimentary section. Radiolarians are common to abundant and generally well preserved in the Quaternary to Miocene sediments (Wise *et al.* 1989). The integrated biostratigraphical and magnetostratigraphical data previously used (palaeomagnetics: Heider *et al.* 1992; diatoms: Harwood & Maruyama 1992; calcareous nannofossils: none; radiolarians: Lazarus 1992) and calibrated to the GPTS of Berggren *et al.* (1985*a, b*) is shown in Table 36. An age–depth graph for Hole 120-747B using these data is shown in Fig. 51.

Figure 52 compares the polarity chronozones (Heider *et al.* 1992) and diatom zones assigned by previous workers (Harwood & Maruyama 1992) with the chronozonal and diatom zonal assignments in this study. The stratigraphical ranges reflect the placement of FO and LO of stratigraphically useful species as recognized by Harwood & Maruyama (1992) and Lazarus (1990), respectively. The revised chronozones and diatom zones illustrated are based on reassessing the data sets of these workers (Table 37). Comparison of the chronozonal assignments of Heider *et al.* (1992; Table 36) to those of this study (Table 37) indicates that the chronozonal placements are similar.

Figure 53 shows an age–depth graph for Hole 120-747B using the stratigraphical markers determined in this study. The polarity sequence in this hole has intervals of uncertainty (Fig. 52). One hiatus is identified at approximately 28.5 mbsf. The upper boundary of this break is constrained by the FO of *Nitzschia interfrigidaria* (21.80–23.80 mbsf), the FO of *Helotholus vema* (24.4–25.1 mbsf), the FO of *Nitzschia barronii* (26.80–28.10) and the FO of *Nitzschia praeinterfrigidaria* (28.1–28.3 mbsf). The lower boundary is constrained by the top of C4An (29.54 mbsf) and the base of C4An (30.33 mbsf).

Harwood & Maruyama (1992) recognize three hiatuses. The stratigraphical break between 12.30 and 12.8 mbsf is based on the absence of *Thallassiosira vulnifica* in this hole and a single occurence of this species in Hole 747A. The LO of *Thallassiosira kolbei* (12.80–14.30 mbsf) in conjunction with the base of C1n (7.63 mbsf) and the top of C2An.1n (13.6 mbsf) suggest a possibly continuous record. The LO of *Thallassiosira vulnifica* is considered a primary marker and the absence of this species/event in this hole may support placement of a stratigraphical break. The intermediate hiatus between 26.8 and 28.06 mbsf is based on a single occurrence of *Nitzschia barronii*. in this hole compared with a more continuous occurrence of this species in Hole 747A. The placement of this break approximates the hiatus idenified in this study. In the absence of primary statigraphical markers we are unable to evaluate the placement of the hiatus recorded between 31.81 and 36.31 mbsf which Harwood & Maruyama (1992) base on the absence of *Hemidiscus ovalis*.

Table 36. *Composite dataset for ODP Hole 747B containing the diatom, radiolarian, calcareous nannofossil and palaeomagnetic data calibrated to the GPTS of Cande & Kent (1995)*

	Event	Sample (top)	Sample (bottom)	Depth (t)	Depth (b)	Age (t)	Age (b)
LO	*Actinocyclus ingens*	1H-2, 144-145	1H-CC	2.50	2.80	0.67	0.75
FO	*Nitzschia weaveri*	1H-CC	2H-1, 20-21	2.80	3.00	0.75	0.80
LO	*Thalassiosira elliptopora*	1H-CC	2H-1, 20-21	2.80	3.00	0.75	0.80
B	C1n			2.90	2.90	0.78	0.78
B	C2n			7.63	7.63	1.95	1.95
FO	*Thalassiosira oestrupii*	2H-CC	3H-1, 50-51	12.30	12.80	?	?
	Unconformity			***	***	***	***
LO	*Thalassiosira inura*	3H-1, 50-51	3H-2, 50-51	12.80	14.30	?	?
LO	*Thalassiosira kolbei*	3H-1, 50-51	3H-2, 50-51	12.80	14.30	?	?
FO	*Actinocyclus actinochilus*	3H-2, 50-51	3H-5, 147-148	14.30	19.80	?	?
LO	*Hemidiscus karstenii*	3H-2, 50-51	3H-5, 147-148	14.30	19.80	?	?
LO	*Nitzschia interfrigidaria*	3H-2, 50-51	3H-5, 147-148	14.30	19.80	?	?
FO	*Nitzschia kerguelensis*	3H-2, 50-51	3H-5, 147-148	14.30	19.80	?	?
FO	*Thalassiosira elliptopora*	3H-2, 50-51	3H-5, 147-148	14.30	19.80	?	?
LO	*Thalassiosira insigna*	3H-2, 50-51	3H-5, 147-148	14.30	19.80	?	?
FO	*Thalassiosira lentiginosa*	3H-2, 50-51	3H-5, 147-148	14.30	19.80	?	?
T	C2An.1n			17.92	17.92	2.58	2.58
B	C2An.2n			21.32	21.32	3.22	3.22
FO	*Nitzschia interfrigidaria*	3H-CC	4H-2, 50-51	21.80	23.80	3.30	?
LO	*Nitzschia praeinterfrigidria*	3H-CC	4H-2, 50-51	21.80	23.80	3.30	?
FO	*Thalassiosira insigna*	3H-CC	4H-2, 50-51	21.80	23.80	3.30	?
FO	*Thalassiosira kolbei*	3H-CC	4H-2, 50-51	21.80	23.80	3.30	?
LO	*Lampromitra coronata*	3H-CC	4H-2, 32	21.80	23.62	3.30	?
LO	*Nitzschia barronii*	4H-2, 50-51	4H-4, 50-51	23.80	26.80	3.60	?
FO	*Helotholus vema*			24.42	25.12	3.75	?
FO	*Nitzschia barronii*	4H-4, 50-51	4H-5, 50-51	26.80	28.10	4.17	?
FO	*Thalassiosira inura*	4H-4, 50-51	4H-5, 26-27	26.80	28.10	4.17	?
T	C3n.1n			26.84	26.84	4.18	4.18
	Unconformity			***	***	***	***
FO	*Nitzschia praeinterfrigidria*	4H-5, 26-27	4H-5, 50-51	28.10	28.30	?	?
LO	*Denticulopsis dimorpha*	4H-5, 50-51	4H-CC	28.30	31.30	?	?
LO	*Thalassiosira miocenica*	4H-5, 50-51	4H-CC	28.30	31.30	?	?

(*continued*)

Table 36. (*continued*)

	Event	Sample (top)	Sample (bottom)	Depth (t)	Depth (b)	Age (t)	Age (b)
T	C3n.2n			29.54	29.54	4.48	4.48
B	C3n.2n			30.33	30.33	4.62	4.62
FO	*Nitzschia reinholdii*	4H-CC	5H-1, 50-51	31.30	31.80	5.19	5.49
FO	*Thalassiosira miocenica*	4H-CC	5H-1, 50-51	31.30	31.80	5.19	5.49
FO	*Acrosphaera labrata*	4H-CC	5H-1, 32	31.30	31.62	5.19	5.38
LO	*Cycladophora spongothorax*	5H-1, 32	5H-1, 112	31.62	34.42	5.38	6.45
LO	*Asteromphalus kennettii*	5H-1, 50-51	5H-4, 50-51	31.80	36.30	5.49	?
B	C3An.1n			32.90	32.90	6.14	6.14
B	C3An.2n			35.00	35.00	6.57	6.57
	Unconformity			***	***	***	***
T	C4An			36.32	36.32	8.70	8.70
FO	*Asteromphalus kennettii*	5H-5, 50-51	5H-CC	37.80	40.80	8.88	9.77
B	C4An			39.10	39.10	9.03	9.03
T	C5n.1n			40.53	40.53	9.74	9.74
FO	*Thalassiosira torokina*	5H-CC	6H-1, 67-68	40.80	41.50	9.77	9.87
LO	*Actinomma golownini*	6H-4, 113	6H-5, 32	46.43	47.12	10.53	10.62
FO	*Acrosphaera australis*	6H-7, 16	6H-CC	47.93	49.43	10.73	10.93
FO	*Eucyrtidium pseudoinflatum*	6H-6, 32	6H-6, 113	48.62	49.43	10.82	10.93
B	C5n.2n			49.61	49.61	10.95	10.95
FO	*Cycladophora spongothorax*	6H-7, 16	6H-CC	49.96	50.30	10.99	?
LO	*Denticulopsis praedimorpha*	6H-7, 20-23	6H-CC	50.00	50.30	?	?

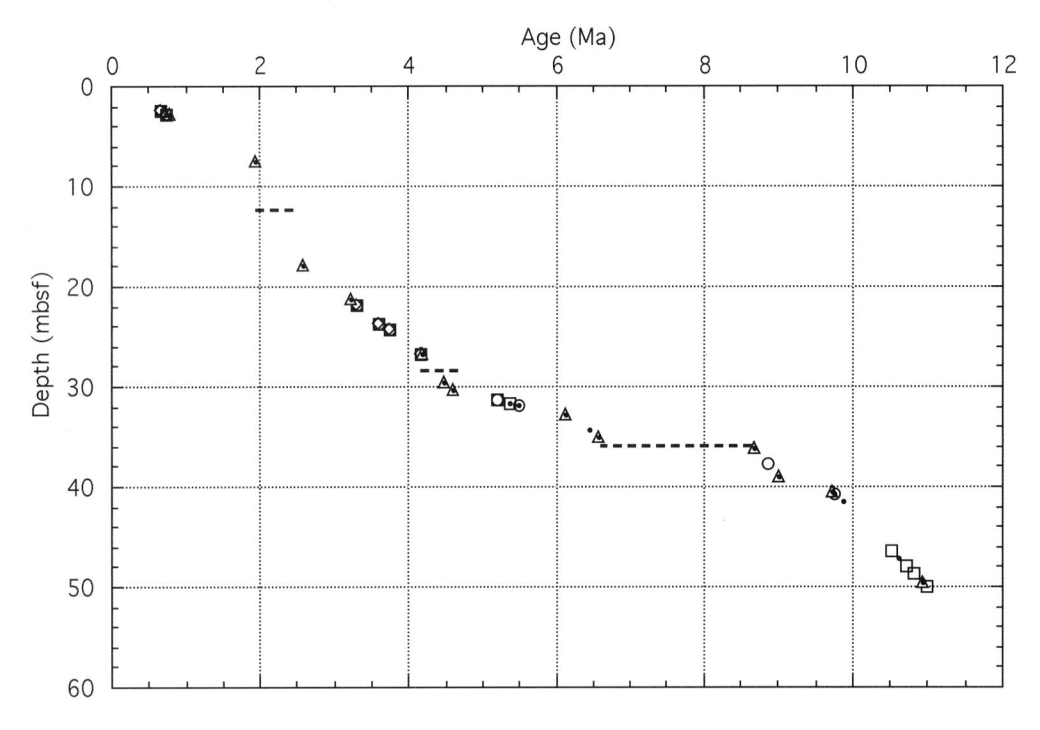

Data sets are integrated from the work of numerous authors (see Table 2).

Fig. 51. An age–depth plot for DSDP Hole 747B utilizing the previously used biostratigraphical datums and polarity reversal interpretation calibrated to the GPTS of Cande & Kent (1995). See Fig. 5 for legend of symbols.

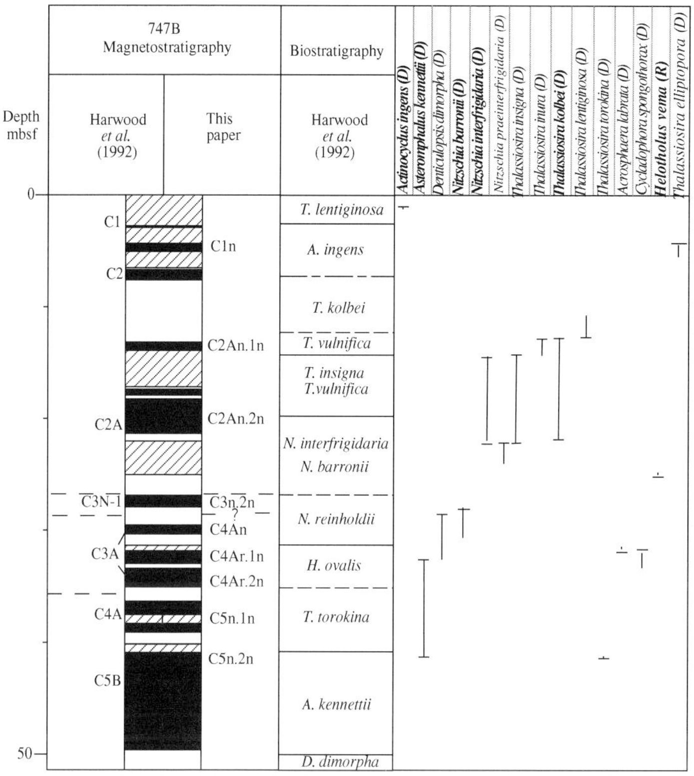

HOLE 747B

Fig. 52. Comparison of the ODP Hole 747B polarity chronozones and diatom zonal assignments by previous workers with the chronozonal assignments of this study. The stratigraphical ranges of selected stratigraphical markers are based on the original data sets (see text for references). D, diatoms; R, radiolaria; N, calcareous nannofossils. Emboldened events indicate primary biostratigraphical events recognized in this study. The polarity record shown represents the original interpretation calibrated to the GPTS of Cande & Kent (1995) and the interpretation from this study correlated with the GPTS of Cande & Kent (1995). Diagonal lines indicate intervals of uncertainty.

Table 37. *Composite dataset for ODP Hole 747B containing the diatom, radiolarian, calcareous nannofossil and palaeomagnetic data calibrated to the GPTS of Cande & Kent (1995) using the biochronological interpretation based on this study*

	Event	Sample (top)	Sample (bottom)	Depth (t)	Depth (b)	Age (t)	Age (b)
LO	*Actinocyclus ingens*	1H-2, 144-145	1H-CC	2.50	2.80	0.26	0.29
FO	*Nitzschia weaveri*	1H-CC	2H-1, 20-21	2.80	3.00	0.29	0.31
LO	*Thalassiosira elliptopora*	1H-CC	2H-1, 20-21	2.80	3.00	0.29	0.31
B	C1n			7.63	7.63	0.78	0.78
FO	*Thalassiosira oestrupii*	2H-CC	3H-1, 50-51	12.30	12.80	2.30	2.46
LO	*Thalassiosira inura*	3H-1, 50-51	3H-2, 50-51	12.80	14.30	2.46	2.70
LO	*Thalassiosira kolbei*	3H-1, 50-51	3H-2, 50-51	12.80	14.30	2.46	2.70
T	C2An.1n			13.16	13.16	2.58	2.58
FO	*Actinocyclus actinochilus*	3H-2, 50-51	3H-5, 147-148	14.30	19.80	2.70	3.17
LO	*Hemidiscus karstenii*	3H-2, 50-51	3H-5, 147-148	14.30	19.80	2.70	3.17
LO	*Nitzschia interfrigidaria*	3H-2, 50-51	3H-5, 147-148	14.30	19.80	2.70	3.17
FO	*Nitzschia kerguelensis*	3H-2, 50-51	3H-5, 147-148	14.30	19.80	2.70	3.17
FO	*Thalassiosira elliptopora*	3H-2, 50-51	3H-5, 147-148	14.30	19.80	2.70	3.17
LO	*Thalassiosira insigna*	3H-2, 50-51	3H-5, 147-148	14.30	19.80	2.70	3.17
FO	*Thalassiosra lentiginosa*	3H-2, 50-51	3H-5, 147-148	14.30	19.80	2.70	3.17
B	C2An.1n			17.40	17.40	3.04	3.04
T	C2An.2n			17.92	17.92	3.11	3.11
B	C2An.2n			21.32	21.32	3.22	3.22
LO	*Lampromitra coronata*	3H-CC	4H-2, 32	21.80	23.62	?	?
FO	*Nitzschia interfrigidaria*	3H-CC	4H-2, 50-51	21.80	23.80	?	?
LO	*Nitzschia praeinterfrigidaria*	3H-CC	4H-2, 50-51	21.80	23.80	?	?
FO	*Thalassiosira insigna*	3H-CC	4H-2, 50-51	21.80	23.80	?	?
FO	*Thalassiosira kolbei*	3H-CC	4H-2, 50-51	21.80	23.80	?	?
LO	*Nitzschia barronii*	4H-2, 50-51	4H-4, 50-51	23.80	26.80	?	?
FO	*Helotholus vema*	4H-2, 112	4H-3, 32	24.42	25.12	?	?
	Unconformity (?)			***	***	***	***
FO	*Nitzschia barronii*	4H-4, 50-51	4H-5, 50-51	26.80	28.10	4.47	?
FO	*Thalassiosira inura*	4H-4, 50-51	4H-5, 26-27	26.80	28.10	4.47	?
T	C3n.2n			26.84	26.84	4.48	4.48
FO	*Nitzschia praeinterfrigidaria*	4H-5, 26-27	4H-5, 50-51	28.10	28.30	?	?
	Unconformity			***	***	***	***
LO	*Denticulopsis dimorpha*	4H-5, 50-51	4H-CC	28.30	31.30	?	?
LO	*Thalassiosira miocenica*	4H-5, 50-51	4H-CC	28.30	31.30	?	?
T	C4An			29.54	29.54	8.70	8.70
B	C4An			30.33	30.33	9.03	9.03
FO	*Nitzschia reinholdii*	4H-CC	5H-1, 50-51	31.30	31.80	9.14	9.14

(*continued*)

Table 37. (*continued*)

	Event	Sample (top)	Sample (bottom)	Depth (t)	Depth (b)	Age (t)	Age (b)
FO	*Thalassiosira miocenica*	4H-CC	5H-1, 50-51	31.30	31.80	9.14	9.14
FO	*Acrosphaera labrata*	4H-CC	5H-1, 32	31.30	31.62	9.14	9.17
LO	*Cycladophora spongothorax*	5H-1, 32	5H-1, 112	31.62	34.42	9.17	9.62
LO	*Asteromphalus kennettii*	5H-1, 50-51	5H-4, 50-51	31.80	36.30	9.19	9.74
B	C4Ar.1n			32.90	32.90	9.31	9.31
T	C4Ar.2n			33.40	33.40	9.58	9.58
B	C4Ar.2n			35.00	35.00	9.64	9.64
T	C5n.1n			36.32	36.32	9.74	9.74
FO	*Asteromphalus kennettii*	5H-5, 50-51	5H-CC	37.80	40.80	9.81	9.95
B	C5n.1n			39.10	39.10	9.88	9.88
T	C5n.2n			40.53	40.53	9.92	9.92
FO	*Thalassiosira torokina*	5H-CC	6H-1, 67-68	40.80	41.50	9.95	10.03
LO	*Actinomma golownini*	6H-4, 113	6H-5, 32	46.43	47.12	10.59	10.67
FO	*Acrosphaera australis*	6H-7,16	6H-CC	47.93	49.43	10.76	10.93
FO	*Eucyrtidium pseudoinflatum*	6H-6, 32	6H-6, 113	48.62	49.43	10.84	10.93
B	C5n.2n			49.61	49.61	10.95	10.95
FO	*Cycladophora spongothorax*	6H-7, 16	6H-CC	49.96	50.30	10.99	?
LO	*Denticulopsis praedimorpha*	6H-7, 20-23	6H-CC	50.00	50.30	?	?

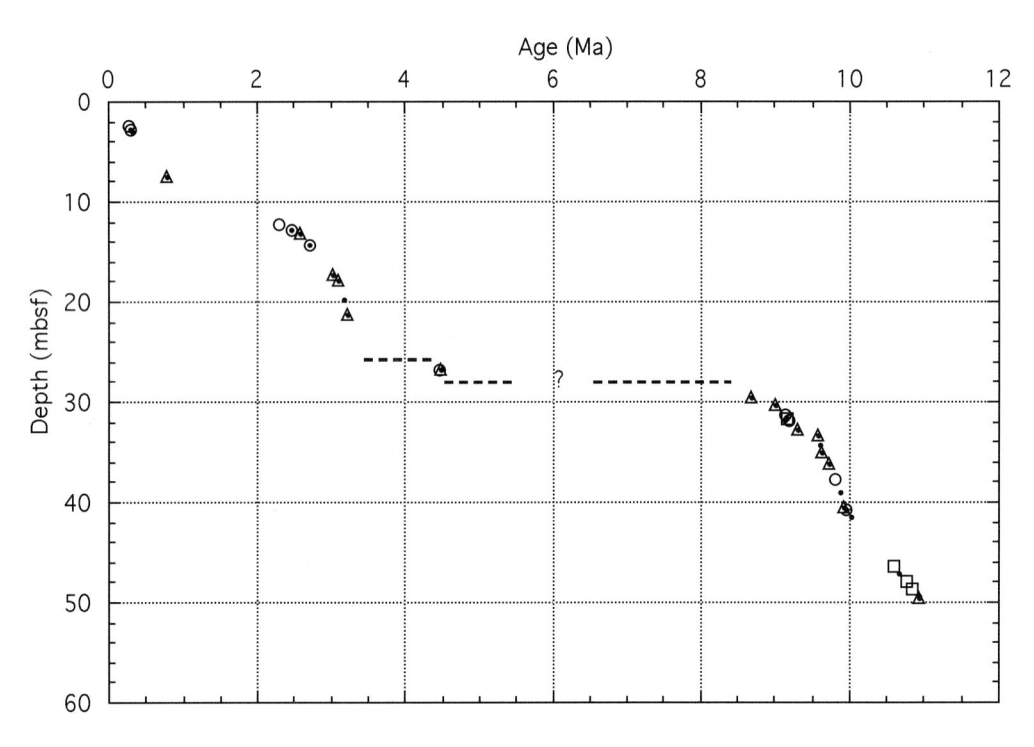

Data sets are integrated from the work of numerous authors (see Table 2).

Fig. 53. An age–depth plot for DSDP Hole 747B utilizing the biostratigraphical and polarity reversal events identified in this study and calibrated to the GPTS of Cande & Kent (1995). See Fig. 5 for legend of symbols.

Hole 748B

Twenty-five cores were recovered from a total penetration of 225.1 mbsf representing a recovery of 84% of the cored interval. Sediments recovered consist of Quaternary to Pliocene diatom ooze, upper Miocene to middle Eocene calcareous nannofossils ooze, and middle Eocene chalk with chert. A palaeomagnetic polarity record was obtained for the Quaternary to middle Eocene portion of the section (Wise *et al.* 1989; Inokuchi & Heider 1992). Abundant and well-preserved diatoms occur in Quaternary to lower Oligocene sediments. Eocene diatoms are poorly preserved. Calcareous nannofossils are generally absent in Quaternary and Pliocene sediments (Wei & Wise 1992*a*). Miocene and upper Eocene calcareous nannofossils are abundant and moderately preserved (Wei & Wise 1992*a*; Aubry 1992). Radiolarians were observed in samples examined throughout the Quaternary to middle Miocene sediments and significant reworking is recorded in upper Pliocene sediments (Wise *et al.* 1989). The integrated biostratigraphical and magnetostratigraphical data previously used (palaeomagnetics: Inokuchi & Heider 1992; Harwood *et al.* 1992; diatoms: Harwood & Maruyama 1992; calcareous nannofossils: Wei & Wise 1992; Wei 1992; radiolarians: Abelmann 1992; Lazarus 1992) and calibrated to the GPTS of Berggren *et al.* (1985*a, b*) is shown in Table 38. An age–depth graph for Hole 120-748B using this data is shown in Fig. 54.

Figure 55 compares the polarity chronozones and diatom zones assigned by previous workers (Inokuchi & Heider 1992; Harwood & Maruyama 1992) with the chronozonal and diatom zonal assignments used in this study. In addition, this figure illustrates the stratigraphical ranges of diatoms, radiolarians and calcareous nannofossils. The stratigraphical ranges reflect the placement of FO and LO of stratigraphically useful species as recognized by Harwood & Maruyama (1992), Abelmann (1992), Lazarus (1992) and Wei & Wise (1992), respectively. The revised chronozones and diatom zones illustrated are based on reassessing the data sets of these workers (Table 38). Comparison of the chronozonal assignments of Wise *et al.* (1989) and Inokuchi & Heider (1992) with those of this study (Table 39) indicate only minimal modifications based either on differences in GPTS or in the species selected for biostratigraphical control. There are significant modifications concerning the continuity of the sequence.

Figure 56 shows an age–depth graph for Hole 120-748B using the stratigraphical markers determined in this study. Four hiatuses are identified. The youngest occurs between 2.8 and 3.0 mbsf based on then joint LO of *Thallassiosira kolbei* and *Thallassiosira vulnifica*. The intermediate hiatus occurs between 10.10 and 11.60 mbsf and is constrained by the FO of *Nitzschi barronii* and *Nitzschia praeinterfrigidaria* at a depth of 10.00–10.30, the FO of *Nitzschia interfrigidaria* (8.00–8.40 mbsf) and the base of C3n.1n (6.95 mbsf). The two oldest unconformities occur at approximately 38 and 40 mbsf respectively. Placement of these two breaks is constrained by a continuous succession of magnetic reversals to C5r.2n (37.6 mbsf), above the interval, and the FO of *Crucidenticula kanayae* (41.6–43.10 mbsf) below the interval. The primary markers, the FO of *Denticulopsis hustedtii* the FO of *Nitzschia denticuloides* and the FO of *Denticulopsis praedimorpha* between 38.10 and 38.6 mbsf separate these two unconformities.

This study suggests that the stratigraphical section may be more continuous than previously recognized. Harwood & Maruyama (1992) identified 13 hiatuses. The youngest occurs at 0–0.10 mbsf. A second hiatus 2.80 and 3.00 mbsf is based on the co-occurrence of the LO of *Thallassiosira kolbei, Thallassiosira vulnifica* and *Thallassiosira insigna*. We concur with the placement of a hiatus at 2.80 mbsf. An additional hiatus is placed between 10.03 and 10.28 mbsf. This coincides with the placement of the second hiatus identified in this study. Harwood & Maruyama (1992) place a hiatus at 19.57–21.07 mbsf using a criterion similar to that at Hole 747A, namely the absence of a gap between the last common occurrence of *Denticulopsis dimorpha* and the FO of *Asteromphalus kennettii*. The two hiatuses recognized by Harwood & Maruyama (1992) between 38.10 and 38.57 mbsf, and 38.57 and 40.07 mbsf approximate the two older hiatuses recognized in this study. These workers define a hiatus at 57.1–57.50 mbsf based on the stratigraphical relationship between *Rossiella symmetrica*, *Azpeitia gombosi* and *Thalassiosira fraga*. In this study the chronostratigraphy of the FO of *Thalassiosira fraga* is shown to be a potentially useful marker, and there is minimal data available to assess the chronostratigraphical reliability of the other two species. Harwood & Maruyama (1992) place four additional hiatuses at 60.57–63.57, 68.57–70.07, 70.07–71.57 and 79.57–81.07 mbsf. This series of hiatuses occur within a polarity sequence which is interpreted to be continuous from the base C6Bn.2n (62.20) to the base of C15n at 119.07. Although no primary events are identified in ths interval the LO of *Reticulofenestra oamaruensis* (115.84–115.94) is used to constrain the base of the hole. Harwood & Maruyama (1992) recognize two additional hiatuses at 89.07–90.57 and 107.26–108.08 mbsf based on the stratigraphical ranges of *Rocella vigilans* var. A and *Triceratium polymorphus*. The succession of diatom markers as used by these workers is complete throughout this entire stratigraphical interval and we question the placement of these hiatuses.

Table 38. *Composite dataset for ODP Hole 748B containing the diatom, radiolarian, calcareous nannofossil and palaeomagnetic data calibrated to the GPTS of Cande & Kent (1995)*

	Event	Sample (top)	Sample (bottom)	Depth (t)	Depth (b)	Age (t)	Age (b)
	Unconformity			***	***	***	***
B	C1n			1.30	1.30	0.78	0.78
LO	*Triceraspyris antarctica*	2H-2, 45	2H-3, 0	2.05	3.10	?	?
LO	*Prunopyle titan*	2H-2, 45	2H-3, 0	2.05	3.10	?	?
LO	*Thalassiosira vulnifica*	2H-2, 120	2H-2, 140	2.07	2.80	?	?
LO	*Nitzschia barronii*	2H-2, 47-48	2H-2, 100	2.10	2.80	?	?
LO	*Nitzschia interfrigidaria*	2H-2, 120	2H-2, 140	2.80	3.00	?	?
FO	*Thalassiosira elliptopora*	2H-2, 120	2H-2, 140	2.80	3.00	?	?
LO	*Thalassiosira insigna*	2H-2, 120	2H-2, 140	2.80	3.00	?	?
LO	*Thalassiosira inura*	2H-2, 120	2H-2, 140	2.80	3.00	?	?
LO	*Thalassiosira kolbei*	2H-2, 120	2H-2, 140	2.80	3.00	?	?
	Unconformity			***	***	***	***
LO	*Cycladophora pliocenica*	2H-3, 0	2H-3, 150	3.10	4.60	?	?
LO	*Eucyrtidium calvertense*	2H-3, 0	2H-3, 0	3.10	4.60	?	?
LO	*Desmospyris spongiosa*	2H-3, 0	2H-3, 0	3.10	4.60	?	?
LO	*Helotholus vema*	2H-3, 0	2H-3, 0	3.10	4.60	?	?
FO	*Cycladophora davisiana*	2H-3, 0	2H-3, 0	3.10	4.60	?	?

(*continued*)

Table 38. (continued)

	Event	Sample (top)	Sample (bottom)	Depth (t)	Depth (b)	Age (t)	Age (b)
FO	Thalassiosira insigna	2H-3, 100	2H-3, 140	4.10	4.50	?	?
FO	Thalassiosira vulnifica	2H-3, 100	2H-3, 140	4.10	4.50	?	?
LO	Hemidiscus karstenii	2H-4, 140	2H-5, 20	6.00	6.30	?	?
LO	Simonseniella barboi	2H-4, 140	2H-5, 20	6.30	6.60	?	?
LO	Nitzschia praeinterfrigidria	2H-5, 20	2H-5, 80	6.30	6.90	?	?
FO	Helotholus vema	2H-5, 45	2H-6, 45	6.55	8.55	?	?
FO	Thalassiosira torokina	2H-5, 80	2H-5, 140	6.90	7.50	?	?
B	C2An3n			6.95	6.95	3.58	3.58
FO	Nitzschia interfrigidaria	2H-6, 40	2H-6, 80	8.00	8.40	?	?
LO	Lampromitra coronata	2H-6, 45	2H-6, 45	8.05	10.60	?	?
LO	Navicula wisei	2H-5, 140	2H-6, 140	8.40	8.80	?	?
FO	Navicula wisei	2H-6, 120	2H-7, 40	8.80	9.60	?	?
FO	Rouxia heteropolara	3H-1, 3	3H-1, 25	9.60	9.90	?	?
FO	Thalassiosira kolbei	3H-1, 3	3H-1, 25	9.60	9.90	?	?
FO	Nitzschia barronii	3H-1, 43	3H-1, 68	10.00	10.30	?	?
FO	Nitzschia praeinterfrigidria	3H-1, 43	3H-1, 68	10.00	10.30	?	?
FO	Nitzschia reinholdii	3H-1, 43	3H-1, 68	10.00	10.30	?	?
FO	Thalassiosira complicata	3H-1, 43	3H-1, 68	10.00	10.30	?	?
FO	Thalassiosira inura	3H-1, 43	3H-1, 68	10.00	10.30	?	?
FO	Thalassiosira oestrupii	3H-1, 43	3H-1, 68	10.00	10.30	?	?
LO	Asteromphalus kennettii	3H-1, 43	3H-1, 68	10.03	10.30	?	?
LO	Denticulopsis dimorpha	3H-1, 68	3H-1, 110	10.10	10.30	?	?
	Unconformity			***	***	***	***
LO	Neobrunia miriabilis	3H-2, 47-48	3H-3, 47-48	11.60	13.10	?	?
T	C4Ar.2n			13.80	13.80	9.23	9.23
T	C5n.1n			17.25	17.25	9.74	9.74
FO	Asteromphalus kennettii	3H-CC	4H-1, 47-48	19.10	19.60	?	?
LO	Cycladophora humerus	3H-CC	4H-1, 45	19.10	19.55	?	?
FO	Acrosphaera australis	4H-1, 45	4H-2, 45	19.55	21.05	?	?
FO	Eucyrtidium pseudoinflatum	4H-1, 45	4H-2, 45	19.55	21.05	?	?
FO	Cycladophora spongothorax	4H-1, 45	4H-2, 45	19.55	21.05	?	?
LO	Nitzschia denticuloides	4H-1, 47-48	4H-2, 47-48	19.60	20.10	?	?
	Unconformity			***	***	***	***
LO	Calcidiscus macintyrei			21.18	22.68	?	?
B	C5n.2n			25.10	25.10	10.95	10.95
B	C5r.1n			26.90	26.90	11.10	11.10
LO	Denticulopsis praedimorpha	5H-3, 47-48	5H-4, 47-48	32.10	33.60	11.33	11.40
T	C5r.2n			35.40	35.40	11.48	11.48
FO	Denticulopsis lauta	5H-6, 47-48	5H-7, 47-48	36.60	38.10	11.51	11.54
FO	Hemidiscus ovalis	5H-6, 47-48	5H-7, 47-48	36.60	38.10	11.51	11.54
LO	Crucidenticula kanayae	5H-6, 47-48	5H-7, 47-48	36.60	38.10	11.51	11.54
FO	Calcidiscus macintyrei	5-H6, 58-62	5-H7, 10-11	36.70	37.62	11.51	11.54
B	C5r.2n			37.60	37.60	11.53	11.53
LO	Coscinodiscus lewisianus	5H-CC	6H-1, 47-48	38.10	38.60	?	?
	Unconformity			***	***	***	***
FO	Actinocyclus ingens	5H-CC	6H-1, 47-48	38.10	38.60	?	?
FO	Denticulopsis dimorpha	5H-CC	6H-1, 47-48	38.10	38.60	?	?
FO	Denticulopsis hustedtii	5H-CC	6H-1, 47-48	38.10	38.60	?	?
FO	Denticulopsis praedimorpha	5H-CC	6H-1, 47-49	38.10	38.60	?	?
FO	Nitzschia denticuloides	5H-CC	6H-1, 47-49	38.10	38.60	?	?
LO	Nitzschia grossepunctata	5H-CC	6H-1, 47-49	38.10	38.60	?	?
FO	Dendrospyris megalocephalis	5H-CC	6H-1, 65	38.10	38.75	?	?
FO	Actinomma golownini	5H-CC	6H-1, 65	38.10	38.75	?	?
	Unconformity			***	***	***	***
FO	Denticulopsis maccolumnii	6H-1, 47-48	6H-2, 47-48	38.60	40.10	?	?
FO	Nitzschia grossepunctata	6H-1, 47-48	6H-2, 47-48	38.60	40.10	?	?
LO	Nitzschia maleinterpretaria	6H-1, 47-48	6H-2, 47-48	38.60	40.10	?	?
	Unconformity			***	***	***	***
T	C5Bn.2n			39.80	39.80	15.03	15.03
	Unconformity			***	***	***	***
LO	Denticulopsis maccolumnii	5H-CC	6H-1, 47-49	40.07	41.57	?	?
LO	Thalassiosira fraga	6H-2, 47-48	6H-3, 47-48	40.10	41.60	?	?
FO	Crucidenticula kanayae	6H-3, 47-48	6H-4, 47-48	41.60	43.10	?	?
FO	Calcidiscus leptoporus	6-H3, 58-62	6-H4, 58-62	41.68	43.68	?	?
B	C5Dn			42.00	42.00	17.62	17.62
LO	Reticulofenestra pseudoumbilica	6-H5, 58-62	6-H6, 58-62	44.70	46.70	18.02	18.29
T	C5En			46.50	46.50	18.28	18.28
B	C5En			53.69	53.69	18.78	18.78
LO	Thalassiosira spumellaroides	7H-5, 47-48	7H-6, 47-48	54.07	55.57	18.85	?

Table 38. (*continued*)

	Event	Sample (top)	Sample (bottom)	Depth (t)	Depth (b)	Age (t)	Age (b)
T	C6n			55.20	55.20	19.05	19.05
	Unconformity			***	***	***	***
FO	*Nitzschia maleinterpretaria*	7H-CC	8H-1, 47-49	57.10	57.60	?	?
T	C6A.2n			58.90	58.90	21.00	21.00
FO	*Thalassiosira fraga*	8H-2, 47-48	8H-3, 47-48	59.10	60.60	?	?
FO	*Thalassiosira spumellaroides*	8H-7, 47-48	8H-CC	60.57	63.57	?	?
LO	*Bogorovia veniamini*	8H-3, 47-48	8H-5, 47-48	60.57	63.57	?	?
LO	*Thalassiosira primalabiata*	8H-3, 47-48	8H-5, 47-48	60.57	63.57	?	?
	Unconformity			***	***	***	***
B	C6Bn.1n			62.20	62.20	22.75	22.75
B	C6Cn.1n			64.00	64.00	23.54	23.54
LO	*Reticulofenestra bisecta*	8H-7, 55-59	9H-1, 10-11	66.65	66.70	?	?
LO	*Rocella vigilans*	9H-2, 47-48	9H-3, 47-48	68.60	70.10	?	?
LO	*Lisitzina ornata*	9H-2, 47-48	9H-3, 47-48	68.60	70.10	?	?
T	C7n.1n			69.60	69.60	24.73	24.73
	Unconformity			***	***	***	***
FO	*Rocella gelida*	9H-3, 47-48	9H-6, 47-48	70.10	71.60	?	?
	Unconformity			***	***	***	***
T	C8n.1n			71.10	71.10	25.82	25.82
LO	*Chiasmolithus altus*	9H-5, 10-11	9H-6, 10-11	72.20	72.70	25.93	?
B	C8n.1n			72.40	72.40	25.95	25.95
	Unconformity			***	***	***	***
B	C11n.2n			98.80	98.80	30.10	30.10
FO	*Rocella vigilans*	12H-4, 47-49	12H-5, 47-49	100.07	101.57	30.29	30.52
T	C12n			101.29	101.29	30.48	30.48
B	C12n			104.90	104.90	30.94	30.94
FO	*Cyclicargolithus abisectus*	13H-1, 10-11	13H-2, 10-11	104.70	106.20	30.89	31.25
LO	*Isthmolithus recurvus*	13H-4, 10-11	13H-5, 10-11	109.80	113.30	32.11	32.94
T	C13n			113.80	113.80	33.06	33.06
B	C13n			114.70	114.70	33.55	33.55
LO	*Reticulofenestra oamaruensis*	14H-2, 24-26	14H-3, 24-26	115.84	115.95	33.74	33.76
T	C16n.1n			125.25	125.25	35.34	35.34

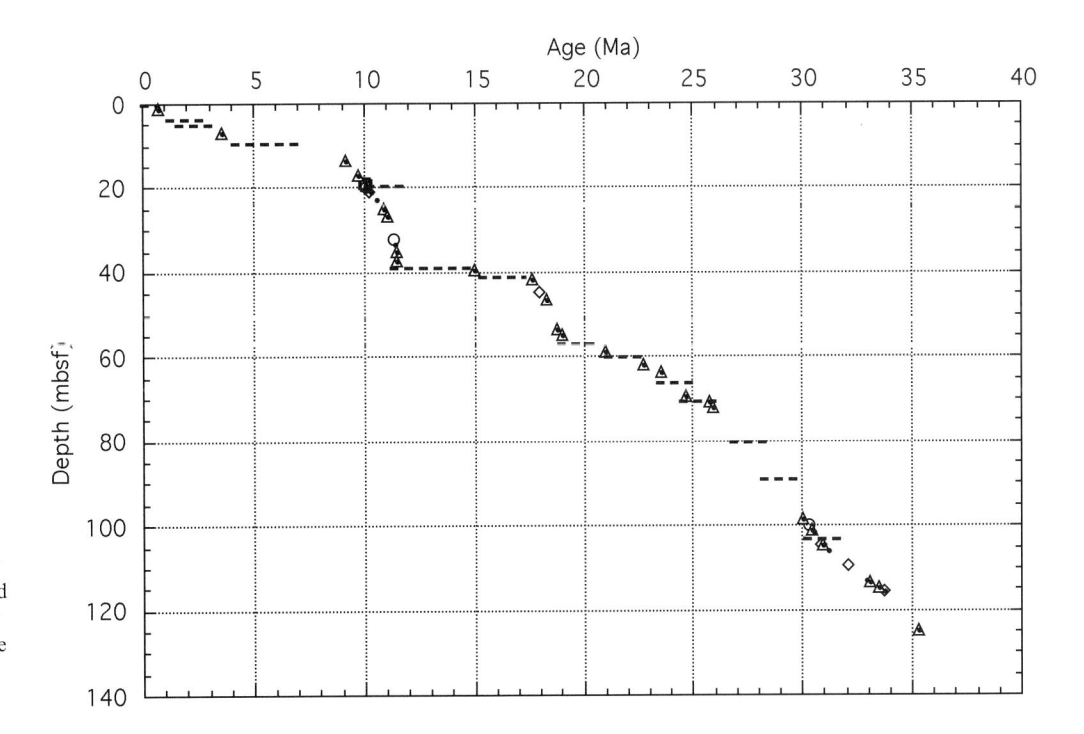

Data sets are integrated from the work of numerous authors (see Table 2).

Fig. 54. An age–depth plot for DSDP Hole 748B utilizing the previously used biostratigraphical datums and polarity reversal interpretation calibrated to the GPTS of Cande & Kent (1995). See Fig. 5 for legend of symbols.

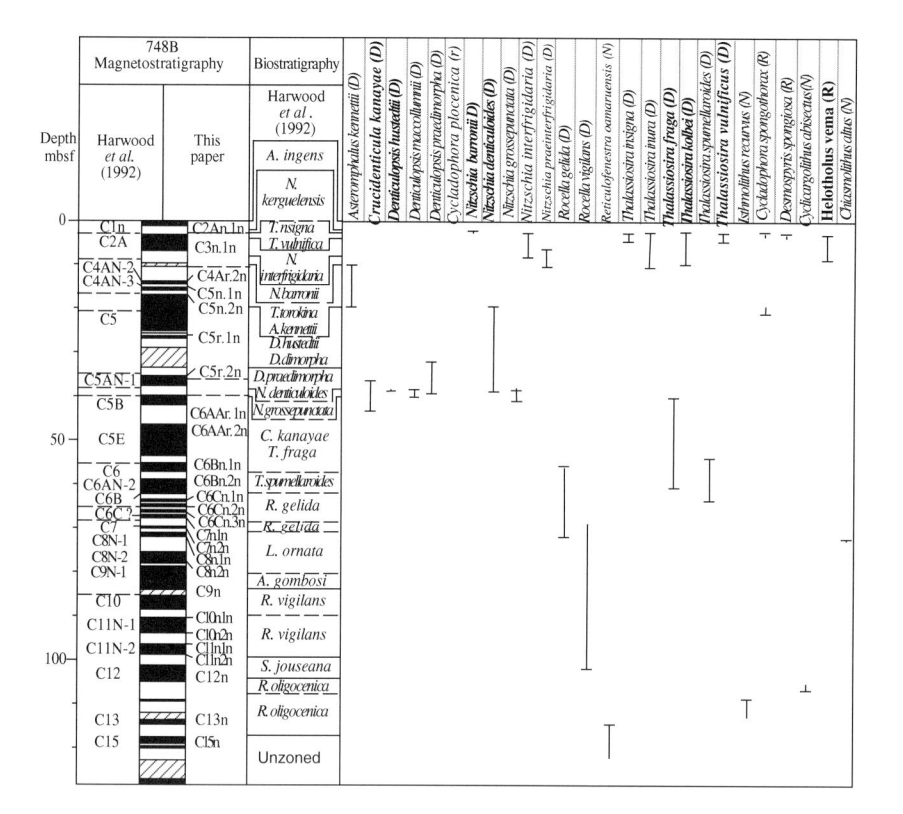

Fig. 55. Comparison of the ODP Hole 748B polarity chronozones and diatom zonal assignments by previous workers with the chronozonal assignments of this study. The stratigraphical ranges of selected stratigraphical markers are based on the original data sets (see text for references). D, diatoms; R, radiolaria; N, calcareous nannofossils. Emboldened events indicate primary biostratigraphical events recognized in this study. The polarity record shown represents the original interpretation calibrated to the GPTS of Cande & Kent (1995) and the interpretation from this study correlated with the GPTS of Cande & Kent (1995). Diagonal lines indicate intervals of uncertainty.

Table 39. *Composite dataset for ODP Hole 748B containing the diatom, radiolarian, calcareous nannofossil and palaeomagnetic data calibrated to the GPTS of Cande & Kent (1995) using the biochronological interpretation based on this study*

	Event	Sample (top)	Sample (bottom)	Depth (t)	Depth (b)	Age (t)	Age (b)
	Unconformity			***	***	***	***
B	C1n			1.30	1.30	0.78	0.78
LO	Triceraspyris antarctica	2H-2, 45	2H-3, 0	2.05	3.10	1.62	2.81
LO	Prunopyle titan	2H-2, 45	2H-3, 0	2.05	3.10	1.62	2.81
LO	Nitzschia barronii	2H-2, 47-48	2H-2, 100	2.10	2.80	1.68	2.47
LO	Nitzschia interfrigidaria	2H-2, 120	2H-2, 140	2.80	3.00	2.47	2.69
LO	Thalassiosira kolbei	2H-2, 120	2H-2, 140	2.80	3.00	2.47	2.69
LO	Thalassiosira inura	2H-2, 120	2H-2, 140	2.80	3.00	2.47	2.69
LO	Thalassiosira insigna	2H-2, 120	2H-2, 140	2.80	3.00	2.47	2.69
LO	Thalassiosira vulnifica	2H-2, 120	2H-2, 140	2.80	3.00	2.47	2.69
T	C2An.1n			2.90	2.90	2.58	2.58
FO	Thalassiosira insigna	2H-2, 120	2H-2, 140	4.10	4.50	3.09	3.26
FO	Thalassiosira vulnifica	2H-2, 120	2H-2, 140	4.10	4.50	3.09	3.26
LO	Cycladophora pliocenica	2H-3, 0	2H-3, 150	3.10	4.60	2.66	3.30
LO	Eucyrtidium calvertense	2H-3, 0	2H-3, 0	3.10	4.60	2.66	3.30
LO	Desmospyris spongiosa	2H-3, 0	2H-3, 0	3.10	4.60	2.66	3.30
LO	Helotholus vema	2H-3, 0	2H-3, 0	3.10	4.60	2.66	3.30
FO	Cycladophora davisiana	2H-3, 0	2H-3, 0	3.10	4.60	2.66	3.30
LO	Hemidiscus karstenii	2H-4, 140	2H-5, 20	6.00	6.30	3.89	4.02
LO	Simonseniella barboi	2H-4, 140	2H-5, 20	6.30	6.60	4.02	4.15
LO	Nitzschia praeinterfrigidaria	2H-5, 20	2H-5, 80	6.30	6.90	4.02	4.27
FO	Helotholus vema	2H-5, 45	2H-6, 45	6.55	8.05	4.12	4.75
FO	Thalassiosira torokina	2H-5, 80	2H-5, 140	6.90	7.50	4.27	?
B	C3n.1n			6.95	6.95	4.29	4.29
FO	Nitzschia interfrigidaria	2H-6, 40	2H-6, 80	8.00	8.40	?	?
LO	Lampromitra coronata	2H-6, 45	2H-6, 45	8.05	9.60	?	?
LO	Navicula wisei	2H-5, 140	2H-6, 140	8.40	8.80	?	?
FO	Navicula wisei	2H-6, 120	2H-7, 40	8.80	9.60	?	?
FO	Rouxia heteropolara	3H-1, 3	3H-1, 25	9.60	9.90	?	?
FO	Thalassiosira kolbei	3H-1, 3	3H-1, 25	9.60	9.90	?	?
FO	Nitzchia barronii	3H-1, 43	3H-1, 68	10.00	10.30	?	?
FO	Nitzchia praeinterfrigidaria	3H-1, 43	3H-1, 68	10.00	10.30	?	?
FO	Thalassiosira complicata	3H-1, 43	3H-1, 68	10.00	10.30	?	?
FO	Thalassiosira inura	3H-1, 43	3H-1, 68	10.00	10.30	?	?
FO	Thalassiosira oestrupii	3H-1, 43	3H-1, 68	10.00	10.30	?	?
LO	Asteromphalus kennettii	3H-1, 43	3H-1, 67	10.03	10.30	?	?
LO	Denticulopsis dimorpha	3H-1, 68	3H-1, 110	10.10	10.30	?	?

Table 39. (continued)

	Event	Sample (top)	Sample (bottom)	Depth (t)	Depth (b)	Age (t)	Age (b)
	Unconformity			***	***	***	***
LO	Neobrunia miriabilis	3H-2, 47-48	3H-3, 47-48	11.60	13.10	?	?
T	C4Ar.2n			13.80	13.80	9.58	9.58
B	C4Ar.2n			14.57	14.57	9.64	9.64
T	C5n.1n			14.80	14.80	9.74	9.74
B	C5n.1n			16.10	16.10	9.88	9.88
T	C5n.2n			16.98	16.98	9.92	9.92
FO	Asteromphalus kennettii	3H-CC	4H-1, 47-48	19.10	19.60	10.19	10.25
LO	Cycladophora humerus	3H-CC	4H-1, 45	19.10	19.55	10.19	10.25
FO	Acrosphaera australis	4H-1, 45	4H-2, 45	19.55	21.05	10.25	10.44
FO	Eucyrtidium pseudoinflatum	4H-1, 45	4H-2, 45	19.55	21.05	10.25	10.44
FO	Cycladophora spongothorax	4H-1, 45	4H-2, 45	19.55	21.05	10.25	10.44
LO	Nitzschia denticuloides	4H-1, 47-48	4H-2, 47-48	19.60	20.10	10.25	10.32
FO	Eucyrtidium pseudoinflatum	4H-2, 45	4H-3, 45	21.05	22.55	10.44	10.63
LO	Calcidiscus macintyrei			21.18	22.68	10.45	10.64
B	C5n.2n			25.10	25.10	10.95	10.95
T	C5r.1n			25.60	25.60	11.05	11.05
B	C5r.1n			26.90	26.90	11.10	11.10
LO	Denticulopsis praedimorpha	5H-3, 47-48	5H-4, 47-48	32.10	33.60	11.33	11.40
T	C5r.2n			35.40	35.40	11.48	11.48
FO	Denticulopsis lauta	5H-6, 47-48	5H-7, 47-48	36.60	38.10	11.48	?
FO	Hemidiscus ovalis	5H-6, 47-48	5H-7, 47-48	36.60	38.10	11.48	?
LO	Crucidenticula kanayae	5H-6, 47-48	5H-7, 47-48	36.60	38.10	11.48	?
FO	Calcidiscus macintyrei	5-H6, 58-62	5-H7, 10-11	36.70	37.62	11.49	11.51
B	C5r.2n			37.60	37.60	11.53	11.53
	Unconformity			***	***	***	***
LO	Coscinodiscus lewisianus	5H-CC	6H-1, 47-48	38.10	38.57	?	?
FO	Actinocyclus ingens	5H-CC	6H-1, 47-48	38.10	38.60	?	?
FO	Denticulopsis dimorpha	5H-CC	6H-1, 47-48	38.10	38.60	?	?
FO	Denticulopsis hustedtii	5H-CC	6H-1, 47-48	38.10	38.60	?	?
LO	Denticulopsis maccolumnii	5H-CC	6H-1, 47-49	38.10	38.60	?	?
FO	Denticulopsis praedimorpha	5H-CC	6H-1, 47-49	38.10	38.60	?	?
FO	Nitzschia denticuloides	5H-CC	6H-1, 47-49	38.10	38.60	?	?
LO	Nitzschia grossepunctata	5H-CC	6H-1, 47-49	38.10	38.60	?	?
FO	Dendrospyris megalocephalis	5H-CC	6H-1, 65	38.10	38.75	?	?
FO	Actinomma golownini	5H-CC	6H-1, 65	38.10	38.75	?	?
LO	Nitzschia maleinterpretaria	6H-1, 47-48	6H-2, 47-48	38.60	40.10	?	?
FO	Nitzschia grossepunctata	6H-1, 47-48	6H-2, 47-48	38.60	40.10	?	?
B	Magnetic reversal :- unidentified			39.80	39.80	?	?
FO	Denticulopsis maccollumnii	6H-1, 47-48	6H-2, 47-48	40.07	41.57	?	?
	Unconformity			***	***	***	***
LO	Thalassiosira fraga			40.10	41.60	?	?
FO	Crucidenticula kanayae	6H-3, 47-48	6H-4, 47-48	41.60	43.10	?	?
FO	Calcidiscus leptoporus	6H-3, 58-62	6H-4, 58-62	41.68	43.68	?	?
B	C6AAr.1n			42.00	42.00	22.25	22.25
LO	Reticulofenestra pseudoumbilica	6H-5, 58-62	6H-6, 58-62	44.70	46.70	22.50	22.46
T	C6AAr.2n			46.50	46.50	22.46	22.46
B	C6AAr.2n			53.69	53.69	22.49	22.49
LO	Thalassiosira spumellaroides	7H-5, 47-48	7H-6, 47-48	54.07	55.57	22.52	22.62
T	C6Bn.1n			55.20	55.20	22.59	22.59
LO	Rocella gelida	7H-6, 47-48	7H-CC	55.70	57.10	22.63	22.73
FO	Nitzschia maleinterpretaria	8H-1, 47-48	8H-2, 47-48	57.10	57.60	22.73	22.76
B	C6Bn.1n			57.40	57.40	22.75	22.75
T	C6Bn.2n			58.90	58.90	22.80	22.80
FO	Thalassiosira fraga	8H-2, 47-48	8H-3, 47-48	59.10	60.60	22.82	22.93
FO	Thalassiosira spumellaroides	8H-7, 47-48	8H-CC	60.57	63.57	22.94	23.38
LO	Bogorovia veniamini	8H-7, 47-48	8H-CC	60.57	63.57	22.94	23.38
LO	Thalassiosira primalabiata	8H-7, 47-48	8H-CC	60.57	63.57	22.94	23.38
B	C6Bn.2n			62.20	62.20	23.07	23.07
T	C6Cn.1n			63.48	63.48	23.35	23.35
B	C6Cn.1n			64.00	64.00	23.54	23.54
T	C6Cn.2n			64.50	64.50	23.68	23.68
B	C6Cn.2n			65.20	65.20	23.80	23.80
T	C6Cn.3n			65.70	65.70	24.00	24.00
B	C6Cn.3n			66.35	66.35	24.12	24.12
LO	Reticulofenestra bisecta	8H-7, 55-59	9H-1, 10-11	66.65	66.70	24.53	24.59
T	C7n.1n			66.80	66.80	24.73	24.73
B	C7n.1n			67.70	67.70	24.78	24.78

(continued)

Table 39. (*continued*)

	Event	Sample (top)	Sample (bottom)	Depth (t)	Depth (b)	Age (t)	Age (b)
T	C7n.2n			69.60	69.60	24.84	24.84
B	C7n.2n			70.07	70.07	25.18	25.18
FO	*Rocella gelida*	9H-3, 47-48	9H-4, 47-48	70.10	71.60	25.20	25.87
T	C8n.1n			71.10	71.10	25.82	25.82
LO	*Chiasmolithus altus*	9H-5, 10-11	9H-6, 10-11	72.20	72.70	25.93	25.95
B	C8n.1n			72.40	72.40	25.95	25.95
T	C8n.2n			75.44	75.44	25.99	25.99
B	C8n.2n			78.06	78.06	26.55	26.55
T	C9n			78.71	78.71	27.03	27.03
B	C9n			88.50	88.50	27.97	27.97
T	C10n.1n			90.40	90.40	28.28	28.28
B	C10n.2n			93.90	93.90	28.75	28.75
T	C11n.1n			96.40	96.40	29.40	29.40
B	C11n.2n			98.80	98.80	30.10	30.10
FO	*Rocella vigilans*	12H-4, 47-49	12H-5, 47-49	100.07	101.57	30.29	30.52
T	C12n			101.29	101.29	30.48	30.48
B	C12n			104.90	104.90	30.94	30.94
FO	*Cyclicargolithus abisectus*	13H-1, 10-11	13H-2, 10-11	104.70	106.20	30.89	31.25
LO	*Isthmolithus recurvus*	13H-4, 10-11	13H-5, 10-11	109.80	113.30	32.11	32.94
T	C13n			113.80	113.80	33.06	33.06
B	C13n			114.70	114.70	33.55	33.55
LO	*Reticulofenestra oamaruensis*	14H-2, 24-26	14H-3, 24-26	115.84	115.95	33.99	34.03
T	C15n			117.60	117.60	34.66	34.66
B	C15n			119.07	119.07	34.94	34.94

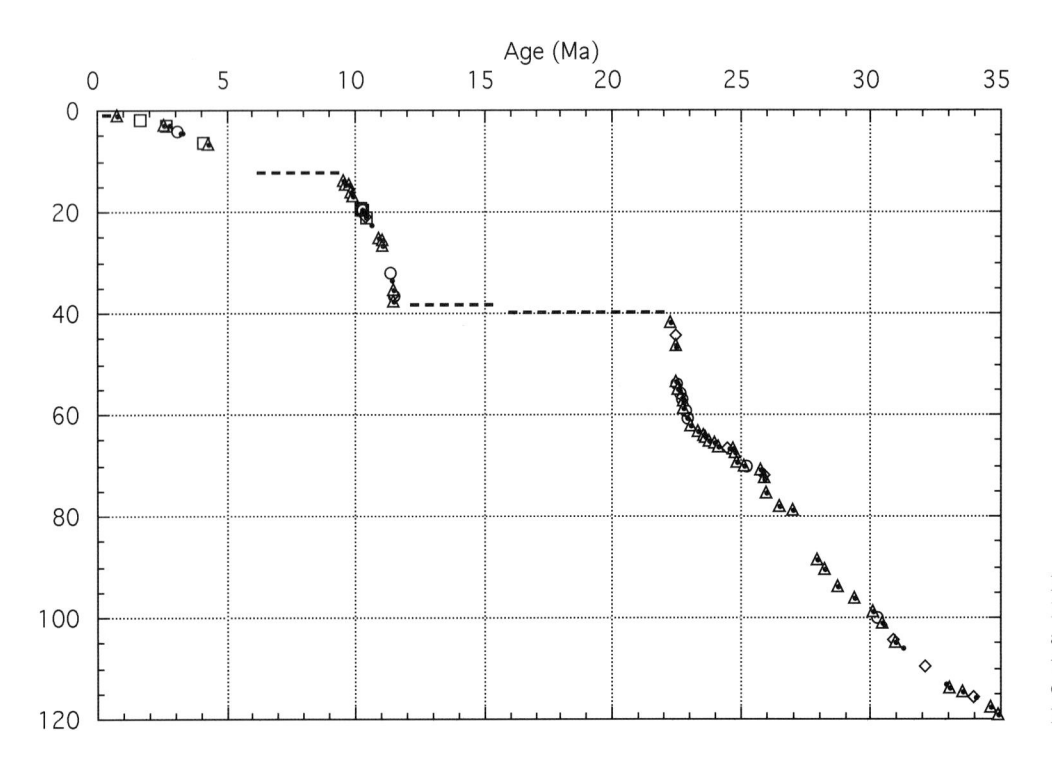

Data sets are integrated from the work of numerous authors (see Table 2).

Fig. 56. An age–depth plot for DSDP Hole 748B utilizing the biostratigraphical and polarity reversal events identified in this study and calibrated to the GPTS of Cande & Kent (1995). See Fig. 5 for legend of symbols.

Hole 751A

Eighteen cores were recovered from a total penetration of 166.2 mbsf representing a recovery of 98% of the cored interval. Sediments recovered consist of Quaternary and lower Pliocene diatom ooze with chert and Miocene calcareous nannofossil ooze (Wise *et al.* 1989). A partial palaeomagnetic polarity record was obtained from the sequence. The interval from 2–25 and 70–98 mbsf is difficult to interpret as a result of the lack of data and scatter (Wise *et al.* 1989; Heider *et al.* 1992). All three microfossil groups were recorded from this hole. Diatoms are generally common and well preserved. Two intervals, one in the upper middle Miocene and the other in the lower Miocene contain specimens exhibiting poor preservation (Wise *et al.* 1989). Calcareous nannofossils are absent or rare in the Quaternary and Pliocene and are abundant with moderate preservation for the Miocene (Wei & Wise 1992*a*). Radiolarians are abundant and well preserved for the Quaternary to middle Miocene and abundant and poorly preserved in the lower Miocene (Wise *et al.* 1989). The integrated biostratigraphical and magnetostratigraphical data previously used (palaeomagnetics: Heider *et al.* 1992; Harwood *et al.* 1992; diatoms: Harwood & Maruyama, 1992; calcareous nannofossils: none; radiolarians: Abelmann 1992; Lazarus 1992) calibrated to the GPTS of Berggren *et al.* (1985*a,b*) are shown in Table 40. An age–depth graph for Hole 120-751A using this data is shown in Fig. 57.

Figure 58 compares the polarity chronozones and diatom zones assigned by previous workers (Heider *et al.* 1992; Harwood *et al.* 1992; diatoms: Harwood & Maruyama 1992) with the chronozonal and diatom zonal assignments used in this study. In addition, this figure illustrates the stratigraphical ranges of diatoms and radiolarians. The stratigraphical ranges reflect the placement of FO and LO of stratigraphically useful species as recognized by Abelmann (1992) and Lazarus (1992), respectively. The revised chronozones and diatom zones illustrated are based on reassessing the data sets of these workers (Table 41). Comparison of the chronozonal assignments of Harwood *et al.* (1992) show minimal differences reflecting the GPTS used in each study.

Figure 59 shows an age–depth graph for Hole 120-51A using the stratigraphical markers determined in this study. One hiatus is placed in the section between approximately 39 and 42 mbsf. The upper boundary of this break is constrained by the FO of *Nitzsciha praeinterfrigidaria* at 34.80–35.80 mbsf and the top of C3n.3n at 36.5 mbsf. The lower boundary is constrained by a continuous polarity sequence from the base of C4An at 41 mbsf to the base of C5Cn.2n at 134.35 mbsf. Two primary events the FO of *Nitzschia denticuloides* and *Denticulopsis hustedtii* occur within this sequence.

Harwood & Maruyama (1992) recognize seven hiatuses within the sequence. The youngest is placed between 0 and 0.1 mbsf. The second hiatus is placed between 1.05 and 1.47 mbsf. A third hiatus between 5.71 and 6.30 mbsf is based on the ranges of *Thallassiosira vulnifica*, *Thallassiosira insigna* and *Thallassiosira kolbei*. We cannot assess the placement of this hiatus given the current data set. A fourth hiatus occurs between 34.8 and 35.75 mbsf based on the ranges of *Nizschia barronii*, *Thallassiosira oestrupii* and *Asteromphalus kennettii*. Based on our interpretation of the data we use the FO of *Nizschia barronii* as a primary stratigraphical marker and question the placement of this hiatus. The fourth hiatus of Harwood & Maruyama (1992) between 42.7 and 43.75 mbsf approximates the placement of our hiatus at 40 mbsf. The hiatus between 103.75 and 105.25 mbsf is based on the range of *Nitzschia grossepunctata*. This hiatus is placed at a depth interval well constrained by our study with both biostratigraphical and palaeomagnetic data. The oldest unconformity between 129.25 and 130.75 mbsf corresponds to an interval with a continuous polarity sequence constrained by the first occurrence of *Crucidenticula kanayae* (137.70–138.80 mbsf)

Table 40. *Composite dataset for ODP Hole 751A containing the diatom, radiolarian, calcareous nannofossil and palaeomagnetic data calibrated to the GPTS of Cande & Kent (1995)*

	Event	Sample (top)	Sample (bottom)	Depth (t)	Depth (b)	Age (t)	Age (b)
	Unconformity			***	***	***	***
LO	*Nitzschia barronii*	1H-1, 105-106	1H-1, 145-146	1.10	1.50	?	?
	Unconformity			***	***	***	***
LO	*Thalassiosira elliptopora*	1H-1, 145-146	1H-3, 10-11	1.50	3.30	?	?
LO	*Antarctissa cylindrica*	1H-1, 150	1H-2, 106	1.50	2.56	?	?
LO	*Triceraspyris antarctica*	1H-2, 106	1H-3, 45	2.56	3.45	?	?
LO	*Cycladophora pliocenica*	2H-1, 98	2H-2, 98	5.68	7.18	?	?
LO	*Eucyrtidium calvertense*	2H-1, 98	2H-2, 98	5.68	7.18	?	?
LO	*Helotholus vema*	2H-1, 98	2H-2, 98	5.68	7.18	?	?
LO	*Thalassiosira insigna*	2H-1, 105-107	2H-2, 10-11	5.70	6.30	?	?
LO	*Thalassiosira vulnifica*	2H-1, 105-107	2H-2, 10-11	5.70	6.30	?	?
	Unconformity			***	***	***	***
FO	*Actinocyclus actinochilus*	2H-2, 10-12	2H-2, 105-106	6.30	7.30	?	?
LO	*Nitzschia interfrigidaria*	2H-3, 10-12	2H-3, 105-106	6.30	7.30	?	?
FO	*Nitzschia kerguelensis*	2H-2, 10-11	2H-2, 105-106	6.30	7.30	?	?
LO	*Nitzschia weaveri*	2H-2, 10-11	2H-2, 105-106	6.30	7.30	?	?
LO	*Simonseniella barboi*	2H-2, 10-11	2H-2, 105-106	6.30	7.30	?	?
LO	*Thalassiosira inura*	2H-2, 10-11	2H-2, 105-106	6.30	7.30	?	?
LO	*Thalassiosira kolbei*	2H-2, 10-11	2H-2, 105-106	6.30	7.30	?	?
LO	*Desmospyris spongiosa*	2H-2, 98	2H-3, 98	7.18	8.68	?	?
FO	*Cycladophora davisiana*	2H-2, 98	2H-3, 98	7.18	8.68	?	?
LO	*Thalassiosira complicata*	2H-3, 10-11	2H-3, 105-106	7.80	8.80	?	?
FO	*Thalassiosira vulnifica*	2H-3, 10-11	2H-3, 105-106	7.80	8.80	?	?
LO	*Nitzschia praeinterfrigidaria*	2H-3, 10-12	2H-3, 105-106	7.81	8.76	?	?
FO	*Nitzschia weaveri*	2H-3, 105-106	2H-4, 10-11	8.80	9.30	?	?
FO	*Thalassiosira elliptopora*	2H-4, 145-146	2H-CC	9.30	14.20	?	?
FO	*Thalassiosira insigna*	2H-CC	3H-1, 47-48	14.20	15.30	?	?
LO	*Navicula wisei*	3H-1, 47-48	3H-1, 105-106	14.70	15.30	?	?
LO	*Lampromitra coronata*	2H-CC	3H-1, 98	14.80	15.80	?	?
FO	*Thalassiosira torokina*	3H-2, 10-11	3H-2, 105-106	15.80	16.80	?	?

(*continued*)

Table 40. (continued)

	Event	Sample (top)	Sample (bottom)	Depth (t)	Depth (b)	Age (t)	Age (b)
LO	*Prunopyle titan*	3H-3, 98	3H-4, 98	18.18	19.68	?	?
FO	*Nitzschia interfrigidaria*	3H-4, 105-106	3H-5, 10-11	19.80	20.30	?	?
LO	*Denticulopsis dimorpha*	3H-6, 10-11	3H-CC	21.80	23.70	?	?
FO	*Nitzschia barronii*	3H-CC	4H-2, 105-106	23.70	26.30	?	?
FO	*Thalassiosira kolbei*	4H-2, 105-106	4H-3, 10-11	26.30	26.80	?	?
FO	*Navicula wisei*	4H-4, 10-11	4H-4, 105-106	28.30	29.30	?	?
FO	*Actinocyclus ingens*	4H-CC	5H-1, 10-11	33.20	33.30	?	?
FO	*Rouxia heteropolara*	5H-1, 10-11	5H-1, 105-106	3.30	34.30	?	?
T	C3n.3n			34.50	34.50	4.89	4.89
	Unconformity			***	***	***	***
FO	*Thalassiosira oestrupii*	5H-2, 10-11	5H-2, 105-106	34.80	35.80	?	?
FO	*Thalassiosira inura*	5H-2, 105-106	5H-3, 10-11	34.80	35.80	?	?
FO	*Nitzschia praeinterfrigidaria*			34.80	35.80	?	?
T	C3An.2n			36.50	36.50	6.27	6.27
LO	*Asteromphalus kennettii*	5H-4, 10-11	5H-4, 105-106	37.80	38.75	6.42	6.47
LO	*Thalassiosira miocenica*	5H-4, 105-106	5H-5, 10-11	38.80	39.30	6.47	6.50
LO	*Reticulofenestra hesslandii*	5H-5, 5-7	5H-6, 5-7	39.25	40.75	6.49	6.55
FO	*Hemidiscus ovalis*	5H-5, 10-11	5H-CC	39.30	42.70	6.49	6.65
FO	*Nitzschia praecurta*	5H-5, 10-11	5H-CC	39.30	42.70	6.49	6.65
FO	*Nitzschia reinholdii*	5H-5, 10-11	5H-CC	39.30	42.70	6.49	6.65
FO	*Thalassiosira miocenica*	5H-5, 10-11	5H-CC	39.30	42.70	6.49	6.65
B	C3An.2n			41.00	41.00	6.57	6.57
	Unconformity			***	***	***	***
FO	*Acrosphaera labrata*	5H-5, 10-11	6H-1, 53	42.70	43.23	?	?
T	C4Ar.1n			43.60	43.60	9.23	9.23
LO	*Cycladophora spongothorax*	6H-1, 98	6H-2, 50	43.68	44.70	9.23	9.30
B	C4Ar.1n			44.90	44.90	9.31	9.31
T	C4Ar.2n			45.20	45.20	9.58	9.58
FO	*Asteromphalus kennettii*	6H-4, 105-106	6H-5, 105-106	48.30	49.60	9.63	9.69
B	C4Ar.2n			48.60	48.60	9.64	9.64
T	C5n.1n			50.50	50.50	9.74	9.74
T	C5n.2n			51.88	51.88	9.92	9.92
LO	*Reticulofenestra pseudoumbilica*	7H-4, 5-7	7H-5, 5-7	56.80	58.30	10.14	10.21
FO	*Acrosphaera australis*	8H-3, 98	8H-4, 98	65.68	67.18	10.54	10.61
LO	*Cycladophora humerus*	8H-CC	9H-1, 98	71.20	72.28	10.79	10.84
FO	*Eucyrtidium pseudoinflatum*	8H-CC	9H-1, 98	71.20	72.28	10.79	10.84
LO	*Actinomma golownini*	9H-2, 98	9H-3, 98	73.68	75.18	10.90	10.97
LO	*Denticulopsis praedimorpha*	11H-2, 105-107	11H3, 105-107	92.80	94.30	11.77	11.84
LO	*Nitzschia denticuloides*	11H-2, 105-107	11H3, 105-107	94.25	95.75	11.83	11.90
FO	*Cycladophora spongothorax*	11H-6, 98	11H-CC	98.68	99.70	12.03	12.08
B	C5An.1n			99.70	99.70	12.08	12.08
FO	*Denticulopsis dimorpha*	12H-1, 105-107	12H-2, 105-107	100.80	102.30	12.15	?
LO	*Actinocyclus ingens v. nodus*	12H-1, 105-107	12H-2, 105-107	101.00	102.30	12.17	?
T	C5An.2n			101.20	101.20	12.18	12.18
B	C5An.2n			102.26	102.26	12.40	12.40
LO	*Denticulopsis maccollumnii*	12H-2, 105-107	12H-3, 105-107	102.30	103.80	?	?
T	C5Ar.2n			103.75	103.75	12.78	12.78
FO	*Denticulopsis praedimorpha*	12H-3, 105-107	12H-4, 105-107	103.75	105.25	?	?
LO	*Nitzschia grossepunctata*	12H-3, 105-107	12H-4, 105-107	103.75	105.25	?	?
	Unconformity			***	***	***	***
T	C5AAn			104.22	104.22	12.99	12.99
B	C5ABn			108.20	108.20	13.51	13.51
FO	*Actinomma golownini*	12H-CC 13H-CC	109.20	118.70	13.66	14.40	FO
	Nitzschia denticuloides	12H-CC	13H-1, 105-107	109.30	110.30	13.67	13.84
T	C5ACn			109.48	109.48	13.70	13.70
B	C5ACn			111.75	111.75	14.08	14.08
FO	*Denticulopsis hustedtii*	13H-2, 105-107	13H-2, 105-107	111.80	113.30	14.08	14.15
T	C5ADn			113.84	113.84	14.18	14.18
FO	*Actinocyclus ingens v. nodus*	13H-5, 105-107	13H-6, 105-107	116.30	117.80	14.29	14.36
FO	*Crucidenticula nicobarica*	13H-CC	14H-1, 105-107	118.80	119.80	14.41	14.45
LO	*Synedra jouseana*			121.25	122.75	14.52	14.59
B	C5ADn			123.30	123.30	14.61	14.61
T	C5Bn.1n			127.20	127.20	14.80	14.80
FO	*Denticulopsis lauta*	14H-CC	15H-1, 105-107	128.20	129.30	?	?
	Unconformity			***	***	***	***
FO	*Nitzschia grossepunctata*	15H-1, 105-107	15H-2, 105-107	129.30	130.80	?	?
B	C5Bn.2n			129.51	129.51	15.16	15.16
T	C5Cn.1n			132.56	132.56	16.01	16.01
FO	*Calcidiscus macintyrei*	15H-2, 5-6	15H-3, 55-56	132.75	137.72	16.03	16.53

Table 40. (*continued*)

	Event	Sample (top)	Sample (bottom)	Depth (t)	Depth (b)	Age (t)	Age (b)
B	C5Cn.1n			134.35	134.35	16.29	16.29
T	C5Cn.2n			136.02	136.02	16.33	16.33
LO	*Nitzschia maleinterpretaria*	15H-6, 105-107	15H-CC	136.80	137.70	16.42	16.53
FO	*Denticulopsis maccollumnii*	15H-CC	16H-1, 105-107	137.70	138.80	16.53	16.66
LO	*Crucidenticula kanayae*	15H-CC	16H-1, 105-107	137.70	138.80	16.53	16.66
FO	*Calcidiscus leptoporus*	16H-7, 2-4	17H-1, 22-25	146.70	147.40	17.58	17.67
B	C5Dn			147.00	147.00	17.62	17.62
FO	*Crucidenticula kanayae*	16H-CC	17H-2, 105-107	147.20	149.80	17.64	17.94
FO	*Nitzschia maleinterpretaria*	16H-CC	17H-2, 105-107	147.20	149.80	17.64	17.94
T	C5En			152.75	152.75	18.28	18.28
LO	*Thalassiosira fraga*	17H-6, 105-107	17H-CC	165.30	166.20	?	?

Data sets are integrated from the work of numerous authors (see Table 2).

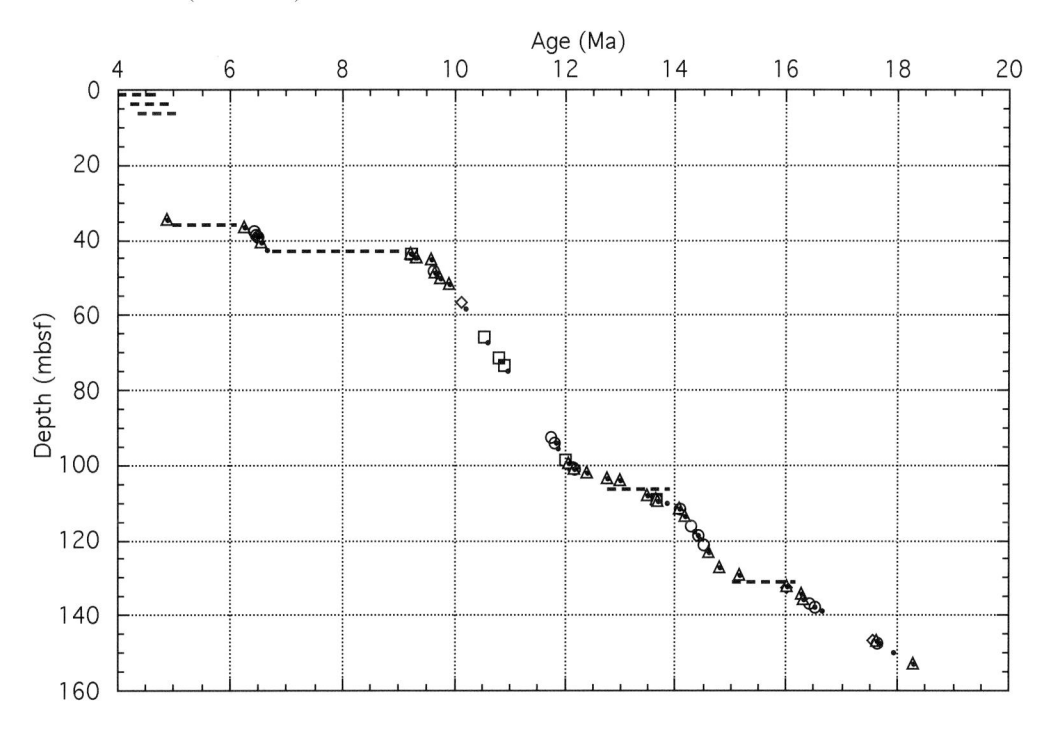

Fig. 57. An age–depth plot for DSDP Hole 751A utilizing the previously used biostratigraphical datums and polarity reversal interpretation calibrated to the GPTS of Cande & Kent (1995). See Fig. 5 for legend of symbols.

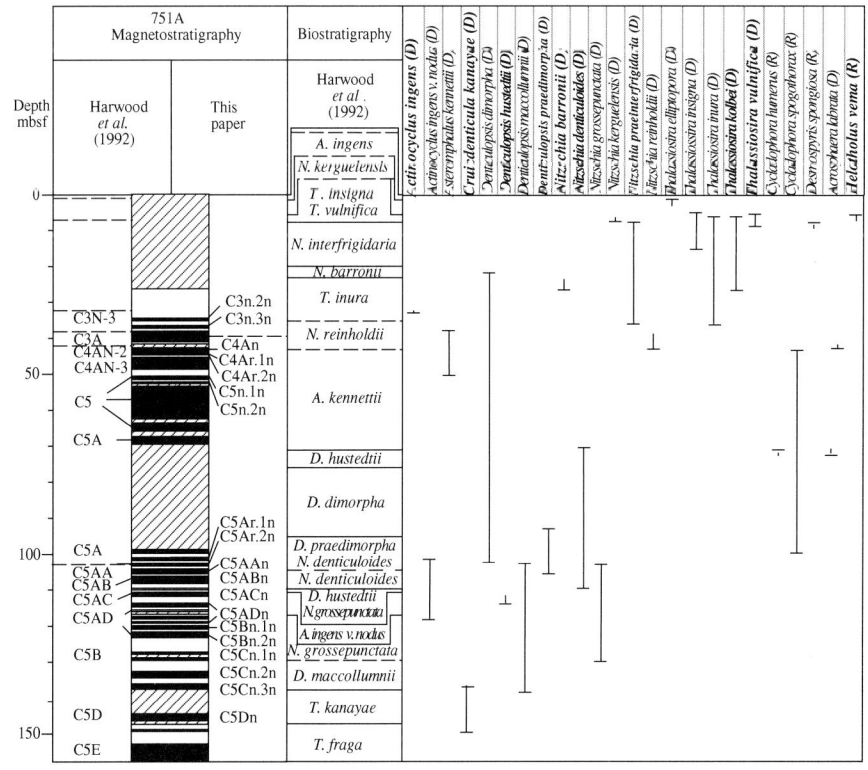

Fig. 58. Comparison of the ODP Hole 751A polarity chronozones and diatom zonal assignments by previous workers with the chronozonal assignments of this study. The stratigraphical ranges of selected stratigraphical marker are based on the original data sets (see text for references). D, diatoms; R, radiolaria; N, calcareous nannofossils. Emboldened events indicate primary biostratigraphical events recognized in this study. The polarity record shown represents the original interpretation calibrated to the GPTS of Cande & Kent (1995) and the interpretation from this study correlated with the GPTS of Cande & Kent (1995). Diagonal lines indicate intervals of no core recovery, dashed lines indicate intervals of uncertainty.

Table 41. *Composite dataset for ODP Hole 751A containing the diatom, radiolarian, calcareous nannofossil and palaeomagnetic data calibrated to the GPTS of Cande & Kent (1995) using the biochronological interpretation based on this study*

	Event	Sample (top)	Sample (bottom)	Depth (t)	Depth (b)	Age (t)	Age (b)
LO	*Nitzschia barronii*	1H-1, 105-106	1H-1, 145-146	1.10	1.50	?	?
LO	*Thalassiosira elliptopora*	1H-1, 145-146	1H-3, 10-11	1.50	3.30	?	?
LO	*Antarctissa cylindrica*	1H-1, 150	1H-2, 106	1.50	2.56	?	?
LO	*Triceraspyris antarctica*	1H-2, 106	1H-3, 45	2.56	3.45	?	?
LO	*Cycladophora pliocenica*	2H-1, 98	2H-2, 98	5.68	7.18	?	?
LO	*Eucyrtidium calvertense*	2H-1, 98	2H-2, 98	5.68	7.18	?	?
LO	*Helotholus vema*	2H-1, 98	2H-2, 98	5.68	7.18	?	?
LO	*Thalassiosira insigna*	2H-1, 105-107	2H-2, 10-11	5.70	6.30	?	?
LO	*Thalassiosira vulnifica*	2H-1, 105-107	2H-2, 10-11	5.70	6.30	?	?
FO	*Actinocyclus actinochilus*	2H-2, 10-12	2H-2, 105-106	6.30	7.30	?	?
LO	*Nitzschia interfrigidaria*	2H-3, 10-12	2H-3, 105-106	6.30	7.30	?	?
FO	*Nitzschia kerguelensis*	2H-2, 10-11	2H-2, 105-106	6.30	7.30	?	?
LO	*Nitzschia weaveri*	2H-2, 10-11	2H-2, 105-106	6.30	7.30	?	?
LO	*Simonseniella barboi*	2H-2, 10-11	2H-2, 105-106	6.30	7.30	?	?
LO	*Thalassiosira inura*	2H-2, 10-11	2H-2, 105-106	6.30	7.30	?	?
LO	*Thalassiosira kolbei*	2H-2, 10-11	2H-2, 105-106	6.30	7.30	?	?
LO	*Desmospyris spongiosa*	2H-2, 98	2H-3, 98	7.18	8.68	?	?
FO	*Cycladophora davisiana*	2H-2, 98	2H-3, 98	7.18	8.68	?	?
LO	*Thalassiosira complicata*	2H-3, 10-11	2H-3, 105-106	7.80	8.80	?	?
FO	*Thalassiosira vulnifica*	2H-3, 10-11	2H-3, 105-106	7.80	8.80	?	?
LO	*Nitzschia praeinterfrigidaria*	2H-3, 10-12	2H-3, 105-106	7.81	8.76	?	?
FO	*Nitzschia weaveri*	2H-3, 105-106	2H-4, 10-11	8.80	9.30	?	?
FO	*Thalassiosira elliptopora*	2H-4, 145-146	2H-CC	9.30	14.20	?	?
FO	*Thalassiosira insigna*	2H-CC	3H-1, 47-48	14.20	15.30	?	?
LO	*Navicula wisei*	3H-1, 47-48	3H-1, 105-106	14.70	15.30	?	?
LO	*Lampromitra coronata*	2H-CC	3H-1, 98	14.80	15.80	?	?
FO	*Thalassiosira torokina*	3H-2, 10-11	3H-2, 105-106	15.80	16.80	?	?
LO	*Prunopyle titan*	3H-3, 98	3H-4, 98	18.18	19.68	?	?
FO	*Nitzschia interfrigidaria*	3H-4, 105-106	3H-5, 10-11	19.80	20.30	?	?
LO	*Denticulopsis dimorpha*	3H-6, 10-11	3H-CC	21.80	23.70	?	?
FO	*Nitzschia barronii*	3H-CC	4H-2, 105-106	23.70	26.30	?	?
FO	*Thalassiosira kolbei*	4H-2, 105-106	4H-3, 10-11	26.30	26.80	?	?
FO	*Navicula wisei*	4H-4, 10-11	4H-4, 105-106	28.30	29.30	?	?
FO	*Actinocyclus ingens*	4H-CC	5H-1, 10-11	33.20	33.30	?	?
FO	*Rouxia heteropolara*	5H-1, 10-11	5H-1, 105-106	33.30	34.30	?	?
T	C3n.2n			34.50	34.50	?	?
FO	*Thalassiosira oestrupii*	5H-2, 10-11	5H-2, 105-106	34.80	35.80	?	?
FO	*Thalassiosira inura*	5H-2, 105-106	5H-3, 10-11	34.80	35.80	?	?
FO	*Nitzschia praeinterfrigidaria*	5H-2, 105-106	5H-3, 10-11	34.80	35.80	?	?
B	C3n.2n			35.20	35.20	4.62	4.62
T	C3n.3n			36.50	36.50	4.80	4.80
LO	*Asteromphalus kennettii*	5H-4, 10-11	5H-4, 105-106	37.80	38.75	?	?
LO	*Thalassiosira miocenica*	5H-4, 105-106	5H-5, 10-11	38.80	39.30	?	?
LO	*Reticulofenestra hesslandii*	5H-5, 5-7	5H-6, 5-7	39.25	40.75	?	?
	Unconformity			***	***	***	***
FO	*Hemidiscus ovalis*	5H-5, 10-11	5H-CC	39.30	42.70	?	?
FO	*Nitzschia praecurta*	5H-5, 10-11	5H-CC	39.30	42.70	?	?
FO	*Nitzschia reinholdii*	5H-5, 10-11	5H-CC	39.30	42.70	?	?
FO	*Thalassiosira miocenica*	5H-5, 10-11	5H-CC	39.30	42.70	?	?
B	C4An			41.00	41.00	9.03	9.03
FO	*Acrosphaera labrata*	5H-CC	6H-1, 53	42.70	43.23	9.21	9.27
B	C4Ar.1n			43.60	43.60	9.31	9.31
LO	*Cycladophora spongothorax*	6H-1, 98	6H-2, 50	43.68	44.70	9.32	9.57
T	C4Ar.2n			45.20	45.20	9.58	9.58
FO	*Asteromphalus kennettii*	6H-4, 105-106	6H-5, 105-106	48.30	49.60	9.63	9.69
B	C4Ar.2n			48.60	48.60	9.64	9.64
T	C5n.1n			50.50	50.50	9.74	9.74
B	C5n.1n			51.88	51.88	9.88	9.88
T	C5n.2n			51.88	51.88	9.92	9.92
LO	*Reticulofenestra pseudoumbilica*	7H-4, 5-7	7H-5, 5-7	56.80	58.30	10.18	10.25
FO	*Acrosphaera australis*	8H-3, 98	8H-4, 98	65.68	67.18	10.64	10.71
LO	*Cycladophora humerus*	8H-CC	9H-1, 98	71.20	72.28	10.92	10.98
FO	*Eucyrtidium pseudoinflatum*	8H-CC	9H-1, 98	71.20	72.28	10.92	10.98
LO	*Actinomma golownini*	9H-2, 98	9H-3, 98	73.68	75.18	11.05	11.13
LO	*Denticulopsis praedimorpha*	11H-2, 105-107	11H3, 105-107	92.80	94.30	12.04	12.12
LO	*Nitzschia denticuloides*	11H-2, 105-107	11H3, 105-107	92.80	94.30	12.04	12.12
FO	*Cycladophora spongothorax*	11H-6, 98	11H-CC	98.68	99.70	12.35	12.40
B	C5An.2n			99.70	99.70	12.40	12.40
FO	*Denticulopsis dimorpha*	12H-1, 105-107	12H-2, 105-107	100.80	102.30	12.61	12.78
LO	*Actinocyclus ingens v. nodus*	12H-1, 105-107	12H-2, 105-107	101.00	102.30	12.64	12.78

Table 41. (*continued*)

	Event	Sample (top)	Sample (bottom)	Depth (t)	Depth (b)	Age (t)	Age (b)
T	C5Ar.1n			101.20	101.20	12.68	12.68
B	C5Ar.1n			101.71	101.71	12.71	12.71
T	C5Ar.2n			102.26	102.26	12.78	12.78
LO	*Denticulopsis maccollumnii*	12H-2, 105-107	12H-3, 105-107	102.30	103.80	12.78	12.83
B	C5Ar.2n			103.72	103.72	12.82	12.82
FO	*Denticulopsis praedimorpha*	12H-3, 105-107	12H-4, 105-107	103.75	105.25	12.83	13.11
LO	*Nitzschia grossepunctata*	12H-3, 105-107	12H-4, 105-107	103.75	105.25	12.83	13.11
T	C5AAn			104.22	104.22	12.99	12.99
B	C5AAn			105.50	105.50	13.14	13.14
T	C5ABn			105.76	105.76	13.30	13.30
B	C5ABn			108.20	108.20	13.51	13.51
FO	*Actinomma golownini*	12H-CC	13H-CC	109.20	118.70	13.66	14.57
FO	*Nitzschia denticuloides*	12H-CC	13H-1, 105-107	109.30	110.30	13.67	13.84
T	C5ACn			109.48	109.48	13.70	13.70
B	C5ACn			111.75	111.75	14.08	14.08
FO	*Denticulopsis hustedtii*	13H-2, 105-107	13H-2, 105-107	111.80	113.30	14.08	14.15
T	C5ADn			113.84	113.84	14.18	14.18
FO	*Actinocyclus ingens v. nodus*	13H-5, 105-107	13H-6, 105-107	116.30	117.80	14.38	14.50
FO	*Crucidenticula nicobarica*	13H-CC	14H-1, 105-107	118.80	119.80	14.58	14.67
B	C5ADn			119.21	119.21	14.61	14.61
T	C5Bn.1n			119.47	119.47	14.80	14.80
B	C5Bn.1n			120.90	120.90	14.89	14.89
T	C5Bn.2n			121.28	121.28	15.03	15.03
B	C5Bn.2n			123.00	123.00	15.16	15.16
T	C5Cn.1n			127.20	127.20	16.01	16.01
FO	*Denticulopsis lauta*	14H-CC	15H-1, 105-107	128.20	129.30	16.13	16.26
FO	*Nitzschia grossepunctata*	15H-1, 105-107	15H-2, 105-107	129.30	130.80	16.26	16.31
B	C5Cn.1n			129.51	129.51	16.29	16.29
T	C5Cn.2n			132.56	132.56	16.33	16.33
FO	*Calcidiscus macintyrei*	15H-2, 5-6	15H-3, 55-56	132.75	137.72	16.35	16.72
B	C5Cn.2n			134.35	134.35	16.49	16.49
T	C5Cn.3n			136.02	136.02	16.56	16.56
LO	*Nitzschia maleinterpretaria*	15H-6, 105-107	15H-CC	136.80	137.70	16.63	16.72
FO	*Denticulopsis maccollumnii*	15H-CC	16H-1, 105-107	137.70	138.80	16.72	16.83
LO	*Crucidenticula kanayae*	15H-CC	16H-1, 105-107	137.70	138.80	16.72	16.83
FO	*Calcidiscus leptoporus*	16H-7, 2-4	17H-1, 22-25	146.70	147.40	17.58	17.65
B	C5Dn			147.10	147.10	17.62	17.62
FO	*Crucidenticula kanayae*	16H-CC	17H-2, 105-107	147.20	149.80	17.63	17.94
FO	*Nitzschia maleinterpretaria*	16H-CC	17H-2, 105-107	147.20	149.80	17.63	17.94
T	C5En			152.75	152.75	18.28	18.28
LO	*Thalassiosira fraga*	17H-6, 105-107	17H-CC	165.30	166.20	?	?

Data sets are integrated from the work of numerous authors (see Table 2).

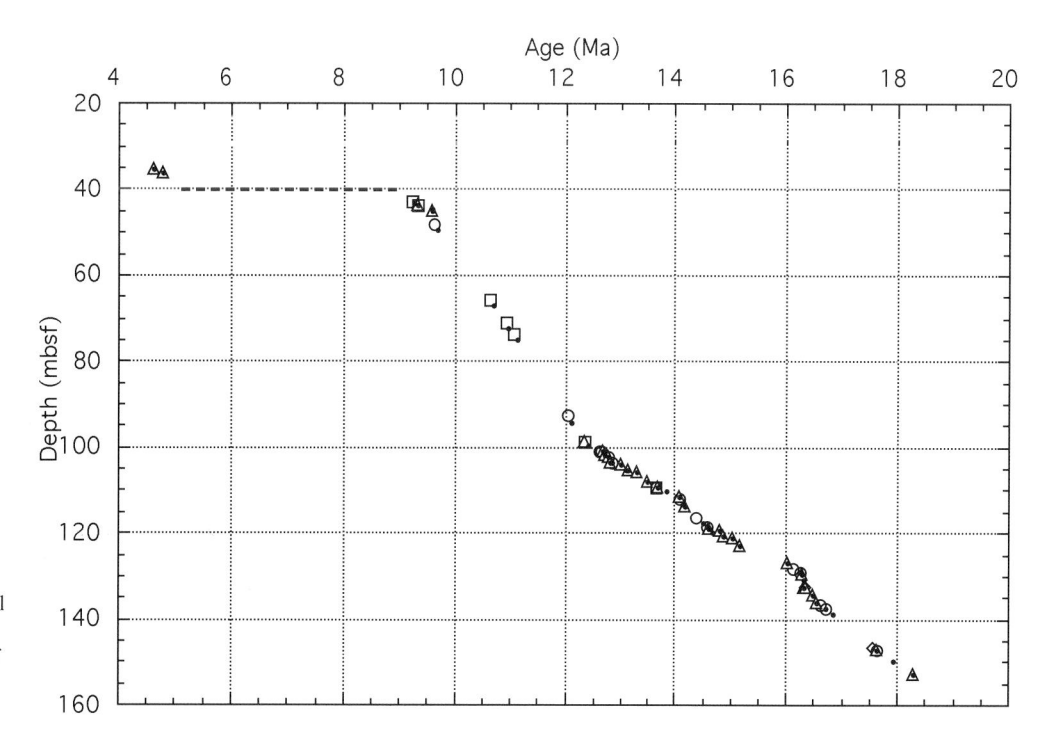

Fig. 59. An age–depth plot for DSDP Hole 751A utilizing the biostratigraphical and polarity reversal events identified in this study and calibrated to the GPTS of Cande & Kent (1995). See Fig. 5 for legend of symbols.

Discussion

Ages for the original 173 events calibrated to the Cande & Kent (1995) GPTS are presented in Table 3. These assigned ages have been revised based on the iterative process used in this study and the results are shown in Table 42. All events shown in Table 42 have been used previously as biostratigraphical markers in the Southern Ocean. It is impossible to evaluate the stratigraphical usefulness of the 37 datums that occur at only one site. Of the 136 datums that occur at more than one site, 15 are calcareous nannofossils, 27 are radiolarians, and 94 are diatoms. Of these events six are exclusive to the South Atlantic sector and nine to the Indian Ocean sector of the Southern Ocean. All 136 datums datums occur within intervals where a reliable polarity record was obtained and it is possible to evaluate their value as stratigraphical markers. Age constraints for each datum are influenced by the original sampling constraints in a given hole. Most datums are constrained within an age range of less than 0.3 million years. In several holes, however, age constraints exceeds 1.0 million years. This accuracy is largely an artefact of sampling and can only be reduced by additional sampling. The information provided by these less-constrained events is often still useful when compared to the age estimates of the same event in the remaining holes. Such a comparison often provides an indication to the datums reliability throughout the Southern Ocean.

The number of primary stratigraphical events exhibiting chronological consistency throughout the Southern Ocean is surprisingly few and suggests that the biochronological resolution which can be achieved for the Southern Ocean is coarser than previously believed using the currently available data. For example, more than forty datums previously defined are currently being used as biostratigraphical indicators for the Pliocene and Pleistocene. Results from this study suggest that, at most, six of these events have consistent stratigraphical ranges and can be used to correlate throughout the Southern Ocean. An additional eight events may prove biostratigraphically useful with a better understanding of their spatial and temporal distribution. The remaining biostratigraphical events are considered unreliable given the diachronous ages of each specific event. Similarly, 61 events are currently used as biostratigraphical indicators for the Miocene. This study suggests that only seven primary events are useful for the Miocene to Late Eocene interval.

Thirteen primary datums are defined as control points for the Southern Ocean biostratigraphy and an additional twenty datums are identified as potentially useful (Table 42; Table 43). Primary datums show a generally consistent occurrence in most holes. In cases where an estimated age for a datum in one hole is anomalous to its age in other holes we consider the event as useful until further studies are completed.

In compiling the primary events we have identified an additional 20 datums that have potential stratigraphical usefulness (Table 43). These events are not promoted as primary indicators for the following reasons: the paucity of data, the occurrence of events limited to two holes, poorly constrained events with age range distributions greater than 0.25 million years and the inconsistent cluster of age constraints for a specific event.

Pliocene–Quaternary

Six primary biostratigraphical markers consisting of five diatom events and one radiolarian event were identified within the Pliocene of the Southern Ocean (Table 42). Nine potentially useful events (five diatom and one radiolarian) are also identified within this interval. The paucity of primary events results from the inconsistent occurrence of the majority of marker species previously determined as stratigraphically useful or limited data. The calculation of age ranges for first occurrences and last occurrences is constrained by sampling intervals, the availability of magnetic data within critical intervals and by the occurrence of stratigraphical breaks which result in unreliable age determinations. The further recognition of potential markers however, may provide local or regionally reliable datums. The limited number of events (diatoms and radiolarians) and the elimination of calcareous nannofossil datums raises doubts concerning the reliability of previous biostratigraphical interpretations for the Pliocene of the Southern Ocean. The primary and potential events defined for this time interval are discussed below.

LO *Thalassiosira kolbei* (diatom). This event is recorded by previous workers in twelve of the holes used in this study (Table 42). Age constraints, which are calculated for this event in eleven sites, vary in each hole and reflect differences in the sampling density. Given the sampling constraint this event is contemporaneous at eight of the 11 sites. In six holes this event is constrained by less than 0.3 million years and at one site by 0.15 million years. The LO of *Thalassiosira kolbei* approximates or occurs directly below the base of Chron C2n at eight of the 11 holes. In three holes (704B, 747B and 748B) this event is older. The similarity in age estimates for this event from holes in the Atlantic Ocean (Leg 113) and the Indian Ocean (Leg 119) sectors of the Southern Ocean suggest that this event is a useful primary biostratigraphical indicator.

LO *Thallassiosira vulnifica* (diatom). This event is recorded in ten holes (Table 42). In eight of these holes it falls within our constraint of 0.25 million years. This event is best constrained in two holes, Hole 514 (2.3–2.40 Ma) and Hole 699A (2.43–2.49 Ma). At five of the other holes the age of this event is younger e.g. at Hole 744B this event has an age between 1.76 and 2.32 Ma. We did not anticipate the usefulness of this species given its susceptibility to dissolution.

FO *Nitzschia praeinterfrigidaria* (diatom). This event is recorded in 11 holes (Table 42). In four of these holes the occurrence of an unconformity prevents the determination of age constraints. In Hole 747A this event is constrained to within 0.3 million years and in six of the remaining seven holes this event is constrained to less than 0.17 million years and to within 0.3 million years at Hole 747A. With the exception of an occurrence in Hole 514 (3.24–3.36 Ma) this event occurs between the base of C3n.2n and the top of C3n.3n and has an estimated age of 4.70 Ma. We consider this species as an excellent primary event.

FO *Nitzschia interfrigidaria* (diatom). This event is recorded in thirteen holes used in this study (Table 42). The occurrence of unconformities in seven holes prevents the calculation of age constraints for this event. In the six remaining holes its constraints are less than 0.3 million years. In two holes, 737A (3.98–4.03 Ma) and 745B (3.71–3.80 Ma) this species is constrained by less than 0.1 million years. In three of the four remaining holes the age ranges for this species fall within our constraint of 0.25 million years. At Hole 695A this event ranges from 3.09 to 3.26 Ma.

FO *Helotholus vema* (radiolarian). This event is recorded in eight holes considered in this study (Table 42). The presence of unconformities in three holes prevents the calculation of age constraints for this event. In five holes this event occurs within an interval of less than 0.3 million years. The event is best constrained at two holes 689B (4.40–4.50 Ma) and 745B (4.57–4.60 Ma) and occurs within Chron C3n.2n. There are no holes where a discrepancy in ages is recorded.

deriveditzschia barronii (diatom). This event is recorded in 12 holes used in this study (Table 42). The presence of unconformities at six holes prevents the calculation of age constraints for this species. In four of the remaining six holes the age range of the species is less than 0.1 million years. The event is best constrained at Hole 696A (4.51–4.52 Ma) and coincides with the FO of *Helotholus vema*. In Hole 695A the event ranges from 4.12 to 4.18 Ma.

Miocene

The paucity of reliable stratigraphical markers identified in this study contrasts distinctly with the voluminous number of events previously identified. The results from this study suggest that five diatom events are useful primary markers (Table 42). An additional nine events (seven diatom and two radiolarian) are identified as potentially useful biostratigraphical indicators.

FO *Denticulopsis praedimorpha* (diatom). This event occurs in seven of the holes examined (Table 42). An unconformity in three holes prevents an age calculation for this event. In the remaining four holes this event is constrained by 0.28 million years and has an age range of 12.83–13.14 Ma in Holes 747A and 751A. In Hole 690B the age of this event ranges from 13.09–13.35 Ma and overlaps the age of this event at the other two holes. In Hole 689B this event has a younger age of 12.51–12.76 Ma.

FO *Nitzschia denticuloides* (diatom). This event is recorded in seven holes used in this study (Table 42). An unconformity prevents calculation of age constraints in Holes 699A. In the six remaining holes the age range of this event is constrained by 0.38 million years. The event is best constrained in Holes 689B (13.43–13.54 Ma) and Hole 747A (13.57–13.66 Ma). An older age range of 15.21–15.55 Ma is recorded in Hole 744B.

FO *Denticulopsis hustedtii* (diatom). This event is recorded in eight holes (Table 42). The occurrence of unconformities in three sites precludes the calculation of age ranges for this event. The age range of this event is best constrained in Hole 751A between 14.08 and 14.15 Ma. In two Holes 689B and 747A the age ranges for this event partially overlap the range recorded in Hole 751A. In Holes 690B and 744B the age ranges for the event fall within our constraint of 0.25 million years.

LO *Synedra jouseana* (diatom). This event occurs in five holes (Table 42). The occurrence of unconformities at three holes prevents determination of an age constraint for this event. In Hole 689B the age of this event ranges from 14.51 to 14.57 Ma and at Hole 747A from 14.49 to 14.80 Ma.

FO *Crucidenticula kanayae* (diatom). This event is recorded in seven holes used in this study (Table 42). An unconformity in Hole 748B prevents an age estimate for this event. The age range is less than 0.1 million years in three holes (689B, 690B and 751A). In five holes this event has an age range between 17.42 and 17.69 Ma and in Hole 704B the age range, between 16.56 and 16.81 Ma, is significantly younger.

Oligocene–late Eocene

The definition of events within this interval is hindered by limited core recovery and/or the paucity of palaeomagnetic data. Two primary events are recognized within this interval for the Southern Ocean (Table 42). These consist of the LO and FO of two calcareous nannofossil species respectively. This study confirms the usefulness of calcareous nannofossil events recognized by Wei & Wise (1990, 1992a,b), Wei (1990; 1992) and Wei et al. (1992).

FO *Isthmolithus recurvus* (calcareous nannofossil). This event is recorded in three holes (Table 42) and the estimate of the total age range for this event occurs between 35.73 and 36.15 Ma. The age of this event approximates 36 Ma.

LO *Reticulofenestra reticulata* (calcareous nannofossil). This event is recorded in three holes (Table 42) and the estimate of the total age range for this event occurs between 35.73 and 36.15 Ma. The age of this event coincides with that of the LO of *Isthmolithus recurvus* at approximately 36 Ma.

Potential events:

In compiling the primary events we have identified an additional 20 datums that have potential stratigraphical usefulness (Table 43). These events are not promoted as primary indicators for the following reasons: the paucity of data, the occurrence of events limited to two holes, poorly constrained events with age range distributions greater than 0.25 million years and the inconsistent cluster of age constraints for a specific event. Five examples are given below.

LO *Chiasmolithus altus* has an age constraint, in Holes 744A, 747A, and 748B, between 25.93 and 26.11 Ma, suggesting that this event is chronostratigraphically useful for these holes. In Hole 699A this event ranges from 26.54 to 26.80 Ma and in Hole 689B this event has an age younger than 29.43 Ma. These age ranges indicate the questionable reliability of this event and confirm the the results of Wei (1992).

LO *Coscinodiscus rhombicus* is constrained between 16.66 and 16.72 Ma in Hole 689B and has an age constraint between 16.81 and 16.99 Ma in Hole 704B. Age constraints are unavailable for other sites. We therefore elect not to use this event as a primary marker.

FO *Acrosphaera labrata* occurs in four holes. This event has an age constraint of 9.32–9.46 Ma (Hole 689B), 8.46–8.82 Ma (Hole 747A), 9.14–9.17 Ma (747B) and 9.21–9.27 Ma (Hole 751A) Although these data show that this event is useful for two holes and potentially for a third hole we cannot asses the usefulness of this event until further data become available.

LO *Denticulopsis hustedtii* occurs in four holes. This event has an age contraint of 4.28–4.37 Ma (695A), 4.38–4.49 Ma (699A) and 4.18–5.24 Ma (Hole 704B) suggesting that this event is potentially a primary event. The age constraint of this event in Hole 697B (2.61–2.82 Ma), raises questions concerning the reliability of this event as a primary marker.

LO *Antarctissa cylindrica* is well constrained at two holes (745B and 747A) with an age constraint between 0.56 and 0.65 Ma. In Holes 690B and 695A this event has an age range between 0.00 and 2.27 Ma.

A comparison between the Miocene diatom datum levels for the Southern Ocean (this study) and the North Pacific (Barron and Gladenkov, 1995) shows that the first occurrences of *Crucidenticula kanayae* and *Denticulopsis praedimorpha* provide useful datum in

…

Table 42. *Calculated ages for previously used biostratigraphic events using a calibration to the GPTS of Cande & Kent (1995) and the biochronological and polarity interpretation of this study*

Event	514t	514b	689Bt	689Bb	690Bt	690Bb	695At	695Ab	696At	696Ab	697Bt	697Bb	699At	699Ab	701At	701Ab
LO *Pterocanium trilobium*																
LO *Triceraspyris antarctica*																
LO *Pseudoemiliana lacunosa*																
FO *Emiliana huxleyii*																
LO *Thalassiosira elliptopora*	0.72	0.84											0.55	0.66		
LO *Hemidiscus karstenii*													0.00	0.03	0.23	0.26
LO *Simonseniella barboi*	1.82	1.83			3.19	3.32					3.86	4.05	1.68	1.92	1.48	1.58
FO *Nitzschia miocenica*																
FO *Nitzschia kerguelensis*	2.40	2.42	?	?	0.00	0.81	1.17	1.70	5.13	5.65						
LO *Actinocyclus ingens*	0.59	0.72	?	?	0.81	1.98	0.20	0.64	?	?	0.79	1.10	0.54	0.55	0.81	0.83
LO *Thalassiosira kolbei*	**1.92**	**2.15**	?	?	**1.98**	**2.27**					**1.74**	**1.95**	**1.99**	**2.14**	**1.73**	**1.98**
LO *Thalassiosira inura*			?	?	1.98	2.27	2.94	3.08	?	?	2.82	2.94				
LO *Thalassiosira vulnifica*	**2.37**	**2.40**					**0.64**	**1.95**			**2.17**	**2.39**	**2.43**	**2.49**	**2.12**	**2.26**
LO *Thalassiosira insigna*							2.00	2.14			2.94	3.08	2.68	2.72	2.47	?
LO *Nitzschia interfrigidaria*	2.91	2.93	?	?	1.98	2.27	2.15	2.61			3.08	3.11	2.72	2.72	2.47	?
LO *Denticulopsis dimorpha*													?	?		
LO *Nitzschia barronii*			?	?	1.98	2.27	2.58	2.61	?	?	1.10	1.25				
LO *Helicosphaera sellii*																
LO *Discoaster browerii*																
FO *Thalassiosira elliptopora*													4.86	4.89		
LO *Rouxia diploneides*																
LO *Nitzschia cylindrica*																
FO *Actinocyclus actinochilus*																
LO *Antarctissa cylindrica*			?	?	0.00	2.26	0.00	2.27								
LO *Cycladophora pliocenica*			?	?	0.00	2.26	0.00	2.27								
LO *Eucyrtidium calvertense*			?	?	0.00	2.26										
LO *Desmospyris spongiosa*	2.42	2.50	?	?	0.00	2.26	2.00	2.15			2.39	2.82	2.44	3.26	2.12	2.46
LO *Helotholus vema*	2.42	2.50	?	?	0.00	2.26	2.00	2.15			2.39	2.82	2.44	3.26	2.12	2.46
FO *Cycladophora davisiana*	2.59	2.63	?	?	0.00	2.26					3.08	3.34			2.12	2.46
LO *Nitzschia weaveri*	2.74	2.76					2.58	2.61					3.43	3.70	?	?
LO *Nitzschia reinholdii*	2.91	2.93											2.53	3.70		
LO *Cosmiodiscus intersectus*							4.48	?								
LO *Rouxia heteropolara*																
LO *Thalassiosira torokina*																
LO *Rouxia californica*																
LO *Stephanopyxis grunowii*																
LO *Nitzschia praeinterfrigidaria*			?	?	3.29	3.68	2.14	2.15	4.52	4.64	3.50	3.65	3.43	3.56		
FO *Stephanopyxis grunowii*																
FO *Thalassiosira lentiginosa*	2.91	2.93			3.68	3.81					3.40	3.50	4.86	4.89		
FO *Thalassiosira kolbei*	3.24	3.36	4.69	4.75							2.61	2.82				
FO *Thalassiosira inura*			4.84	4.89	?	?			6.18	6.77	?	?				
FO *Thalassiosira vulnifica*							2.15	2.61			2.61	2.82				
FO *Nitzschia praeinterfrigidaria*	**3.24**	**3.36**			?	?	**4.57**	?	**4.65**	**4.76**			**5.57**	**5.74**		
FO *Navicula wisei*																
FO *Nitzschia weaveri*	3.37	3.44					4.23	4.27					3.43	3.70		
LO *Lampromitra coronata*																
FO *Nitzschia interfrigidaria*	?	?	?	?	**3.81**	**3.98**	**3.09**	**3.26**			**4.05**	**4.35**	**3.82**	**4.11**		
LO *Prunopyle titan*			3.53	3.92	4.10	4.60	4.28	4.37					3.26	4.64		
LO *Nitzschia fossilis*																
LO *Rhizosolenia miocenica*																
FO *Helotholus vema*			**4.40**	**4.50**	**4.10**	**4.60**										
FO *Nitzschia barronii*			**4.46**	**4.55**	**4.52**	?	**4.02**	**4.18**	**4.51**	**4.52**	?	?				
LO *Thalassiosira complicata*									4.52	4.52	3.52	3.53				
LO *Denticulopsis hustedtii*							4.28	4.37	?	?	2.61	2.82	4.38	4.49		
FO *Thalassiosira complicata*																
LO *Neobrunia miraibilis*																
FO *Nitzschia praecurta*									?	?	?	?				
FO *Nitzschia angulata*													4.35	4.37		
FO *Thalassiosira oestrupii*									?	?	?	?				
LO *Thalassiosira burckliana*																
LO *Thalassiosira miocenica*																
LO *Hemidiscus cuneiformis*													4.47	4.63		
LO *Amphymenium challengerae*																
LO *Stichocorys peregrina*																
LO *Nitzschia marina*																

DISCUSSION

704Bt	704Bb	737At	737Ab	744At	744Ab	744Bt	744Bb	745Bt	745Bb	746At	746Ab	747At	747Ab	747Bt	747Bb	748Bt	748Bb	751At	751Ab
								0.00	0.09			0.56	0.64						
0.00	0.15											1.97	1.99			1.62	2.81	?	?
												0.11	1.25						
						0.89	1.76	0.79	1.27			0.57	0.64	0.29	0.31			?	?
0.00	0.07	3.81	3.84			1.76	2.32	0.26	0.44	7.99	8.01			2.70	3.17	3.89	4.02		
2.12	2.13	3.68	3.76			2.32	3.46	1.57	1.75			1.65	1.99			4.02	4.15	?	?
										6.70	6.74								
		?	?			1.76	2.32	2.12	2.18			2.19	2.79	2.70	3.17			?	?
0.58	0.64	7.17	7.35			0.89	1.76	0.44	0.61	8.29	8.79	0.25	0.35	0.26	0.29				
2.30	**2.34**					**1.76**	**2.32**	**1.75**	**1.97**			**1.99**	**2.19**	**2.46**	**2.70**	**2.47**	**2.69**	**?**	**?**
												0.46	0.57	2.46	2.70	2.47	2.69	?	?
3.00	**3.13**					**1.76**	**2.32**	**2.23**	**2.33**			**2.19**	**3.22**			**2.47**	**2.69**	**?**	**?**
4.42	4.57	3.28	3.60			1.76	2.32	2.18	2.23			2.19	3.22	2.70	3.17	2.47	2.69	?	?
		?	?			1.76	2.32	1.97	2.12			2.19	2.79	2.70	3.27	2.47	2.69	?	?
9.33	9.48					1.76	2.32	0.61	0.79	6.60	6.65	4.46	4.76	?	?	?	?	?	?
												0.78	0.88	?	?	1.68	2.47	?	?
2.47	2.65																		
2.96	3.13																		
						2.32	3.46	1.57	1.75			0.88	0.99	2.70	3.17			?	?
												?	?						
								2.23	2.33										
												2.19	2.79	2.70	3.17			?	?
								0.61	0.65			0.56	0.64					?	?
						1.32	1.69	1.57	1.58			1.67	1.73			2.66	3.33	?	?
						1.76	3.46	0.61	0.62			1.97	1.99			2.66	3.33	?	?
		3.20	3.33			1.76	3.46	2.50	2.53			2.17	2.77			2.66	3.33	?	?
						1.76	3.46	2.48	2.50			2.17	2.77			2.66	3.33	?	?
								2.66	2.72			3.28	?			2.66	3.33	?	?
		?	?			3.46	?	1.75	1.97			2.19	2.79	0.29	0.31			?	?
4.42	4.57							2.71	2.83										
4.52	4.57																		
								2.18	2.23										
		4.77	4.78					2.12	2.18										
								2.23	2.43										
								2.71	2.83										
		?	?			2.32	3.46	2.71	2.83			?	?	?	?	4.02	4.27	?	?
		4.08	4.13			1.76	2.32					4.33	4.43	2.70	3.17				
		3.12	3.33			2.32	3.46	2.90	2.96			4.33	4.43	?	?	?	?	?	?
		4.72	4.77			2.32	3.46	4.80	4.89			8.84	9.03	?	?	?	?	4.54	4.70
3.31	3.32					2.32	3.46	2.71	2.83			3.22	3.29			3.09	3.26	?	?
		4.70	**4.72**			**?**	**?**	**4.72**	**4.87**			**4.46**	**4.76**	**?**	**?**	**?**	**?**	**4.54**	**4.70**
																?	?	?	?
3.48	3.66	?	?			?	?	3.06	3.13			?	?					?	?
												?	?	?	?	?	?	?	?
		3.98	**4.03**			**?**	**?**	**3.71**	**3.80**			**?**	**?**	**?**	**?**	**?**	**?**	**?**	**?**
						1.76	3.46	3.44	3.52			3.28	?			1.62	2.81	?	?
4.57	4.63							3.95	4.08										
								3.95	4.08										
		4.32	**4.66**					**4.57**	**4.60**			**4.43**	**4.69**	**?**	**?**	**4.12**			
		4.32	**4.62**			**?**	**?**	**4.22**	**4.27**			**4.33**	**4.69**	**?**	**?**	**?**	**?**	**?**	**?**
												?	?			?	?	?	?
4.18	5.42																		
												4.46	4.76						
												4.76	?			?	?		
												4.76	?					?	?
4.05	4.18																		
4.95	5.06	5.01	5.08			3.46	?	5.38	5.56			4.33	4.33	2.30	2.60	?	?	4.54	4.70
		6.92	7.07							6.83	6.88								
4.18	4.25	5.41	5.70			?	?	?	?							?	?	?	?
4.18	4.25							1.57	1.75										
4.73	5.75																		
4.73	5.75	5.07	5.12															?	?
4.57	4.62																		

(continued)

Table 42. (*continued*)

Event	514t	514b	689Bt	689Bb	690Bt	690Bb	695At	695Ab	696At	696Ab	697Bt	697Bb	699At	699Ab	701At	701Ab
LO *Thalassiosira praeconvexa*																
LO *Thalassiosira convexa v. aspinosa*																
LO *Lamprocyclas aegles*																
FO *Thalassiosira praeconvexa*																
LO *Nitzschia porteri*																
FO *Lamprocyclas aegles*																
LO *Diartus hughesi*																
FO *Stichocorys peregrina*																
FO *Nitzschia reinholdii*			4.67	4.69									?	?		
FO *Thalassiosira insigna*	3.20	3.26					2.65	2.80								
FO *Nitzschia marina*																
LO *Denticulopsis lauta*													4.90	5.19		
FO *Navicula wisei*																
FO *Thalassiosira inura*																
FO *Rouxia heteropolara*									6.18	6.77	?	?				
LO *Cycladophora spongothorax*			8.28	8.31	?	?										
LO *Hemiaulus triangularis*																
FO *Simonseniella barboi*													?	?		
LO *Raphidodiscus marylandicus*																
FO *Cestodiscus peplum*																
LO *Thalassiosira convexa*																
FO *Thalassiosira convexa v. aspinosa*																
LO *Trinacria excavata*																
FO *Hemidiscus karstenii*													?	?		
LO *Asteromphalus kennettii*			8.55	9.25	9.27	9.31										
LO *Reticulofenestra pseudoumbilica*																
FO *Amphymenium challengerae*																
FO *Thalassiosira burckliana*																
FO *Hemidiscus ovalis*																
LO *Actinocyclus ingens v. nodus*			9.25	9.29												
LO *Acrosphaera australis*																
FO *Cosmiodiscus intersectus*			9.31	9.35	9.58	9.63					?	?				
FO *Acrosphaera labrata*			9.32	9.46												
LO *Reticulofenestra gelida*			9.60	9.92	12.08	12.85										
FO *Acrosphaera australis*			10.16	10.25												
LO *Calcidiscus macintyrei*																
FO *Thalassiosira torokina*													?	?		
LO *Cycladophora humerus*			10.16	10.25	10.22	10.57										
FO *Eucyrtidium pseudoinflatum*			10.41	10.47	10.00	10.58										
FO *Nitzschia fossilis*																
FO *Diartus hughesi*																
FO *Hemidiscus cuneiformis*	2.60	2.62											?	?		
FO *Thalassiosira miocenica*																
LO *Denticulopsis dimorpha*			3.95	4.02	11.98	12.86										
LO *Actinomma golownini*			11.86	11.87	10.00	10.58							?	?		
FO *Asteromphalus kennettii*			10.23	10.31	10.17	10.29										
FO *Actinocyclus fryxellae*																
LO *Cyrtocapsella japonica*																
LO *Denticulopsis praedimorpha*					10.88	11.02										
LO *Nitzschia denticuloides*			11.80	11.92	11.46	11.52							?	?		
LO *Reticulofenestra hesslandii*			11.91	11.93	12.08	12.85										
FO *Reticulofenestra hesslandii*																
FO *Cyrtocapsella japonica*																
FO *Denticulopsis dimorpha*			11.91	11.92	12.91	12.96										
LO *Dendrospyris megalocephalis*			11.88	11.92												
FO *Cycladophora spongothorax*			11.90	11.91	12.03	12.96										
LO *Nitzschia grossepunctata*			12.15	12.51	13.68	14.02										
FO *Denticulopsis praedimorpha*			**12.51**	**12.76**	**13.09**	**13.35**										
LO *Crucidenticula nicobarica*			11.95	12.14	12.85	12.91										
FO *Actinomma golownini*			13.06	13.28	13.41	13.75										
FO *Nitzschia denticuloides*			**13.43**	**13.54**	**13.68**	**14.02**							?	?		
FO *Antarctissa deflandrei*			13.38	13.43	13.75	14.51										
LO *Denticulopsis maccollumnii*			13.99	14.24	14.24	14.44										
FO *Denticulopsis hustedtii*			**13.99**	**14.24**	**14.24**	**14.44**					?	?				
FO *Cycladophora humerus*			14.13	?	13.75	14.51										

DISCUSSION

704Bt	704Bb	737At	737Ab	744At	744Ab	744Bt	744Bb	745Bt	745Bb	746At	746Ab	747At	747Ab	747Bt	747Bb	748Bt	748Bb	751At	751Ab
4.77	4.89					?	?												
6.02	6.19																		
6.20	6.24																		
6.97	7.01									6.60	6.65								
7.71	7.83																		
7.80	8.05																		
8.12	8.14																		
8.12	8.14															5.14			?	?
8.54	8.73	?	?			?	?							9.11	?	3.09	3.26	?	?
8.54	8.73							2.90	2.96					?					
8.54	8.73									8.12	8.18								
7.09	7.12					10.78	10.89									?	?	?	?
								?	?							?	?	?	?
		6.36	6.40					3.95	4.08			4.46	4.76						
										9.41	9.44	9.23	9.25	9.17	9.62			9.32	9.70
												4.76	?						
11.39	11.66			20.69	?														
?	?																		
								5.98	6.09										
								6.09	6.18										
								6.09	6.18										
										?	?								
						?	?			9.50	?	4.76	?	9.19	9.74	?	?	?	?
3.83	4.01					10.42	10.52					16.51	16.75			22.46	22.50	10.18	10.25
6.20	6.24																	?	?
										9.50	?								
												9.25	9.27			11.48	?	?	?
11.06	12.95					14.28	14.56			6.60	6.65	12.88	13.14					12.64	12.78
						10.50	10.70												
8.54	8.74	?	?							8.65	8.75								
												8.46	8.82	9.14	9.17			9.21	9.27
														10.76	10.93	10.25	10.44	10.64	10.71
2.96	3.13					10.52	10.62					9.72	9.80			10.45	10.64		
						?	?			8.52	8.56	9.03	9.24	9.95	10.03	4.27	?	?	?
				8.23	8.29					8.23	8.29					10.19	10.25	10.92	10.98
						16.45	17.20			?	?	10.08	10.37	10.84	10.93	10.44	10.63	10.92	10.98
9.48	9.59																		
9.83	9.84																		
9.87	9.90					10.58	10.69					10.38	10.66						
9.97	10.16	6.56	6.67			?	?			?	?			9.11	9.14			?	?
10.15	10.30			4.46	4.76	2.32	3.46			6.60	6.65			?	?	?	?	?	?
						10.79	11.12					10.37	10.94	10.59	10.67			11.05	11.13
10.42	10.52					10.25	10.35			?	?			9.81	9.95	10.19	10.25	9.63	9.69
		?	?			10.99	11.05												
10.48	10.54																		
						11.38	11.44					11.67	11.81			11.33	11.40	12.04	12.12
						?	?					11.81	11.98			10.25	10.32	12.04	12.12
						17.55	17.92					2.82	3.56					?	?
												11.93	12.18						
10.99	11.05																		
11.06	12.95					?	?	0.79	1.27			12.88	13.14			?	?	12.61	12.78
						11.32	?					11.66	11.76	?	?	10.25	10.44	12.35	12.40
13.11	13.64					?	?					13.14	13.57			?	?	12.83	13.11
						?	?					**12.88**	**13.14**	?	?	?	?	**12.83**	**13.11**
12.95	13.11																		
						?	?					12.88	13.11			?	?	13.66	14.57
13.64	**14.02**					**15.21**	**15.55**					**13.57**	**13.66**			?	?	**13.67**	**13.84**
13.11	13.64					?	?					14.32	14.49			?	?	12.78	12.83
?	?	?	?			**14.42**	**14.56**					**14.14**	**14.32**			?	?	**14.08**	**14.15**
						14.28	14.42			?	?								

(continued)

Table 42. (*continued*)

Event	514t	514b	689Bt	689Bb	690Bt	690Bb	695At	695Ab	696At	696Ab	697Bt	697Bb	699At	699Ab	701At	701Ab
FO *Dendrospyris megalocephalis*			14.13	?												
LO *Synedra jouseana*			**14.51**	**14.57**	**14.51**	**?**										
FO *Actinocyclus ingens v. nodus*			?	?	14.51	?										
LO *Coscinodiscus lewisianus*			16.61	16.66												
FO *Simonseniella barboi*																
LO *Coscinodiscus rhombicus*			16.66	16.72												
FO *Eucyrtidium punctatum*			16.72	17.33												
LO *Nitzschia maleinterpretaria*			16.85	17.07	14.51	?										
LO *Crucidenticula kanayae*			16.81	17.05	16.91	17.19										
FO *Denticulopsis lauta*																
FO *Nitzschia grossepunctata*			16.85	17.09	17.19	17.42										
FO *Nitzschia maleinterpretaria*					17.19	17.42										
LO *Thalassiosira fraga*			17.09	17.20	17.86	18.59										
FO *Calcidiscus macintyrei*																
FO *Actinocyclus ingens*	2.35	2.37	17.22	17.45	17.19	17.42					3.21	3.34				
FO *Denticulopsis maccollumnii*			17.22	17.45	17.42	17.50										
FO *Calcidiscus leptoporus*																
FO *Crucidenticula kanayae*			**17.45**	**17.55**	**17.42**	**17.50**										
FO *Cycladophora golli regipileus*			18.94	19.00	18.28	19.20										
FO *Crucidenticula nicobarica*			19.01	19.14	19.50	19.70										
FO *Cyrtocapsella tetrapera*			18.94	?	19.55	20.04										
FO *Cyrtocapsella longithorax*			?	?	19.20	19.55										
LO *Thalassiosira aspinosa*																
FO *Coscinodiscus lewisianus*																
LO *Thalassiosira spumellaroides*			15.57	?	19.50	19.55										
FO *Raphidodiscus marylandicus*			?	?	19.99	20.04										
FO *Thalassiosira fraga*			?	?	19.85	19.99										
LO *Chiasmolithus altus*			?	24.93	?	?							26.54	26.80		
LO *Reticulofenestra bisecta*			?	24.93	29.29	29.54							23.83	23.86		
FO *Thalassiosira aspinosa*					19.99	20.04										
LO *Rocella gelida*			?	24.84	19.85	19.99							23.43	23.51		
LO *Rocella schraderi*																
LO *Rocella vigilans*			24.91	25.05	19.99	20.04										
LO *Bogorovia veniamini*													?	?		
LO *Lisitzina ornata*					20.04	?										
FO *Thalassiosira spumellaroides*			24.77	25.05	20.04	?										
FO *Rocella vigilans*			26.07	26.13	26.48	27.06										
FO *Rocella gelida*			26.07	26.13	26.48	27.06							25.33	25.76		
LO *Thalassiosira primalabiata*																
FO *Lisitzina ornata*					27.62	27.78										
LO *Reticulofenestra umbilica*			30.74	31.03	?	?							31.75	31.88		
LO *Isthmolitus recurvus*			31.83	32.35	?	?							32.31	32.39		
LO *Reticulofenestra oamaruensis*			33.60	34.01									33.48	?		
FO *Reticulofenestra oamaruensis*			35.24	35.26												
FO *Isthmolitus recurvus*			**35.82**	**36.01**	**35.73**	**35.92**										
FO *Cyclicargolithus abisectus*					30.77	?							30.99	31.49		
LO *Reticulofenestra reticulata*			**35.82**	**36.01**	**35.73**	**35.92**										
LO *Rhizosolenia oligocenica*																
LO *Synedra jouseana*																

Emboldened events are primary stratigraphical indicators based on this study. ? indicates the occurrence of undated specific biostratigraphical events. Dating these events was prohibited either by the absence of magnetic data and/or the presence of an unconformity.

both regions. The ages of the FO of *Denticulopsis praedimorpha* are nearly isochronous (12.8 Ma in the North Pacific and 12.96 Ma in the Southern Ocean). Thus, using the criteria adopted in this study, *Denticulopsis praedimorpha* would qualify as a primary datum level for correlating between the two regions. The FO of *Crucidenticula kanayae*, 16.9 Ma in the North Pacific, is younger than the age of this event, 17.55 Ma, in the Southern Ocean. We would not consider this event either as a primary or potentially useful event for correlation between the North Pacific and Southern Ocean. Some Miocene diatom datum levels which have been recorded as diachronous in the Southern Ocean, FO *Denticulopsis lauta*, FO *Denticulopsis dimorpha*, LO *Denticulopsis dimorpha*, and FO *Thalassiosira oestrupii*, appear to be approximately isochronous and widely applicable in the North Pacific (Barron and Gladenkov, 1995). These observations confirm our reservations concerning the transfer and application of useful diatom datum levels between the high latitude regions of the northern and southern hemispheres.

Stratigraphical continuity

The occurrence of regionally consanguineous hiatuses is used by palaeoceanographers to infer temporal changes in the characteristics of major deep water masses (Ramsay *et al.* 1994). Burckle & Abrams (1987) reconsidered the diatom zonation which was used to identify and correlate hiatuses in Quaternary and Pliocene (0–5 Ma)

DISCUSSION

704Bt	704Bb	737At	737Ab	744At	744Ab	744Bt	744Bb	745Bt	745Bb	746At	746Ab	747At	747Ab	747Bt	747Bb	748Bt	748Bb	751At	751Ab
						?	?									?	?	?	?
						?	?					**14.49**	**14.80**					?	?
13.11	13.64			15.55	15.89					8.07	8.12	14.49	14.68					14.38	14.50
13.11	13.64			1.76	2.32							13.14	13.57			?	?		
?	?																		
16.81	16.99																		
14.03	?					16.40	16.80					14.46	14.58			?	?	16.63	16.72
12.95	13.11					16.40	16.80									11.48	?	16.72	16.83
						16.95	17.02					15.91	16.01			11.48	?	16.13	16.26
16.52	16.52					16.95	17.02					15.91	16.01			?	?	16.26	16.31
18.66	19.37	25.98	26.23													22.73	22.76	17.63	17.94
?	?					16.80	16.95					18.49	19.13			?	?	?	?
?	16.52					?	?					17.17	17.45			11.49	11.51	16.35	16.72
?	?					15.89	16.40	2.23	2.33			15.91	16.01			?	?	?	?
16.50	16.52					16.95	17.02					16.74	17.28			?	?	16.72	16.83
						?	?					17.67	17.96					17.58	17.65
16.56	**16.81**					**17.49**	**17.69**					**17.52**	**17.65**			?	?	**17.63**	**17.67**
						17.20	?												
17.42	17.47					17.49	17.69											14.58	14.67
11.35	11.83																		
		19.61	19.78																
17.00	17.12																		
17.12	17.42																		
19.86	20.12	19.61	19.78													22.52	22.62		
		20.69	21.34	17.01	17.18							21.98	22.42						
20.12	21.07	19.85	20.02									20.18	20.57			22.82	22.93		
		26.04	26.11									25.96	25.97			25.93	25.95		
		?	?									23.80	24.03			24.53	24.59		
19.74	19.76											?	?					?	?
18.66	19.37	20.54	20.70									22.68	23.50			22.63	22.73		
18.66	19.37	20.01	20.24																
18.66	19.37							5.38	5.56			25.37	25.50						
20.12	21.07	20.69	21.34									21.48	21.22			22.94	23.38		
		24.94	25.80									23.34	23.80						
		20.55	20.70													22.94	23.38		
		30.69	30.83					?	?			26.53	?			30.29	30.52		
?	?	26.23	26.38									25.89	25.92			25.28	26.12		
		?	?:									23.78	24.11			22.94	23.38		
		26.29	26.45									26.53	?						
		31.21	31.39													22.46	22.50		
		31.72	31.93													32.11	32.94		
		33.67	33.75													33.99	34.03		
		35.86	35.92																
		36.03	**36.15**																
		31.39	31.50													30.89	31.25		
		36.03	**36.15**																
		33.56	33.66																
		30.27	30.56																

sequences from piston cores from the Southern Ocean. In this study, which also incorporated palaeomagnetic data, these workers recorded the occurrence of fewer hiatuses than had been previously identified and identified diachronous diatom events. Burckle & Abrams (1987) concluded that the recognition of regionally synchronous hiatuses in the Southern Ocean required both the reevaluation of the magnetostratigraphy and the development of locally and regionally applicable biozonations.

The choice of an age model used to interpret the stratigraphy of drill sites can have a profound influence on the placement and number of hiatuses within a drillsite. Figure 60 shows the distribution of breaks in the stratigraphical sections based on the age model used in this study. The current interpretation indicates that, in general, the sequences are more continuous with fewer breaks than previously recognized (see Fig 3). Comparing Figs 3 and 60 illustrates the noticeable differences, especially the completeness of the records. In particular, the numerous hiatuses previously identified in Holes 690B, 747A, 748B and 751A are a consequence of the biostratigraphical markers and the age model used. Although the current study identifies more continuous sections, the age model used relies to a greater extent on the record of palaeomagnetic reversals with fewer biostratigraphical control points. Therefore this interpretation is based on a coarser stratigraphical resolution which was achieved by eliminating previously used datums that, according to our data sets are less reliable for stratigraphical control. Although the number of

Table 43. *Primary and potentially useful Quaternary, Neogene and late Palaeogene stratigraphical indicators for the Southern ocean based on this study.*

Primary events
LO *Thalassiosira kolbei*
LO *Thalassiosira vulnifica*
FO *Nitzschia praeinterfrigidaria*
FO *Nitzschia interfrigidaria*
FO *Helotholus vema*
FO *Nitzschia barronii*
FO *Denticulopsis praedimorpha*
FO *Nitzschia denticuloides*
FO *Denticulopsis hustedtii*
LO *Synedra jouseana*
FO *Crucidenticula kanayae*
FO *Isthmolithus recurvus*
LO *Reticulofenestra reticulata*

Potential events
LO *Thallasiosira elliptopora*
LO *Hemidiscus karstenii*
LO *Actinocyclus ingens*
LO *Antarctissa cylindrica*
LO *Cycladophora pliocenica*
LO *Desmospyris spongiosa*
LO *Helotholus vema*
FO *Nitzschia weaveri*
LO *Denticulopsis hustedtii*
LO *Thalassiosira burckliana*
LO *Cycladophora spongothorax*
LO *Asteromphalus kennettii*
FO *Acrosphaera labrata*
FO *Asteromphalus kennettii*
FO *Denticulopsis dimorpha*
LO *Crucidenticula nicobarica*
LO *Coscinodiscus rhombicus*
LO *Thalassiosira fraga*
LO *Chiasmolithus altus*
LO *Reticulofenestra oamaurensis*

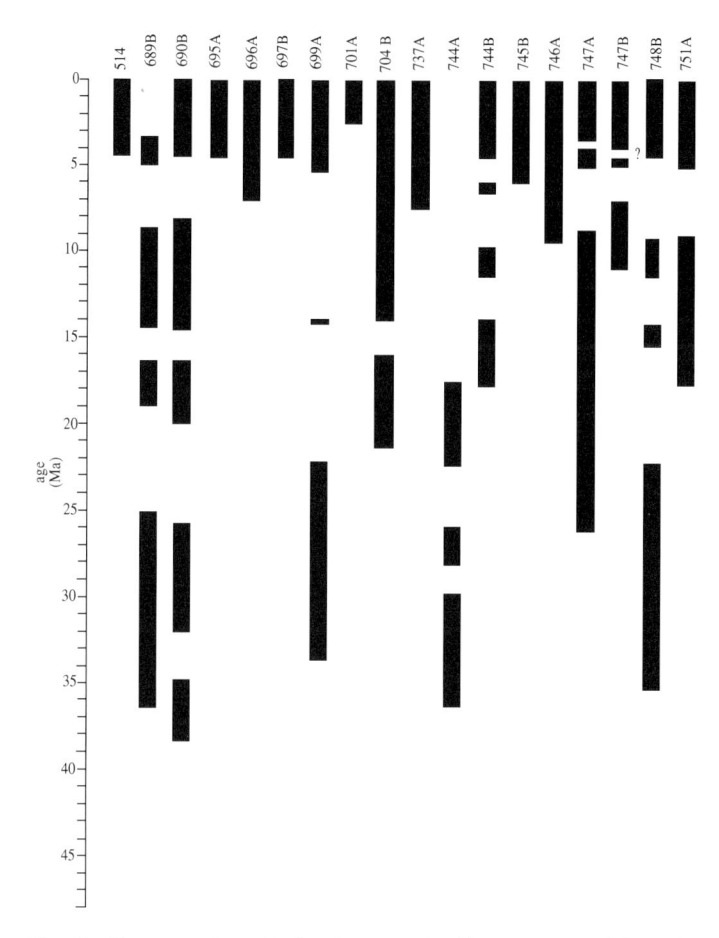

Fig. 60. Chronostratigraphical assignment of sediments recovered from the DSDP and ODP holes used in this study. Solid bars represent intervals containing diatoms, radiolarians or calcareous nannofossils and a polarity record.

biostratigraphical events in the Quaternary and Pliocene provide a reasonable framework the paucity of events in the older stratigraphical sequence limits the available chronostratigraphical resolution when the stratigraphy is constrained only by biostratigraphical events. The chronostratigraphical control is enhanced by using an integrated approach including polarity sequences. It was inevitable that the methodology employed in this study, which included the maintanence of stratigraphical continuity, resulted in the recognition of fewer primary and secondary biostratigraphical markers and fewer unconformities. Thus in some sites e.g. 689B, 690B, 699A and 744A the chronostratigraphy of some intervals is based, in the absence of primary markers, on our reassessment of the polarity sequence. Determination of the polarity stratigraphy for intervals where primary biostratigraphical markers are lacking is based on a coarser stratigraphy that incorporates a broad age assignment of the flora and fauna.

Conclusion

Reinterpretation of the numerous biostratigraphical and palaeomagnetic data sets generated by previous workers for 18 holes in the Southern Ocean have provided the opportunity to reassess the reliability of the current Quaternary–Neogene and late Palaeogene biochronological models. Emphasis, is placed however, on the younger record. This reassessment is focused on Southern Ocean sites that contain both biostratigraphical and polarity reversal records. All data sets were integrated and the polarity reversal records reinterpreted and calibrated to the Geopolarity Timescale of Cande & Kent (1995).

Earlier biochronological models have relied on the assumed isochroneity of specific primary biostratigraphical markers. These markers were used to calibrate the polarity reversal record. In these models 'reliable' primary markers were often identified on the basis of leg-specific results and from secondary and even tertiary correlations. In this study these models were extended by applying an ocean wide approach. The integration of data sets obtained throughout the Southern Ocean provided the first opportunity to assess the accuracy of specific events and determine the reliability of previously identified biochronological models.

The results from this study indicate the paucity of useful biostratigraphical markers and show that the majority of previously used biostratigraphical events (diatoms, radiolarians and calcareous nannofossils) are diachronous throughout the Southern Ocean and less reliable than previously thought. These results suggest that previous palaeoceanographical interpretations will require significant modification.

The biochronology resulting from this study indicates that 13 biostratigraphical events meet the requirements established in this study (age constraint less than 0.25 million years) and are considered as primary biostratigraphical markers. Twenty biostratigraphical events are identified as potential biostratigraphical indicators and additional data are required to further asses their chronostratigraphical potential. The paucity of biostratigraphical events which include diatoms, radiolarians, and calcareous nannofossils has resulted in a much coarser stratigraphy than previously applied.

The results from this study provide an integrated biochronological framework for the Southern Ocean. This framework will continue to be refined as additional stratigraphical sequences are recovered that allow testing the current stratigraphical model and as samples from pre-existing sequences are re-examined to refine the stratigraphical placement of specific species.

References

ABELMANN, A. 1992. Early to middle Miocene radiolarian stratigraphy of the Kerguelen Plateau, Leg 120. *In*: WISE, S. W. JR., SCHLICH, R., PALMER-JULSON, A. & THOMAS, E. (eds) *Proceedings of the Ocean Drilling Program (Scientific Results)*, **120**. College Station, TX (Ocean Drilling Program), 757–784.

—— 1990. Oligocene to middle Miocene radiolarian stratigraphy of Southern high latitudes from Leg 113, sites 689 and 690, Maud Rise. *In*: BARKER, P. F., KENNETT, J. P., O'CONNELL, S. & PISIAS, N. G. (eds) *Proceedings of the Ocean Drilling Program (Scientific Results)*, **113**. College Station, TX (Ocean Drilling Program), 675–708.

ABBOTT, W. H. 1972. *Vertical and lateral patterns of diatomaceous ooze found between Australia and Antarctica*. PhD thesis, University of South Carolina.

AKIBA, F. & YANAGISAWA, Y. 1985. Taxonomy, morphology and phylogeny of the Neogene diatom zonal; marker species in the middle to high latitudes of the North Pacific. *In*: KAGAMI, H., KARIG, D. E., COULBOURN, W. T. *et al.* (eds) *Initial Reports of the Deep Sea Drilling Project*, **87**. Washington (US Govt Printing Office), 483–554.

AUBRY, M. P. 1992. Paleogene calcareous nannofossils from the Kerguelen Plateau, Leg 120. *In*: WISE, S. W. JR., SCHLICH, R., PALMER-JULSON, A. & THOMAS, E. (eds) *Proceedings of the Ocean Drilling Program (Scientific Results)*, **120**. College Station, TX (Ocean Drilling Program), 471–492.

BALDAUF, J. G. & BARRON, J. A. 1991. Diatom biostratigraphy: Kerguelen Plateau and Prydz Bay regions of the Southern Ocean. *In*: BARRON, J., LARSEN, B., BALDAUF, J. G. & ANDERSON, J. (eds) *Proceedings of the Ocean Drilling Program, (Scientific Results)*, **119**. College Station, TX (Ocean Drilling Program), 547–598.

BARKER, P. F., DALZIEL, I. W. D. *et al.* 1976. *Initial Reports of the Deep Sea Drilling Project*, **36**. Washington (US Govt Printing Office).

——, KENNETT, J. P. *et al.* 1988. *Proceedings of the Ocean Drilling Program, (Initial Reports)*, **113**. College Station, TX (Ocean Drilling Program).

——, —— *et al.* 1990. *Proceedings of the Ocean Drilling Program, (Scientific Results)*, **113**. College Station, TX (Ocean Drilling Program).

BARRON, J. A. 1996. Diatom constraints on the problem of the Antarctic Polar Front in the middle part of the Pliocene. *Marine Micropaleontology*, **27**, 195–213.

—— & GLADENKOV, A. Y. 1995. Early Miocene to Pleistocene diatom stratigraphy of Leg 145. *In*: REA, D. K, BASOV, L. A., SCHOLL, D. W. & ALLASN, J. F. (eds) *Proceedings of the Ocean Drilling Program (Scientific Results)*, **145**. College Station, TX (Ocean Drilling Program), 3–19.

——, BALDAUF, J. G., BARRERA, E., CAULET, J. P., HUBER, B. T., KEATING, B. H., LAZARUS, D., SAKAI, H., THIERSTEIN, H. R. & WEI, W. 1991. Biochronologic and magnetochronologic synthesis of Leg 119 sediments from the Kerguelen Plateau and Prydz Bay, Antarctica. *In*: BARRON, J., LARSEN, B., BALDAUF, J. G. & ANDERSON, J. (eds) *Proceedings of the Ocean Drilling Program (Scientific Results)*, **119**. College Station, TX (Ocean Drilling Program), 813–848.

——, LARSEN, B. *et al.* 1989. *Proceedings of the Ocean Drilling Program, (Initial Reports)*, **119**. College Station, TX (Ocean Drilling Program).

——, —— *et al.* 1991. *Proceedings of the Ocean Drilling Program, (Scientific Results)*, **119**. College Station, TX (Ocean Drilling Program).

BERGGREN, W. A., KENT, D. V. & FLYNN, J. J. 1985a. Jurassic to Paleogene: Part 2. Paleogene geochronology and chronostratigraphy. *In*: SNELLING, N. J. (ed.) *The Chronology of the Geological Record*. Geological Society, London, Memoirs, **10**, 141–195.

——, —— & VAN COUVERING, J. A. 1985b. The Neogene: Part 2. Neogene geochronology and chronostratigraphy. *In*: SNELLING, N. J. (ed.) *The Chronology of the Geological Record*. Geological Society, London, Memoirs, **10**, 211–260.

BUKRY, D. 1973a. Low latitude coccolith biostratigraphic zonation. *In*: EDGAR, N. T., SAUNDERS, J. B. *et al.* (eds) *Initial Reports of the Deep Sea Drilling Project*, **15**, Washington (US Govt Printing Office), 685–703.

—— 1973b. Coccolith stratigraphy, eastern Equatorial Pacific, Leg 16 Deep Sea Drilling Program. *In*: VAN ANDEL, T. H., HEATH, G. R. *et al.* (eds) *Initial Reports of the Deep Sea Drilling Project*, **16**. Washington (US Govt Printing Office), 653–711.

—— 1975. Coccolith and silicoflagellate stratigraphy near Antarctica, Deep Sea Drilling Program, Leg 28. *In*: HAYES, D. E., FRAKES, L. A. *et al* (eds) *Initial Reports of the Deep Sea Drilling Project*, **28**. Washington (US Govt Printing Office), 709–715.

BURCKLE, L. H. 1972. Late Cenozoic planktonic diatom zones from the Eastern Equatorial Pacific. *Nova Hedwigia*, **39**, 217.

—— & ABRAMS, N. 1987. Regional late Pliocene: early Pleistocene hiatuses of the Southern Ocean: diatom evidence. *Marine Geology*, **77**, 207–218.

BURNS, D. A. 1975. Nannofossil biostratigraphy for Antarctic sediments, Leg 28, Deep Sea Drilling Program. *In*: HAYES, D. E., FRAKES, L.A. *et al.* (eds) *Initial Reports of the Deep Sea Drilling Project*, **28**. Washington (US Govt Printing Office), 589–598.

CANDE, S. C. & KENT, D. V. 1992. A new geomagnetic polarity time scale for the Late Cretaceous and Cenozoic. *Journal of Geophysical Research*, **97**, 13 917–13 951.

—— & —— 1995. Revised calibration of the geomagnetic polarity time scale for the late Cretaceous and Cenozoic. *Journal of Geophysical Research*, **100**, 6093–6095.

CAULET, J. P. 1982. Faunes de radiolaires et fluctuations climatiques dans les sédiments de l'océan Indien Austral: une nouvelle biozonation. *Bulletin de la Societé Geologique de France*, **24**, 555–562.

—— 1985. Radiolarians from the Southwest Pacific. *In*: KENNETT, J. P, VON DER BORCH, C. C. *et al.* (eds) *Initial Reports of the Deep Sea Drilling Project*, **90**. Washington (US Govt Printing Office), 835–861.

—— 1986. A refined radiolarian biostratigraphy for the Pleistocene of the temperate Indian Ocean. *Marine Micropaleontology*, **11**, 217–229.

—— 1991. Radiolarians from the Kerguelen Plateau, Leg 119. *In*: BARRON, J., LARSEN, B., BALDAUF, J. G., ANDERSON, J. (eds) *Proceedings of the Ocean Drilling Program, (Scientific Results)*, **119**. College Station, TX (Ocean Drilling Program), 513–546.

CHEN, P. H. 1975. Antarctic radiolaria. *In*: HAYES, D. E., FRAKES, L. A. *et al.* (eds) *Initial Reports of the Deep Sea Drilling Project*, **28**, Washington (US Govt Printing Office), 437–513.

CIESIELSKI, P. F. 1983. The Neogene and quaternary diatom biostratigraphy of subantarctic sediments, Deep Sea Drilling Project Leg 71. *In*: LUDWIG, W. J., KRASHENINNIKOV, V. A. *et al.* (eds) *Initial Reports of the Deep Sea Drilling Project*, **71**. Washington (US Govt Printing Office), 635–666.

—— 1986. Middle Miocene to Quaternary diatom biostratigraphy of Deep Sea Drilling Project Site 594, Chatham Rise, Southwest Pacific. *In*: KENNETT, J. P., VON DER BORCH, C. C. *et al.* (eds) *Initial Reports of the Deep Sea Drilling Project*, **90**. Washington (US Govt Printing Office), 863–885.

—— 1991. Relative abundances and ranges of select diatoms and silicoflagellates from Sites 699 and 704, subantarctic South Atlantic. *In*: CIESIELSKI, P. F., KRISTOFFERSEN, Y., CLEMENT, B. & MOORE, T. C. (eds) *Proceedings of the Ocean Drilling Program (Scientific Results)*, **114**. College Station, TX (Ocean Drilling Program), 753–780.

——, KRISTOFFERSEN, Y. *et al.* 1988. *Proceedings of the Ocean Drilling Program (Initial Reports)*, **114**. College Station, TX (Ocean Drilling Program).

——, —— *et al.* 1991. *Proceedings of the Ocean Drilling Program (Scientific Results)*, **114**. College Station, TX (Ocean Drilling Program).

CLEMENT, B. M. & HAILWOOD, E. A. 1991. Magnetostratigraphy of sediments from Sites 701 and 702. *In*: CIESIELSKI, P. F., KRISTOFFERSEN, Y., CLEMENT, B. & MOORE, T. C. (eds) *Proceedings of the Ocean Drilling Program (Scientific Results)*, **114**. College Station, TX (Ocean Drilling Program), 359–366.

——, KENT, D. V. & OPDYKE, N. D. 1996. A synthesis of magnetostratigraphic results from Pliocene–Pleistocene sediments cored using the hydraulic piston corer. *Paleoceanography*, **11**, 299–308.

CRUX, J. A. 1991. Calcareous nannofossils recovered by Leg 114 in the subantarctic South Atlantic Ocean. *In*: CIESIELSKI, P. F., KRISTOFFERSEN, Y., CLEMENT, B. & MOORE, T. C. (eds) *Proceedings of the Ocean Drilling Program (Scientific Results)*, **114**. College Station, TX (Ocean Drilling Program), 155–178.

DONAHUE, J. G. 1970. Pleistocene diatoms as climatic indicators in North Pacific sediments. *In*: HAYS, J. D. (ed.) *Geological Investigations of the North Pacific*. Geological Society of America, Memoirs, **126**, 121–138.

EDWARDS, A. R. 1971. A calcareous nannoplankton zonation of the New Zealand Paleogene. *In*: FARINACCI, A. (ed.) *Proceedings II Planktonic Conference*. Technoscienza, **1**, 381–489.

FENNER, J. M. 1984. Eocene–Oligocene planktonic diatom stratigraphy in the low latitudes and the high southern latitudes. *Micropaleontology*, **30**, 319–342.

—— 1985. Late Cretaceous to Oligocene planktonic diatoms. *In*: BOLLI, H. M., SAUNDERS, J. B. & PERCH-NIELSEN, K. (eds) *Plankton Stratigraphy*. Cambridge University Press, 713–762.

—— 1991. Late Pliocene–Quaternary quantitative diatom stratigraphy in the Atlantic sector of the Southern Ocean. *In*: CIESIELSKI, P. F., KRISTOFFERSEN, Y., CLEMENT, B. & MOORE, T. C. (eds) *Proceedings of the Ocean Drilling Program (Scientific Results)*, **114**. College Station, TX (Ocean Drilling Program), 97–122.

FUNNELL, B. M & RIEDEL, W. R. (eds) 1971. *The Micropalaeontology of Oceans*. Proceedings of the symposium held in Cambridge from 10 to 17 September under the title Micropalaeontology of Marine Bottom Sediments. Cambridge University Press.

FRYXELL, G. A., SIMS, P. A. & WATKINS, T. P. 1986. Azpeitia (Bacillariophyceae): related genera and promorphology. *Systematic Botany Monographs*, **13**, 1–74.

GARD, G. & CRUX, J. A. 1991. Preliminary results from Hole 704A: Arctic-Antarctic correlation through nannofossil biochronology. *In*: CIESIELSKI, P. F., KRISTOFFERSEN, Y., CLEMENT, B. & MOORE, T. C. (eds) *Proceedings of the Ocean Drilling Program (Scientific Results)*, **114**. Ocean Drilling Program, College Station, TX, 193–200.

GERSONDE, R. & BURCKLE, H. 1990. Neogene diatom biostratigraphy of ODP Leg 113, Weddell Sea (Antarctic Ocean). *In*: BARKER, P. F., KENNETT, J. P., O'CONNELL, S. & PISIAS, N. G. (eds) *Proceedings of the Ocean Drilling Program (Scientific Results)*, **113**. College Station, TX (Ocean Drilling Program), 761–790.

——, ABELMANN, A., BURCKLE, L. H., HAMILTON, N., LAZARUS, D. B., MCCARTNEY, K., O'BRIEN, P., SPIESS, V., STOTT, L. D., WEI, W. & WISE, S. W. JR. 1990. Biostratigraphic synthesis of Neogene siliceous microfossils from the Antarctic Ocean, ODP Leg 113 (Weddell Sea). *In*: BARKER, P. F., KENNETT, J. P., O'CONNELL, S. & PISIAS, N. G. (eds) *Proceedings of the Ocean Drilling Program (Scientific Results)*, **113**. College Station, TX (Ocean Drilling Program), 915–936.

GOMBOS, A. M. JR. 1976. Paleogene and Neogene diatoms from the Falkland Plateau and Malvinas outer basin. *In*: BARKER, P. F., DALZIEL, I. W. D. *et al.* (eds) *Initial Reports of the Deep Sea Drilling Project*, **36**. Washington (US Govt Printing Office), 575–690.

—— & CIESIELSKI, P. F. 1983. Late Eocene to early Miocene diatoms from the Southwest Atlantic. *In*: LUDWIG, W. J., KRASHENINNIKOV, V. A. *et al.* (eds) *Initial Reports of the Deep Sea Drilling Project*, **71**. Washington (US Govt Printing Office), 583–634.

GRADSTEIN, F. M., AGTERBERG, F. P., BROWER, J. C. & SCHWARZACHER, W. S. 1985. *Quantitative Stratigraphy*. Kluwer Academic Publishers.

HAILWOOD, E. A. & CLEMENT, B. M. 1991a. Magnetostratigraphy of Sites 699 and 700, East Georgia Basin. *In*: CIESIELSKI, P. F., KRISTOFFERSEN, Y., CLEMENT, B. & MOORE, T. C. (eds) *Proceedings of the Ocean Drilling Program (Scientific Results)*, **114**. College Station, TX (Ocean Drilling Program), 337–358.

—— & ——1991b. Magnetostratigraphy of Sites 703 and 704, Meteor Rise, Southeastern South Atlantic. In: CIESIELSKI, P. F., KRISTOFFERSEN, Y., CLEMENT, B. & MOORE, T. C. (eds) *Proceedings of the Ocean Drilling Program (Scientific Results)*, **114**. College Station, TX (Ocean Drilling Program), 367–386.

HAQ, B. U. 1976. Coccoliths in cores from the Bellinghausen Abyssal Plain and Antarctic continental rise. In: HOLLISTER, C. D., CRADDOCK, C. et al. (eds) *Initial Reports of the Deep Sea Drilling Project*, **35**. US Govt Printing Office, Washington, 557–561.

——, WORSLEY, T. R., BURCKLE, L. H., DOUGLAS, R. G., KEIGWIN, L. D. JR., OPDYKE, L. D., SAVIN, S. M., SOMMER, M. A., VINCENT, E. & WOODRUFF, F. 1980. The late Miocene marine carbon-isotopic shift and the synchroneity of some phytoplanktonic biostratigraphic datums. *Geology*, **8**, 427–431.

HARWOOD, D. M. & MARUYAMA, T. 1992. Middle Eocene to Pleistocene diatom biostratigraphy of Southern Ocean sediments from the Kerguelen Plateau, Leg 120. In: WISE, S. W. JR., SCHLICH, R., PALMER-JULSON, A. & THOMAS, E. (eds) *Proceedings of the Ocean Drilling Program (Scientific Results)*, **120**. College Station, TX (Ocean Drilling Program), 683–734.

——, LAZARUS, D. B., ABELMANN, A., AUBRY, M. P., BERGGREN, W. A., HEIDER, F., INOKUCHI, H., MARUYAMA, I., MCCARTNEY, K., WEI, W. & WISE, S. W. JR. 1992. Neogene integrated magnetobiostratigraphy of the Central Kerguelen Plateau, Leg 120. In: WISE, S. W. JR., SCHLICH, R., PALMER-JULSON, A. & THOMAS, E. (eds) *Proceedings of the Ocean Drilling Program (Scientific Results)*, **120**. College Station, TX (Ocean Drilling Program), 1031–1052.

HAYES, D. E., FRAKES, L. A. et al. 1975. *Initial Reports of the Deep Sea Drilling Project*, **28**. Washington (US Govt Printing Office).

HAYS, J. D. 1965. Radiolaria and late Tertiary and Quaternary history of Antarctic Seas. In: LLANO, G. A. (ed.) *Biology of Antarctic Seas* II. American Geophysical Union Antarctic Research Series, **5**, 125–184.

—— & OPDYKE, N. 1967. Antarctic radiolaria, magnetic reversals, and climatic change. *Science*, **158**, 1001–1011.

HEIDER, F., LEITNER, B. & INOKUCHI, H. 1992. High Southern latitude magnetostratigraphy and rock magnetic properties of sediments from Sites 747, 749, and 751. In: WISE, S. W. JR., SCHLICH, R., PALMER-JULSON, A. & THOMAS, E. (eds) *Proceedings of the Ocean Drilling Program (Scientific Results)*, **120**. College Station, TX (Ocean Drilling Program), 225–246.

HOLLISTER, C. D., CRADDOCK, C. et al. 1976. *Initial Reports of the Deep Sea Drilling Project*, **35**, Washington (US Govt Printing Office).

INOKUCHI, H. & HEIDER, F. 1992. Magnetostratigraphy of sediments from Sites 748 and 750, Leg 120. In: WISE, S. W. JR., SCHLICH, R., PALMER-JULSON, A. & THOMAS, E. (eds) *Proceedings of the Ocean Drilling Program (Scientific Results)*, **120**. College Station, TX (Ocean Drilling Program), 247–254.

JOUSÉ, A. P., KOROLEVA, G. S. & NAGAEVA, G. A. 1963. Stratigraphical and palaeontological investigations in the Indian section of the Southern Ocean. *Okeanol. Issled*, **8**, 137–161.

KEANY, J. 1979. Early Pliocene radiolarian taxonomy and biostratigraphy in the Antarctic region. *Micropaleontology*, **25**, 50–74.

KEATING, B. H. & SAKAI, H. 1991. Magnetostratigraphic studies of sediments from Site 744, southern Kerguelen Plateau. In: BARRON, J., LARSEN, B., BALDAUF, J. G. & ANDERSON, J. (eds) *Proceedings of the Ocean Drilling Program (Scientific Results)*, **119**. College Station, TX (Ocean Drilling Program), 771–794.

KOIZUMI, I. 1972. Marine diatom flora of the Pliocene Tatsunokuchi Formation in Fukushima Prefecture, Northeast Japan. *Transactions and Proceedings of the Paleontological Society of Japan, New Series*, **86**, 340.

——1973. The late Cenozoic diatoms of sites 183–193, Leg 19 Deep Sea Drilling Program. In: CREAGER, J. S., SCHOLL, D. W. et al. (eds) *Initial Reports of the Deep Sea Drilling Project*, **19**, Washington (US Govt Printing Office), 805–838.

——1977. Diatom biostratigraphy in the North pacific region. In: *Proceedings of the First International Congress on Pacific Neogene Stratigraphy*. Tokyo (Kaiyo Shuppan), 235–253.

LA BRECQUE, J. L., HSÜ, K. J. et al. 1983. DSDP Leg 73: Contributions to Palaeogene stratigraphy in nomenclature, chronology and sedimentation rates. *Palaeoecology, Palaeoclimatology, Palaeoecology*, **42**, 91–125.

——, KENT, D. V. & CANDE, S. C. 1977. Revised magnetic polarity time scale for late Cretaceous and Cenozoic time. *Geology*, **5**, 330–335.

LAZARUS, D. B. 1990. Middle Miocene to recent radiolarians from the Weddell Sea, Antarctica, ODP Leg 113. In: BARKER, P. F., KENNETT, J. P., O'CONNELL, S. & PISIAS, N. G. (eds) *Proceedings of the Ocean Drilling Program (Scientific Results)*, **113**. College Station, TX (Ocean Drilling Program), 709–728.

——1992. Antarctic Neogene radiolarians from the Kerguelen Plateau, Legs 119 and 120. In: WISE, S. W. JR., SCHLICH, R., PALMER-JULSON, A. & THOMAS, E. (eds) *Proceedings of the Ocean Drilling Program (Scientific Results)*, **120**. College Station, TX (Ocean Drilling Program), 785–810.

LEDBETTER, M. T. 1983. Magnetostratigraphy of middle-upper Miocene and upper middle Eocene sections in Hole 512. In: LUDWIG, W. J., KRASHENINNIKOV, V. A. et al. (eds) *Initial Reports of the Deep Sea Drilling Project*, **71**. Washington (US Govt Printing Office), 1093–1096.

LING, H. Y. 1991. Cretaceous (Maestrichtian) radiolarians: Leg 114. In: CIESIELSKI, P. F., KRISTOFFERSEN, Y. et al. (eds) *Proceedings of the Ocean Drilling Program (Scientific Results)*, **114**. College Station, TX (Ocean Drilling Program), 317–324.

LUDWIG, W. J., KRASHENINNIKOV, V.A. et al. 1983. *Initial Reports of the Deep Sea Drilling Project*, **71**. Washington (US Govt Printing Office).

MADILE, M. & MONECHI, S. 1991. Late Eocene to early Oligocene calcareous nannofossil assemblages from Sites 699 and 703, Subantarctic South Atlantic Ocean. In: CIESIELSKI, P. F., KRISTOFFERSEN, Y., CLEMENT, B. & MOORE, T. C. (eds) *Proceedings of the Ocean Drilling Program (Scientific Results)*, **114**. College Station, TX (Ocean Drilling Program), 179–192.

MARTINI, E. 1970. Standard Paleogene calcareous nannoplankton zonation. *Nature*, **226**, 560–561.

—— & WORSLEY, T. 1970. Standard Neogene calcareous nannofossil zonation. *Nature*, **225**, 289–290.

MCCOLLUM, D. W. 1975. Diatom stratigraphy of the Southern Ocean. In: HAYES, D. E., FRAKES, L. A. et al. *Initial Reports of the Deep Sea Drilling Project*, **28**. Washington (US Govt Printing Office), 515–571.

MANKINEN, E. A. & COX, A. 1988. Paleomagnetic investigation of some volcanic rocks from the McMurdo volcanic province, Antarctica. *Journal of Geophysical Research*, **93**, 11 599–11 612.

OKADA, H. & BUKRY, D. 1980. Supplementary modification and introduction of code numbers to the 'low latitude coccolith biostratigraphic zonation' (Bukry 1973, 1975). *Marine Micropaleontology*, **5**, 321–325.

OPDYKE, N. D., BURCKLE, L. H. & TODD, A. 1974. The extension of the magnetic time scale in sediments of the central Pacific Ocean. *Earth and Planetary Science Letters*, **22**, 300.

——, GLASS, B., HAYS, J. D. & FOSTER, J. 1966. Paleomagnetic study of Antarctic deep-sea cores. *Science*, **154**, 349–357.

POSPICHAL, J. & WISE, S. W. 1990. Paleocene to middle Eocene calcareous nannofossils of ODP sites 689 and 690, Maud Rise, Weddell Sea. In: BARKER, P. F., KENNETT, J. P., O'CONNELL, S. & PISIAS, N. G. (eds) *Proceedings of the Ocean Drilling Program (Scientific Results)*, **113**. College Station, TX (Ocean Drilling Program), 613–638.

RAMSAY, A. T. S., SYKES, T. J. S. & KIDD, R. B. 1994. Waxing (and waning) lyricalon hiatuses: Eocene–Quaternary Indian Ocean hiatuses as proxy indicators of water mass production. *Paleoceanography*, **9**, 857–977.

ROMAIN, A. J. 1979. Lineages in early Paleogene calcareous nannoplankton. *Utrecht Micropalaeontological Bulletin*, **22**, 1–231.

SALLOWAY, J. C. 1983. Paleomagnetism of sediments from Deep Sea Drilling Project Leg 71. In: LUDWIG, W. J., KRASHENINNIKOV, V. A. et al. (eds) *Initial Reports of the Deep Sea Drilling Project*, **71**. Washington (US Govt Printing Office), 1073–1092.

SAKAI, H. & KEATING, B. 1991. Paleomagnetism of Leg 119-Holes 737A, 738C, 742A, 745B, and 746A. In: BARRON, J., LARSEN, B., BALDAUF, J. G. & ANDERSON, J. (eds) *Proceedings of the Ocean Drilling Program (Scientific Results)*, **119**. College Station, TX (Ocean Drilling Program), 751–770.

SCHLICH, R., WISE, S. W. JR. et al. 1989, *Proceedings of the Ocean Drilling Project, Initial reports*, **120**. Ocean Drilling Program, College Station, TX.

SCHRADER, H. J. 1973. Cenozoic diatoms from the Northeast Pacific, Leg 18. In: KULM, L. D., VON HUENE, R. et al. *Initial Reports of the Deep Sea Drilling Project*, **18**. Washington (US Govt Printing Office), 673–797.

——1976. Cenozoic planktonic diatom biostratigraphy of the Southern Pacific Ocean. In: HOLLISTER, C. D., CRADDOCK, C. et al. (eds) *Initial Reports of the Deep Sea Drilling Project*, **35**. Washington (US Govt Printing Office), 605–672.

SPIESS, V. 1990. Cenozoic magnetostratigraphy of Leg 113 drill sites, Maud Rise, Weddell Sea, Antarctica. In: BARKER, P. F., KENNETT, J. P., O'CONNELL, S. & PISIAS, N. G. (eds) *Proceedings of the Ocean Drilling Program (Scientific Results)*, **113**. College Station, TX (Ocean Drilling Program), 261–318.

STRADNER, H. & EDWARDS, A. R. 1968. Electron microscope studies on upper Eocene coccoliths from the Oamaru Diatomite, New Zealand. *Jahrbuch der Geologischen Bundesanstalt, Sonderband*, **13**, 1–66.

TAKEMURA, A. 1992. Radiolarian Paleogene biostratigraphy in the Southern Indian Ocean Leg 120. In: WISE, S. W. JR., SCHLICH, R., PALMER-JULSON, A. & THOMAS, E. (eds) *Proceedings of the Ocean Drilling Program (Scientific Results)*, **120**. College Station, TX (Ocean Drilling Program), 735–756.

THIERSTEIN, H. R., LAZARUS, D. B. & BECKMANN, J. P. 1994. How synchronous are Neogene marine plankton events. *Paleoceanography*, **9**, 739–763.

TAPPAN, H. 1980. *The Paleobiology of Plant Protists*. W. H. Freeman.

VAROL, O. 1989. Paleocene nannofossil biostratigraphy. In: CRUX, J. A. & VAN HECK, S. E (eds) *Nanofossils and their applications*. Chichester, 267–310.

WEAVER, F. M. 1976. Antarctic radiolaria from the southeast Pacific Basin, Deep Sea Drlling Project, Leg 35. In: HOLLISTER, C. D., CRADDOCK, C. et al. (eds) *Initial Reports of the Deep Sea Drilling Project*, **35**. US Govt Printing Office, Washington, 569–601.

——1983. Cenozoic radiolarians from the southwest Atlantic, Falkland Plateau Region, Deep Sea Drilling Project Leg 71. In: LUDWIG, W. J., KRASHENINNIKOV, V. A. et al. (eds) *Initial Reports of the Deep Sea Drilling Project*, **71**. Washington (US Govt Printing Office), 667–686.

—— & GOMBOS, A. M. 1981. Southern high-latitude diatom biostratigraphy. In: WARME, T. E., DOUGLAS, R. G. & WINTERER, E. L. (eds) *The Deep Sea Drilling Project: A Decade of Progress*. Society of Economic Paleontologists and Mineralogists, Special Publications, **32**, 445–470.

WEI, W. 1991. Middle Eocene–lower Miocene calcareous nannofossils magnetobiobronolgy of ODP holes 699A and 703A in the subantarctic South Atlantic. *Marine Micropaleontology*, **18**, 143–165.

——1992. Paleogene chronology of southern ocean drill holes: an update. In: KENNETT, J. P. & WARNKE, D. A. (eds) *The Antarctic paleoenvironment: a perspective on global change*. American Geophysical Union Antarctic Research Series, **56**, 75–96.

—— & THIERSTEIN, H. R. 1991. Upper Cretaceous and Cenozoic calcareous nannofossils of the Kerguelen Plateau (Southern Indian Ocean) and Prydz Bay (East Antarctica). In: BARRON, J., LARSEN, B., BALDAUF, J. G. & ADERSON, J. (eds) *Proceedings of the Ocean Drilling Program (Scientific Results)*, **119**. College Station, TX (Ocean Drilling Program), 467–492.

—— & WISE, S. W. JR. 1990. Middle Eocene to Pleistocene calcareous nannofossils recovered by the Ocean Drilling Program Leg 113 in the Weddell Sea. In: BARKER, P. F., KENNETT, J. P., O'CONNELL, S. & PISIAS, N. G. (eds) *Proceedings of the Ocean Drilling Program (Scientific Results)*, **113**. College Station, TX (Ocean Drilling Program), 639–666.

—— & ——1992a. Oligocene–Pleistocene calcareous nannofossils from Southern Ocean Sites 747, 748, and 751. In: WISE, S. W. JR., SCHLICH, R., PALMER-JULSON, A. & THOMAS, E. (eds) *Proceedings of the Ocean Drilling Program (Scientific Results)*, **120**. College Station, TX (Ocean Drilling Program), 509–522.

—— & ——1992b. Selected Neogene calcareous nannofossil index taxa of the Southern Ocean: biochronology, biometrics, and paleoceanography. In: WISE, S. W. JR., SCHLICH, R., PALMER-JULSON, A. & THOMAS, E. (eds) *Proceedings of the Ocean Drilling Program (Scientific Results)*, **120**. College Station, TX (Ocean Drilling Program), 523–538.

——, VILLA, G. & WISE, S. W. JR. 1992. Paleoceanographic implications of Eocene–Oligocene calcareous nannofossils from sites 711 and 748 in the Indian Ocean. In: WISE, S. W. JR., SCHLICH, R., PALMER-JULSON, A. & THOMAS, E. (eds) *Proceedings of the Ocean Drilling Program (Scientific Results)*, **120**. College Station, TX (Ocean Drilling Program), 979–999.

WISE, S.W. JR. 1983. Mesozoic and Cenozoic calcareous nannofossils recovered by Deep Sea Drilling Project Leg 71 in the Falkland Plateau region, Southwest Antarctic Ocean. In: LUDWIG, W. J., KRASHENINNIKOV, V. A. et al. (eds) *Initial Reports of the Deep Sea Drilling Project*, **71**. Washington (US Govt Printing Office), 481–550.

—— & CONSTANS, R. E. 1976. Mid Eocene planktonic correlations: Northern Italy–Jamaica, W.I. *Transactions of the Gulf Coast Associations of Geological Societies*, **26**, 144–155.

——, SCHLICH, R. et al. 1989. *Proceedings of the Ocean Drilling Program (Initial Reports)*, **120**. College Station, TX (Ocean Drilling Program).

——, —— et al. 1992. *Proceedings of the Ocean Drilling Program (Scientific Results)*, **120**. College Station, TX (Ocean Drilling Program).

—— & WIND, F. H. 1977. Mesozoic and Cenozoic calcareous nannofossils recovered by DSDP Leg 36 drilling of the Falkland Plateau, Southwest Atlantic sector of the Southern Ocean. In: BARKER, P. F., DALZIEL, I. W. D. et al. (eds) *Initial Reports of the Deep Sea Drilling Project*, **36**. Washington (US Govt Printing Office), 269–492. YANAGISAWA, Y. & AKIBA, A. 1985. Taxonomy and phylogeny of the three marine diatom genera *Crucidenticula*, *Denticulopsis* and *Neodenticula*. *Bulletin of the Geological Survey of Japan*, **41**, 197–301.

Index

abundance data 17
Acrosphaera
  australis 8, 11, 110
    Leg 113   27, 30
    Leg 119   72, 74
    Leg 120   92, 94, 96, 99, 102, 104
  labrata 11, 16, 107, 110, 114
    Leg 113   26, 27, 30
    Leg 120   83, 84, 88, 92, 94, 102, 104
Actinocyclas, ingens 86
Actinocyclus
  actinochilus 9, 108
    Leg 119   71, 74
    Leg 120   84, 87, 91, 93, 101, 104
  fryxellae 11, 110
    Leg 119   64, 66, 72, 74
  ingens 9, 13, 108, 112, 114
    Leg 71   23, 25
    Leg 113   26–8, 30–1, 33–5, 37–8, 40, 42, 43, 45, 46, 48
    Leg 114   50, 51, 55, 57–8, 60–1
    Leg 119   64, 66, 71–2, 74–5, 76, 78, 80, 82
    Leg 120   84–5, 87–9, 91, 93, 96, 99, 102, 104
  ingens v. *nodus* 11, 13, 110, 112
    Leg 113   27–8, 30–1, 35, 38
    Leg 114   58, 61
    Leg 119   72, 75, 80, 82
    Leg 120   85, 88, 102, 104–5
Actinomma, tanyacantha 6
Actinomma golownini 11, 110
  Leg 113   27, 30–1, 34, 37
  Leg 119   72, 74
  Leg 120   84–5, 88, 92, 94, 96, 99, 102, 104–5
age models 13, 14–15, 17, 20, 113
  Age Models 1 and 2   21
age–depth graphs 8, 15
  Hole 514   23, *23*, *25*
  Hole 689B   26, *29*, *32*
  Hole 690B   33, *36*, *39*
  Hole 695A   40, *41*, *42*
  Hole 696A   43, *44*, *45*
  Hole 697B   46, *47*, *48*
  Hole 699A   49, *49*, *53*
  Hole 701A   54, *54*, *56*
  Hole 704B   57, *59*, *62*
  Hole 737A   63, *64*, *66*
  Hole 744A   67, *68*, *70*
  Hole 744B   71, *73*, *75*
  Hole 745B   76, *77*, *79*
  Hole 746A   80, *81*, *82*
  Hole 747A   83, *86*, *90*
  Hole 747B   91, *92*, *94*
  Hole 748B   95, *97*, *100*
  Hole 751A   101, *103*, *105*
Amphymenium, challangerae 8, 9, 11, 108, 110
  Leg 114   57–8, 60
Antarctissa
  cylindrica 9, 107, 108, 114
    Leg 113   26, 30, 33, 37, 40, 42
    Leg 119   76, 78
    Leg 120   84, 87, 101, 104
  deflandrei 11, 110
    Leg 113   27, 31, 34, 38
Antartica, Southern Ocean 3, 6–8
Asteromphalus kennettii 8, 11, 110, 114
  Leg 113   26, 27, 30, 34, 37
  Leg 114   58, 61
  Leg 119   8, 72, 74, 80, 82
  Leg 120   83, 84, 88, 92, 94, 95, 96, 98–9, 101, 102, 104
Atlantic sector, Southern Ocean 7, 21
Azpeitia, gombosi 95

biochronological models 1, 3, 6, 8, 12–14, 16, 115
  limitations 13–14
biostratigraphical and magnetostratigraphical integrations
  Hole 514   20, 23, 23–4, 25
  Hole 689B   26, 26–8, 30–2
  Hole 690B   33, 33–5, 37–8
  Hole 695A   40, 40, 42
  Hole 696A   43, 43, 45
  Hole 697B   46, 46, 48
  Hole 699A   59, 50–1, 51–2
  Hole 701A   54, 54, 55
  Hole 704B   57, 57–9, 60–2
  Hole 737A   63, 63–4, 65–6
  Hole 744A   67, 67–8, 69–70
  Hole 744B   71, 71–3, 74–5
  Hole 745B   76, 76–7, 78–9
  Hole 746A   80, 80, 82
  Hole 747A   83–6, 84–6, 87–9
  Hole 747B   91, 91–2, 93–4
  Hole 748B   95, 95–7, 98–100
  Hole 751A   101, 101–3, 104–5
biostratigraphies 3, 6, 7, 8, 16
biostratigraphy
  calibrated to polarity stratigraphies 1
  potential indicators 21, 106, 107, 114, 115
  primary events 20, 21, 106–7, 112, 114, 115
    Miocene 107, 112
    Oligocene–late Eocene 107
    Pliocene–Quaternary 106–7
  zonations 17
Blackites, spinosus 6, 7
Bogorovia veniamini 13, 112
  Leg 114   51–2, 59, 61
  Leg 119   67, 69
  Leg 120   85, 89, 97, 99

calcareous nannofossils 3, 6, 7, 8, 14–15, 16, 20, 106, 107
  primary stratatigraphical events 107
  zonation 7, 8
calcareous zooplankton 16
Calcidiscus
  leptoporus 8, 13, 112
    Leg 119   73, 75
    Leg 120   8, 85, 89, 96, 99, 103, 105
  macintyrei 8, 11, 13, 110, 112
    Leg 114   57–8, 60–1
    Leg 119   72–3, 74–5
    Leg 120   8, 84–5, 88–9, 96, 99, 102, 105
Cestodiscus peplum 11, 110
  Leg 114   58, 61
Chiasmolithus
  oamaruensis 7
  solitus 7
Chiasmolithus altus 13, 107, 112, 114
  Leg 71   6
  Leg 113   7, 28, 31, 35, 38
  Leg 114   7, 51–2
  Leg 119   67, 67, 69
  Leg 120   86, 89, 97, 100
chronostratigraphic assignment of sediments *15*, 114
chronostratigraphical control 20
chronozones 17, 20–1
Clausicoccus, fenestratus 6
coccolith zonation 3, 7
Coscinodiscus
  insignis 7
  lewisianus 3, 13, 112
    Leg 113   28, 31
    Leg 114   58, 61

    Leg 119   71, 74
    Leg 120   85, 88, 96, 99
  rhombicus 13, 107, 112, 114
    Leg 113   26, 28, 31
    Leg 114   58, 61
  vulnificus see *Thalassiosira vulnifica*
  yabei see *Thalassiosira yabei*
Cosmiodiscus intersectus 9, 11, 108, 110
  Leg 113   7, 27, 30, 34, 37, 40, 42, 46, 48
  Leg 114   57–8, 60
  Leg 119   64, 66, 80, 82
Cretaceous sediments 3, 26, 83
Crucidenticula
  kanayae 13, 107, 112, 112, 114
    Leg 113   26, 28, 31, 33, 35, 38
    Leg 114   58, 61
    Leg 119   71, 72–3, 75
    Leg 120   85, 88–9, 95, 96, 99, 101, 103, 105
  nicobarica 3, 11, 13, 110, 112, 114
    Leg 113   27–8, 30–1, 34–5, 37–8
    Leg 114   58, 61
    Leg 119   75
    Leg 120   102, 105
Cycladophora
  davisiana 9, 108
    Leg 113   24, 25, 27, 30, 33, 37, 40, 42, 46, 48
    Leg 114   55
    Leg 119   76, 78
    Leg 120   84, 87, 95, 98, 101, 104
  golli regipileus 13, 112
    Leg 113   28, 31, 35, 38
    Leg 119   72, 75
  humerus 11, 110
    Leg 113   27–8, 30–1, 34, 37–8
    Leg 119   72, 75, 80, 82
    Leg 120   96, 99, 102, 104
  pliocenica 9, 108, 114
    Leg 113   26, 30, 33, 37, 40, 42
    Leg 119   71, 74, 76–7
    Leg 120   84, 87, 95, 98, 101, 104
  spongothorax 11, 110, 114
    Leg 113   26, 27, 30, 34, 37
    Leg 119   72, 74, 80, 82
    Leg 120   83, 85, 88, 92, 94, 96, 99, 102, 104
Cyclicargolithus
  abisectus 13, 112
    Leg 71   6
    Leg 113   35, 38
    Leg 114   51, 52
    Leg 119   68, 70
    Leg 120   97, 100
  floridanus 8
Cyrtocapsella
  japonica 11, 110
    Leg 114   58, 61
  longithorax 13, 112
    Leg 113   28, 31, 35, 38
    Leg 119   67, 69
  tetrapera 13, 112
    Leg 113   28, 31, 35, 38
    Leg 114   58, 61

data processing 20–1
data quality 16–17
datums, first or last species occurrence 16, 17
Dendrospyris megalocephalis 11, 13, 110, 112
  Leg 113   27–8, 30
  Leg 119   72, 75
  Leg 120   96, 99
Denticula hustedtii see *Denticulopsis hustedtii*
Denticula lauta see *Denticulopsis lauta*
Denticula nicobarica see *Crucidenticula nicobarica*

GENERAL INDEX

Denticulopsis
  dimorpha 9, 11, 108, 110, 112, 114
    Leg 113 27, 30, 34, 37
    Leg 114 50, 52, 58, 60–1
    Leg 119 8, 71, 74, 76, 78, 80, 82
    Leg 120 83, 84–5, 86, 88, 91, 93, 95, 96, 98–9, 102, 104
  hustedtii 9, 11, 107, 108, 110, 114
    Leg 35 3
    Leg 113 7, 26, 33, 27, 31, 34, 38, 40, 42, 43, 45, 46, 46, 48
    Leg 114 50–1, 57, 57–8, 60–1
    Leg 119 63, 64, 66, 71, 72, 75
    Leg 120 85, 86, 88, 95, 96, 99, 101, 102, 105
  lauta 11, 13, 110, 112, 112
    Leg 28 3
    Leg 36 6
    Leg 114 50–1, 52, 58, 60–1
    Leg 119 72, 74–5
    Leg 120 85, 86, 89, 96, 99, 102, 105
  maccollunmii 11, 13, 110, 112
    Leg 113 7, 27–8, 31, 34–5, 38
    Leg 114 58, 61
    Leg 119 72, 74–5
    Leg 120 85, 88–9, 96, 99, 102–3, 105
  praedimorpha 10, 107, 110, 112, 114
    Leg 113 7, 26, 27, 30, 33, 34, 37
    Leg 119 71, 72, 74–5
    Leg 120 85, 86, 88, 92, 94, 95, 96, 99, 102, 104–5
Desmospyris spongiosa 9, 108, 114
  Leg 35 3
  Leg 71 23, 24, 25
  Leg 113 26, 26, 30, 33, 37, 40, 42, 46, 48
  Leg 114 50–1, 55
  Leg 119 63, 65, 71, 74, 76, 78
  Leg 120 84, 87, 95, 98, 101, 104
Diartus hughesi 11, 110
  Leg 114 58, 60–1
diatom zonation 3, *4–5*, 6–7, 112
diatom zones/polarity chronozone comparisons
  Hole 514 23, *24*
  Hole 689B 26, *29*
  Hole 690B 33, *36*
  Hole 695A 40, *41*
  Hole 696A 43, *44*
  Hole 697B 46, *47*
  Hole 699A 49, *50*
  Hole 701A 54, *55*
  Hole 704B 57, *59*
  Hole 737A 63, *65*
  Hole 744A 67, *69*
  Hole 744B 71, *73*
  Hole 745B 76, *78*
  Hole 746A 80, *81*
  Hole 747A 83, *87*
  Hole 747B 91, *93*
  Hole 748B 95, *98*
  Hole 751A 101, *103*
diatoms 15, 16, 17, 20, 106
  changes to taxonomic nomenclature 16
  primary stratigraphical events 106, 107
  Southern Ocean and North Pacific datum levels 1, 107–12
  species distribution in Southern Ocean 17
Discoaster, saipanensis 7
Discoaster browerii 9, 108
  Leg 114 57, 60
Drake passage 3
drilling sites 1, *2*, 3–6
  71-512 (DSDP) 6
  71-514 (DSDP) 16, 23–5, 106
  71-516 (DSDP) 7
  113-689B (ODP) 6, 7, 17, 26–32, 106, 107, 114
  113-690B (ODP) 6, 8, 17, 33–9, 107, 113, 114

  113-693 (ODP) 16
  113-695A (ODP) 16, 40–2, 106, 107
  113-696A (ODP) 16, 43–5, 107
  113-697B (ODP) 16, 46–8, 107
  114-699A (ODP) 7, 8, 17, 49–53, 106, 107, 114
  114-700 (ODP) 8
  114-701A (ODP) 7, 16, 17, 54–6
  114-702 (ODP) 7, 8
  114-703 (ODP) 7, 8
  114-704B (ODP) 7, 17, 57–62, 106, 107
  119-737A (ODP) 8, 63–6, 106
  119-738 (ODP) 8
  119-744A (ODP) 7, 8, 67–70, 107, 114
  119-744B (ODP) 7, 8, 71–5, 106, 107
  119-745B (ODP) 8, 17, 76–9, 106, 107
  119-746A (ODP) 8, 16, 80–2
  120-747A (ODP) 83–90, 106, 107, 113
  120-747B (ODP) 91–4, 106, 107
  120-748B (ODP) 7, 8, 95–100, 106, 107, 113
  120-751A (ODP) 101–5, 107, 113
  266 (DSDP) 6
  278 (DSDP) 6
  328 (DSDP) 6
  geographical locations *2*
  latitude and longitude 1
DSDP (Deep Sea Drilling Project) 1, *2*, 3–6, 16, 17, 23–5

Emiliana huxleyii 9, 108
  Leg 120 84, 87
Eocene sediments
  Leg 36 3–6
  Leg 71 6
  Leg 113 7
  Leg 114 7, 49
  Leg 119 8, 63, 67
  Leg 120 8, 83, 95
Eucyrtidium
  calvertense 9, 108
    Leg 113 26, 30, 33, 37
    Leg 119 71, 74, 76, 78
    Leg 120 84, 87, 95, 98, 101, 104
  pseudoinflatum 11, 110
    Leg 113 27, 30, 34, 37
    Leg 119 72, 75, 80, 82
    Leg 120 84, 88, 92, 94, 96, 99, 102, 104
  punctatum 13, 112
    Leg 113 28, 31

Falkland Plateau, Southern Ocean 3, 6
Fasciculithus, involutus 6
FOs (first occurences) 16, 17, 20, 106–7

Geopolarity Timescales (GPTS) 20
  Berggren (1985*a, b*) 6, 17, 18, 20
  Cande and Kent (1995) 9–13, 15, 17, 18, 20–1, 106, 108–13, 115
  comparison of different versions 18
  La Brecque (1983) 17
  Opdyke (1966) 3
Globorotalia, acostaesis 6

Helicosphaera sellii 9, 108
  Leg 114 57, 60
Heliolithus, universus 6
Helotholus vema 9, 106, 107, 108, 114
  Leg 35 3
  Leg 71 23, 24, 25
  Leg 113 26, 26–7, 30, 33, 33–4, 37, 40, 42, 46, 48
  Leg 114 50–1, 55
  Leg 119 63, 63, 65, 71, 74, 76, 76–7, 78–9
  Leg 120 83, 84, 87–8, 91, 91, 93, 95–6, 98, 101, 104

Hemiaulus triangularis 11, 110
Hemidiscus
  cuneiformis 9, 11, 108, 110
    Leg 71 24, 25
    Leg 114 50–1, 51–2, 57–8, 60–1
    Leg 119 72, 74, 76, 78
    Leg 120 84, 88
  karstenii 9, 11, 108, 110, 114
    Leg 114 50, 51–2, 55, 57, 60
    Leg 119 63, 65, 69, 74, 76, 78, 80, 82
    Leg 120 91, 93, 96, 98
  ovalis 11, 110
    Leg 120 84, 88, 91, 96, 99, 102, 104
  triangularis 11
    Leg 120 84, 88
hiatuses *see* unconformities and hiatuses

Indian Ocean sector, Southern Ocean 7–8, 21, 106
Isthmolitus recurvus 13, 26, 107, 112, 114
  Leg 36 6
  Leg 113 7, 28, 32, 33, 35, 38
  Leg 114 7, 51, 52
  Leg 119 67, 68, 70
  Leg 120 97, 100

Kerguelan Plateau, Southern Ocean 7–8

Lamprocyclas aegles 11, 110
  Leg 114 60
Lampromitra coronata 9, 108
  Leg 114 58
  Leg 120 84, 87, 91, 93, 96, 98, 101, 104
Lisitzina ornata 13, 112
  Leg 113 35, 38
  Leg 119 67, 69
  Leg 120 85–6, 89, 97
LOs (last occurences) 16, 17, 20, 106–7
Lychocanium, grande 83

magnetic data sets 17, 20
  difficulties of data collection 17
  polarity reversals 15, 17, 20, 115
magnetochrons
  Bruhnes, Gauss, Gilbert 7, *24*, 55
  Matuyama 7, 17
magnetostratigraphy 3, 8, 16, 20, 71
  see also biostratigraphical and magnetostratigraphical integration
Malavinas Outer Basin 6
Matuyama magnetic epoch 7, 17
methodology 16–21, 20
  iterative process of re-calibration 21, 106
Mid-Atlantic Ridge, Southern Ocean 6
Miocene sediments
  Leg 28 3
  Leg 35 3
  Leg 36 3–6
  Leg 71 6
  Leg 113 6–7, 26, 33, 40
  Leg 114 7, 49, 57
  Leg 119 8, 63, 67, 71, 76, 80
  Leg 120 8, 83, 91, 95, 101

nannofossils *see* calcareous nannofossils
Navicula wisei 9, 11, 108, 110
  Leg 120 96, 98, 101–2, 104
Neobrunia miraibilis 9, 108
  Leg 120 84, 88, 96, 99
Neogene sediments 3, 6–7, 8
Nitzschia
  angulata 9, 108
    Leg 114 50, 51, 57, 60

GENERAL INDEX

barronii 9, 106, 108, 114
   Leg 113 26, 27, 30, 33, 33–4, 37, 40, 40, 42, 43, 43, 45, 46, 46, 48
   Leg 119 63, 63, 65, 71, 72, 74, 76, 77, 79
   Leg 120 83, 84, 87–8, 91, 91, 93, 95, 95–6, 98, 101, 101–2, 104
cylindrica 9, 108
   Leg 119 76, 78
denticuloides 11, 107, 110, 114
   Leg 36 6
   Leg 113 26, 27, 30–1, 33, 34, 37
   Leg 114 49, 51, 52, 57, 58, 61
   Leg 119 71, 72, 74–5
   Leg 120 83, 85, 86, 88, 96, 99, 101, 102, 104–5
fossilis 9, 11, 108, 110
   Leg 114 57–8, 60, 77, 79
grossepunctata 11, 13, 110, 112
   Leg 36 6
   Leg 113 27–8, 30–1, 34–5, 38
   Leg 114 58, 61
   Leg 119 72, 74–5
   Leg 120 85, 86, 88–9, 96, 99, 101, 102, 105
interfrigidaria 9, 106, 108, 114
   Leg 36 5
   Leg 71 23, 24, 25
   Leg 113 7, 27, 30, 33, 33–4, 37, 40, 42, 46, 46, 48
   Leg 114 49, 50, 51, 55
   Leg 119 63, 63, 65, 71, 71–2, 74, 76, 76–7, 78–9
   Leg 120 83, 84, 87–8, 91, 91, 93, 95, 95–6, 98, 101–2, 104
kerguelensis 9, 108
   Leg 71 24, 25
   Leg 113 33, 37, 40, 42, 43, 45
   Leg 119 63, 65, 71, 74, 76, 78
   Leg 120 83, 84, 87, 91, 93, 101, 104
maleinterpretaria 13, 112
   Leg 113 28, 31, 34–5, 38
   Leg 114 58, 61
   Leg 119 67, 69, 72, 75
   Leg 120 96–7, 99, 103, 105
marina 9, 11, 108, 110
   Leg 114 57–8, 60
   Leg 119 80, 82
miocenica 9, 108
   Leg 119 80, 82
porteri 11, 110
   Leg 114 58, 60
praecurta 9, 108
   Leg 113 43, 45, 46, 48
   Leg 120 84, 88, 102, 104
praeinterfrigidaria 9, 106, 108, 114
   Leg 36 6
   Leg 71 23, 24, 25
   Leg 113 27, 30, 34, 37, 40, 40, 42, 43, 43, 45, 46, 48
   Leg 114 49, 50, 51
   Leg 119 63, 63, 65, 71, 71–2, 74, 76, 76–7, 78–9
   Leg 120 83, 84, 88, 91, 91, 93, 95, 96, 98, 101, 101–2, 104
reinholdii 9, 11, 108, 110
   Leg 71 24, 25
   Leg 113 7, 27, 30
   Leg 114 50, 51–2, 57–8, 60
   Leg 119 64, 66, 72, 74, 76, 78
   Leg 120 83, 92, 93, 96, 102, 104
weaveri 9, 108, 114
   Leg 71 24, 25
   Leg 113 40, 42
   Leg 114 50, 51, 55, 57, 60
   Leg 119 63, 65, 72, 74, 76, 78–9
   Leg 120 83, 84, 87, 91, 93, 101, 104

North Pacific 1, 6, 7, 107, 112
NRM (natural remnant magnetisation) 17

ODP (Ocean Drilling Program) 1, *2*, 3
   Leg 113 6–7, 16, 17, 26–48, 106
   Leg 114 7, 16, 17, 49–62
   Leg 119 7–8, 16, 17, 63–82, 106
   Leg 120 8, 16, 17, 83–105
Oligocene sediments
   Leg 28 3
   Leg 35 3
   Leg 36 3–6
   Leg 71 6
   Leg 113 7
   Leg 114 7, 49, 57
   Leg 119 8, 63, 67
   Leg 120 8, 83, 95

Pacific sector, Southern Ocean 3, 106
Palaeocene sediments
   Leg 36 3–6
   Leg 71 6
   Leg 113 7, 26, 33
   Leg 114 7, 49
   Leg 120 8, 83
palaeomagnetism 15, 16, 17–21
palaeontological data sets 16
planktonic foraminifera 8, 14–15, 16
Pliocene sediments
   Leg 28 3
   Leg 35 3
   Leg 36 3–6
   Leg 71 6, 23
   Leg 113 6–7, 33, 40, 43, 46
   Leg 114 7, 49, 54, 57
   Leg 119 9, 63, 67, 71, 76, 80
   Leg 120 8, 83, 91, 95, 101
Polar Frontal Zone 7, 17
polarity calibration 17
polarity sequences (patterns) 8, 17, 20–1, 114
previous work *1*, 3–15
   Leg 71 23
   Leg 113 26, 33, 40, 43, 46
   Leg 114 49, 54, 57
   Leg 119 63, 67, 71, 76, 80
   Leg 120 83, 86, 91, 95, 101
Prunopyle titan 9, 108
   Leg 113 27, 30, 34, 37, 40, 42
   Leg 114 50, 51
   Leg 119 71, 74, 77, 79
   Leg 120 84, 87, 95, 98, 102, 104
Pseudomiliana lacunosa 9, 108
   Leg 114 57, 60
Pterocanium trilobium 9, 108
   Leg 120 84, 87

Quaternary sediments
   Leg 28 3
   Leg 35 3
   Leg 36 3–6
   Leg 71 6, 23
   Leg 113 6–7, 26, 33, 40, 43, 46
   Leg 114 7, 49, 54, 57
   Leg 119 8, 63, 67, 71
   Leg 120 8, 83, 95, 101

radiolarians 15, 16, 20, 106
   primary stratigraphical events 106
Raphidodiscus marylandicus 11, 13, 110, 112
   Leg 113 28, 31, 35, 38
   Leg 114 58, 61
   Leg 119 67, 69, 72, 75
   Leg 120 85, 89

Reticulofenestra
   bisecta 13, 112
      Leg 71 6
      Leg 113 28, 31, 35, 38
      Leg 114 7, 51, 52
      Leg 119 67, 69
      Leg 120 85, 89, 97, 99
   daviesii 6, 7
   gelida 8, 11, 110
      Leg 113 27, 30, 34, 37
   hesslandii 8, 11, 110
      Leg 113 27, 30, 34, 37
      Leg 119 73, 75
      Leg 120 84–5, 87–8, 102, 104
   oamaruensis 13, 112, 114
      Leg 71 6
      Leg 113 7, 28, 32
      Leg 114 49, 51, 52
      Leg 119 68, 70
      Leg 120 95, 97, 100
   perplexa 8
   pseudoumbilica 11, 110
      Leg 114 57, 60
      Leg 119 72, 74
      Leg 120 85, 89, 96, 99, 102, 104
   reticulata 13, 107, 112, 114
      Leg 113 7, 26, 28, 32, 33, 35, 38
      Leg 119 67, 69, 70
   umbilica 13, 112
      Leg 113 7, 28, 31, 35, 38
      Leg 114 51, 52
      Leg 119 68, 70
Rhizosolenia
   barboi see *Simonseniella barboi*
   costata 83
   miocenica 9, 108
      Leg 119 77, 79
   oligocenica 13, 112
      Leg 119 68, 70
Rocella
   gelida 13, 112
      Leg 113 28, 31, 35, 38
      Leg 114 51, 52, 59, 61
      Leg 119 67, 69
      Leg 120 85–6, 89, 97, 99–100
   schraderi 13, 112
      Leg 114 58, 61, 67, 69
   vigilans 13, 112
      Leg 113 28, 31, 35, 38
      Leg 114 59, 61
      Leg 119 68, 70, 77, 79
      Leg 120 85–6, 89, 97, 100
   vigilans var. A 95
Rosiella symmetrica 95
Rouxia
   californica 9, 108
      Leg 119 76, 78
   diploneides 9, 108
      Leg 120 84, 88
   heteropolara 9, 11, 108, 110
      Leg 113 43, 45, 46, 48
      Leg 119 64, 66, 76–7, 78–9
      Leg 120 84, 88, 96, 98, 102, 104

sedimentary facies
   calcareous nannofossil ooze 57, 63, 67, 71, 83, 91, 95, 101
   calcareous ooze 33
      with diatoms and radiolarians 49
      siliceous 57
      with variable siliceous component 26
   chalk 26, 49, 57
      with calcareous nannofossils 83
      with chert 95
   clay 49

sedimentary facies (*continued*)
  claystone 63
  diatom ooze 23, 26, 33, 63, 67, 71, 80, 91, 95
    with ash and mud 54
    with calcareous nannofossils 83
    with chert 101
    with variable clay content 76
    with variable terrigenous content 40
  limestone 63
  mud
    with diatoms 40
    with variable siliceous content 43, 46
  siliceous ooze
    calcareous 57
    with calcareous nannofossils 49
sedimentation rates 20
siliceous zooplankton 16
Simonseniella barboi 9, 11, 13, 108, 110, 112
  Leg 71 6, 23, 25
  Leg 113 7, 34, 37, 46, 48
  Leg 114 7, 50-1, 51-2, 55, 57-8, 60-1
  Leg 119 63, 65, 71, 74, 76, 78
  Leg 120 84, 87, 96, 98, 101, 104
Siphonosphaera, vesuvius 8
site selection 16, 17
South Atlantic sector, Southern Ocean 106
Southern Ocean
  Atlantic sector 7, 21, 106
  biochronological models 1, 3, 8-14
  drilling sites locations 1
  Falkland Plateau 3, 6
  Indian Ocean sector 7-8, 21, 106
  isochroneity with North Pacific events 1, 7, 107-12
  Kerguelan Plateau 7-8
  Mid-Atlantic Ridge 6
  Pacific sector 3, 106
  previous workers 1
  Weddell Sea 6-7, 8
  see also drilling sites
species concepts 16
Stephanopyxis grunowii 9, 108
  Leg 119 76, 78
Stichocorys peregrina 9, 11, 108, 110
  Leg 71 6
  Leg 114 57-8, 60
  Leg 119 64, 66
stratigraphical continuity 20, 21, 112-14
sub-chronozones 17, 21
submagnetochrons (Jaramillo, Olduvai) 24
Synedra jouseana 13, 107, 112, 114
  Leg 113 26, 28, 31, 33, 34, 38

  Leg 119 68, 70, 71, 72, 75
  Leg 120 85, 86, 88, 102

taxonomic nomenclature 16
Thalassiosira
  aspinosa 13, 112
    Leg 113 35, 38
    Leg 114 58-9, 61
  burckliana 9, 11, 108, 110, 114
    Leg 119 64, 66, 80, 82
  compacta 83
  complicata 9, 108
    Leg 113 43, 45, 46, 48
    Leg 120 84, 87-8, 96, 98, 101, 104
  convexa 11, 110
    Leg 119 77, 79
  convexa v. aspinosa 11, 110
    Leg 114 57, 60, 77, 79
  elliptopora 7, 9, 108, 114
    Leg 71 23, 25
    Leg 114 50, 51-2
    Leg 119 71, 74, 76, 78
    Leg 120 84, 87, 91, 93, 95, 101, 104
  fraga 13, 112, 114
    Leg 113 28, 31, 35, 38
    Leg 114 58-9, 61
    Leg 119 67, 69, 72, 75
    Leg 120 85, 89, 95, 96-7, 99, 103, 105
  gracilis 8
  insigna 9, 11, 108, 110
    Leg 71 24, 25
    Leg 113 40, 42, 43, 45, 46, 48
    Leg 114 50, 51, 55, 57-8, 60
    Leg 119 63, 65, 71, 74, 76, 78
    Leg 120 83, 84, 87, 91, 93, 95, 95-6, 98, 101, 101, 104
  inura 7, 9, 11, 108, 110
    Leg 113 27, 30, 33-4, 37, 40, 42, 43, 45, 46, 48
    Leg 119 63, 65, 71, 74, 77, 79
    Leg 120 8, 83, 84, 87-8, 91, 93, 95-6, 98, 101-2, 104
  jacksonii 8
  kolbei 9, 106, 108, 114
    Leg 71 23, 23-4, 25
    Leg 113 7, 26, 27, 30, 33, 37, 46, 48
    Leg 114 7, 49, 50, 51, 55, 57, 57, 60
    Leg 119 63, 65, 71-2, 74, 76, 76, 78-9
    Leg 120 83, 84, 87-8, 91, 93, 95, 95-6, 98, 101, 101-2, 104
  lentiginosa 9, 108
    Leg 71 24, 25
    Leg 113 34, 37, 46, 48

    Leg 114 50, 52
    Leg 119 63, 65, 71, 74
    Leg 120 84, 88, 91, 93
  miocenica 9, 11, 108, 110
    Leg 114 57-8, 60-1
    Leg 119 64, 66, 72, 74, 77, 79, 80, 82
    Leg 120 83, 91-2, 93-4, 102, 104
  oestrupii 9, 108, 112
    Leg 113 7, 27, 30, 43, 45, 46
    Leg 114 57, 60
    Leg 119 63, 66, 72, 77, 79
    Leg 120 83, 84, 88, 91, 93, 96, 98, 101, 102, 104
  praeconvexa 11, 110
    Leg 71 6
    Leg 114 57-8, 60
    Leg 119 72, 74, 80, 82
  primalabiata 13, 112
    Leg 119 67, 69
    Leg 120 85, 89, 97, 99
  spumellaroides 13, 112
    Leg 113 28, 31, 35, 38
    Leg 114 59, 61
    Leg 119 67, 69
    Leg 120 96-7, 99
  torokina 9, 11, 108, 110
    Leg 114 50, 52
    Leg 119 63, 65, 72, 74, 76, 78, 80, 82
    Leg 120 7, 84, 88, 92, 94, 96, 98, 101, 104
  vulnifica 9, 106, 108, 114
    Leg 71 6, 24, 25
    Leg 113 40, 42, 43, 43, 45, 46, 46, 48
    Leg 114 7, 49, 50, 51, 55, 57, 57, 60
    Leg 119 71, 74, 76, 76, 78
    Leg 120 8, 83, 84, 87, 91, 95, 95-6, 98, 101, 101, 104
  yabei 3
Triceraspyris antartica 9, 108
  Leg 119 76, 78
  Leg 120 84, 87, 95, 98, 101, 104
Triceratium, polymorphus 95
Trinacria excavata 11, 110
  Leg 119 77, 79

unconformities and hiatuses 6, 20, 21, 112-13
  Leg 113 26, 33
  Leg 114 49, 57
  Leg 119 63, 67, 71
  Leg 120 83-6, 91, 95, 101

Weddell Sea, Southern Ocean 6-7, 8